Gentrification as a Global Strategy

This book pays homage to Neil Smith's ideas, offering a critical approach and rich collection of insights that draw on Smith's work for inspiration and debate. With interdisciplinary and international contributions from leading experts, the book demonstrates the impact of Smith's ideas on understanding the role of urbanisation in general and gentrification, in particular, in contemporary society. The book demonstrates how gentrification varies significantly from city to city, across different cultural and political-economic regimes, and in terms of the timing of urban transformations.

This collection provides a forum for debate for those working in urban regeneration and citizenship, and those directly affected by the processes and problems arising from gentrification. It will be of interest to students and scholars in urban geography, urban sociology, cultural studies, and wider social and urban theories.

Abel Albet is an Associate Professor at the Department of Geography, Universitat Autònoma de Barcelona, Spain.

Núria Benach is an Associate Professor at the Department of Geography, Universitat de Barcelona, Spain.

Routledge Critical Studies in Urbanism and the City

This series offers a forum for cutting-edge and original research that explores different aspects of the city. Titles within this series critically engage with, question and challenge contemporary theory and concepts to extend current debates and pave the way for new critical perspectives on the city. This series explores a range of social, political, economic, cultural and spatial concepts, offering innovative and vibrant contributions, international perspectives and interdisciplinary engagements with the city from across the social sciences and humanities.

For a full list of titles in this series, please visit www.routledge.com/Routledge-Critical-Studies-in-Urbanism-and-the-City/book-series/RSCUC

Urban Subversion and the Creative City
Oliver Mould

Mega-Event Mobilities
A Critical Analysis
Edited by Noel B. Salazar, Christiane Timmerman, Johan Wets, Luana Gama Gato and Sarah Van den Broucke

Art and the City
Worlding the Discussion through a Critical Artscape
Edited by Julie Ren and Jason Luger

Gentrification as a Global Strategy
Neil Smith and Beyond
Edited by Abel Albet and Núria Benach

Gentrification as a Global Strategy

Neil Smith and Beyond

Edited by
Abel Albet and Núria Benach

Routledge
Taylor & Francis Group
LONDON AND NEW YORK

First published 2018 by Routledge

2 Park Square, Milton Park, Abingdon, Oxfordshire OX14 4RN
52 Vanderbilt Avenue, New York, NY 10017

Routledge is an imprint of the Taylor & Francis Group, an informa business

First issued in paperback 2019

British Library Cataloguing-in-Publication Data
A catalogue record for this book is available from the British Library

Library of Congress Cataloging-in-Publication Data
A catalog record for this book has been requested

ISBN: 978-1-138-23425-3 (hbk)
ISBN: 978-0-367-21912-3 (pbk)

Typeset in Times New Roman
by codeMantra

Contents

Figures and tables

Figures

Tables

Contributors

Leticia Lyra Acioly is a student in the PGAU (Programa de Graduação em Arquitetura e Urbanismo) affiliated to the Escola de Arquitetura e Urbanismo, Universidade Federal Fluminense (Niterói, Brazil), and a scholar in the BIC/UFF Scientific Initiation Program, in the research group *Arquitetura da Violência*. Contact: leticialyra@id.uff.br

Abel Albet is Associate Professor at the Department of Geography of the Universitat Autònoma de Barcelona (Spain). His teaching and research interests are focused on critical geographies, urban studies, and new cultural/political geographies. He has been a member of the steering committee of the International Critical Geography Group. Together with Núria Benach, he is co-leading the [espais crítics] project and co-editing the *Espacios Críticos* book series (Icaria Editorial). Contact: abel.albet@uab.cat

Núria Benach is Associate Professor at the Department of Geography of the Universitat de Barcelona (Spain). Her research interests include urban transformations, spatial inequalities, and the history of radical geography. Together with Abel Albet, she is co-leading the [espais crítics] project and co-editing the *Espacios Críticos* book series (Icaria Editorial). Contact: nuriabenach@ub.edu

Laura Calbet i Elias is an urban researcher and planner. Since 2009, she has been a research assistant and lecturer at the Chair for Urban Design & Development, Department of Urban and Regional Planning of the Technische Universität Berlin (Germany). From 2012 to 2015, she received a fellowship from the German Research Foundation at the Center for Metropolitan Studies in Berlin. Her research explores the financialised patterns of inner-city real estate development. Additional research interests include planning history of Berlin, urban redevelopment in Barcelona and Berlin, public space in planning politics and the critical pedagogy of urban design. She is a member of the editorial board of *sub\urban. zeitschrift für kritische stadtforschung*. Contact: laura.calbetelias@tu-berlin.de

Luis Gustavo Rosadas Campos has a Bachelor's degree in Architecture and Urbanism from Fluminense Federal University (Brazil) (2012), and a Bachelor's degree in Graphic Design from Estácio de Sá University

(2001). He is doing research on methodologies of presentation of architecture and urbanism projects. He is also a student at the PGAU (Programa de Graduação em Arquitetura e Urbanismo) affiliated to the Escola de Arquitetura e Urbanismo, Universidade Federal Fluminense (Niterói, Brazil). Contact: luiscampos@id.uff.br

Greig Charnock is Senior Lecturer in International Politics at the University of Manchester (United Kingdom), where he teaches international political economy, the politics of globalisation and Marxist critical theory. His most recent research has engaged the 'open Marxism' approach with critical theories of the production of space, and in analyses of crisis formation and management in Spain. He is co-author of *The Limits to Capital in Spain*, and has published articles in such journals as *Antipode*, *Society and Space*, the *International Journal of Urban and Regional Research*, *Historical Materialism*, *New Political Economy* and *South Atlantic Quarterly*. Contact: greig.charnock@manchester.ac.uk

Eric Clark is Professor of Human Geography at Lund University and elected member of the Royal Society of Letters at Lund. He has held associate professorships at the University of Oslo and Copenhagen University, professorships at Uppsala University and (since 1998) Lund University, and guest professorship at National Taiwan University. His research interests include political economy of space, land rent, gentrification, financialisation, sustainability, island studies and historical political ecology – consistently drawing inspiration from Neil Smith's work. Clark is managing editor of the *Journal of Urban Affairs* and edited *Geografiska Annaler B: Human Geography* from 1999 to 2008. Contact: eric.clark@keg.lu.se

Anne Clerval (PhD) is Associate Professor of Geography at the University Paris-Est Marne-la-Vallée (France). She has been studying gentrification in Paris since the early 2000's; her work is focused on social division of space, class and race relations, urban public policies and discourses, with both quantitative and qualitative methodologies. She published *Paris sans le peuple. La gentrification de la capitale* [Paris without the people. The gentrification of the capital] in 2013 (La Découverte). Contact: anne.clerval@univ-paris-est.fr

Vickie Cooper is a Lecturer in Social Policy and Criminology and Co-Director of Harm and Evidence Research Collaborative (HERC) at the Open University (United Kingdom). Vickie's research is on homelessness in relation to housing policy and the criminal justice system. Looking at how housing plays a critical role in the incarceration and policing of disadvantaged and marginalized groups, her previous research considers where housing and prisons, hostels, rehabilitation and the uneven geographical dispersal of women who offend. She is currently researching the relationship between welfare reforms and evictions in the post-crash climate and austerity as a violent political project. Contact: victoria.cooper1@open.ac.uk

Deborah Cowen is Associate Professor at the Department of Geography and Planning of the University of Toronto (Canada). Her research explores the role of organized violence in shaping intimacy, space, and citizenship. She is the author of *The Deadly Life of Logistics: Mapping Violence in Global Trade* and *Military Workfare: The Soldier and Social Citizenship in Canada*, and co-editor with Emily Gilbert, of *War, Citizenship, Territory*. Deborah has also been active in community-based research in Toronto addressing the racialization of sub/urban space, and was a collaborator on the National Film Board of Canada's Emmy award winning HIGHRISE project. Deborah serves on the board of the Groundswell Community Justice Trust Fund. Contact: deb.cowen@utoronto.ca

Myrto Dagkouli-Kyriakoglou. After her BA in Economics, where she had the opportunity to attend courses in Economics and Management, she wished to continue to a more social field and thus she attended a Masters in 'Social Environment and Development' and 'Urban and Regional Planning' at the School of Architecture, National Technical University of Athens (Greece). During this time, she was introduced to Urban Sociology, Human Geography and Geopolitics, all of which solidified her future research interests. At the same time, she was working in personal research projects in the fields of Urban Sociology and Human Geography, especially on housing practices, family strategies, ethnography of housing solutions, migration and digital geographies. Contact: myrtokyriadagk@gmail.com

Katarina Despotovic is a documentary photographer and has worked for five years doing extensive ethnographic research and documenting the demolition of Kvillebäcken (Sweden). Contact: despotkat@gmail.com

Sonia Maria Taddei Ferraz has a Bachelor's degree in Architecture and Urbanism from the Federal University of Rio de Janeiro (1968), a Master's degree in Analyse Regionale et Amenagement de l'Espace at University of Paris I, Pantheon Sorbonne (1979), and a PhD in Communication and Culture from the Federal University of Rio de Janeiro (1999). She is a Professor in the PGAU (Programa de Graduação em Arquitetura e Urbanismo) affiliated to the Escola de Arquitetura e Urbanismo, Universidade Federal Fluminense (Niterói, Brazil). Her main field of interest is Theory of Architecture, with emphasis in popular housing. She is Coordinator of the research group *Arquitetura da Violência* (www.uff.br/arqviol), with financial support from *Fundação de Amparo e Pesquisa do Estado do Rio de Janeiro*. The group takes part in the *Rede de Laboratórios de Políticas Públicas do Rio de Janeiro*. Contact: sonia_tferraz@uol.com.br

Luz Marina García Herrera is Professor of Geography at the University of La Laguna (Tenerife, Canary Islands, Spain). Her research interests include the production of urban space and its social and political implications, gentrification and urban public spaces. She is the co-author, with

Fernando Sabaté Bel, of the book *Neil Smith. Gentrificación urbana y desarrollo desigual* (Icaria, 2015). Contact: lmgarcia@ull.edu.es

Francesco Indovina is Professor of Spatial Planning at Department of Architecture, Design and Urban Planning of the University of Sassari (Italy). He also teaches Urban and Territorial Analysis at Istituto Universitario di Architettura di Venezia (Italy). His research interests are in the field of urban dynamics, with a focus on new forms of urbanisation, urban and spatial planning. He has recently published *La metropoli europea. Una prospettiva* (Milan: Franco Angeli, 2014). Contact: indovina@iuav.it

Hamish Kallin is a Teacher in Human Geography at the University of Edinburgh (United Kingdom). He completed his doctoral thesis in summer 2015 on the topic of urban regeneration, gentrification, and the crises of capitalism, focused on two large-scale developments in Edinburgh (Scotland). He teaches widely, and runs a popular undergraduate course on the mutating forms of capitalism and the (re)making of the Scottish landscape. He is co-author (with Tom Slater) of a 2014 paper in *Environment and Planning A* and a (forthcoming) chapter in *Tainted Urban Spaces* (Routledge), both on the links between territorial stigmatisation and gentrification. Contact: h.kallin@ed.ac.uk

Nemoy Lewis is a PhD candidate in the Department of Geography and Planning at Queen's University in Kingston, Ontario (Canada). Nemoy did both his BA and MA degree in Geography at the University of Toronto prior to commencing his doctoral studies in 2012. His doctoral research analyses the ongoing foreclosure crisis and how it has affected the lives of racialized people, low-income families and economically disadvantaged communities in two of the hardest hit cities across the US. He has presented papers at conferences both home and abroad, and has contributed to a book chapter to the forthcoming book entitled *Neoliberal Chicago*. Contact: nemoy.lewis@queensu.ca

Mayra Duarte Lima is a student at the PGAU (Programa de Graduação em Arquitetura e Urbanismo) affiliated to the Escola de Arquitetura e Urbanismo, Universidade Federal Fluminense (Niterói, Brazil). She is a scholar in the BIC/UFF Scientific Initiation Program, in the research group *Arquitetura da Violência*. Contact: mayradlima@gmail.com

Ernesto López-Morales is Associate Professor at the University of Chile and PhD in Urban Planning from the DPU, University College London. He is principal researcher in the *Contested Cities* international network and the Centre for Social Conflict and Cohesion Studies (COES) where he focuses on gentrification, neoliberal urbanism and housing in Chile and Ibero-American cities. He has co-authored two books with Loretta Lees and Hyun Bang named: *Global Gentrifications: Uneven Development and Displacement* (Policy Press, 2015) and *Planetary Gentrification*

(Polity Press, 2016). Some of his recent journal papers are 'Assessing exclusionary displacement through rent gap analysis in the urban redevelopment of inner Santiago, Chile' (*Housing Studies*), 'Gentrification in Chile: A property-led process of dispossession and exclusion in the inner city of Santiago' (*Urban Geography*) and 'Gentrification in the global South' (*City*). Contact: elopez@uchilefau.cl

Hug March (Tarragona, 1983) has a BA and a MA in Environmental Sciences (Universitat Autònoma de Barcelona, Spain) and a PhD in Environmental Sciences (Universitat Autònoma de Barcelona, 2010). Currently he is a senior researcher at the Internet Interdisciplinary Institute (IN3) of the Universitat Oberta de Catalunya (Spain). His research focuses on the urban political ecology of the hydrosocial cycle and the political ecology of urban transformation and alternative socio-ecological transitions. Contact: hug.march@gmail.com

Paula Ramos C. Carvalho de Mendonça is a student at the PGAU (Programa de Graduação em Arquitetura e Urbanismo) affiliated to the Escola de Arquitetura e Urbanismo, Universidade Federal Fluminense (Niterói, Brazil). She is a scholar in the Scientific Initiation Program of the Research Support Foundation of the State of Rio de Janeiro, in the research group *Arquitetura da Violência*. Contact: paularccm@gmail.com

Don Mitchell is a Professor of Cultural Geography in the Department of Social and Economic Geography at Uppsala University (Sweden). With Neil Smith, he is the general editor of *Revolting New York: 350 Years of Riot, Rebellion, Uprising, and Revolution in the Big Apple* (2017). He is currently at work on a book called *Mean Streets: Homelessness, Public Space, and the Limits to Capital.* Contact: dmmitc01@maxwell.syr.edu

Marc Morell is a Social Anthropologist mainly working on the class character of the production of space in market society. To date, he has conducted fieldwork in Catalonia, Majorca and Malta and he has also been employed as researcher and lecturer at the Universitat de les Illes Balears. His most recent publications in English are: *The Making of Heritage* (co-editor, Routledge, 2015), 'When space draws the line on class' (in *Anthropologies of Class*, Cambridge University Press, 2015), and 'Working class heritage without the working class' (in *Heritage, Labour and the Working Classes*, Routledge, 2011). Contact: marc.morell@uib.cat

Oriol Nel·lo is Associate Professor of Urban Geography and Spatial Planning at the Department of Geography of the Universitat Autònoma de Barcelona (Spain). His research interests are metropolitan dynamics, urban segregation, social movements and urban policies. He has been director of the Institute for Metropolitan Studies of Barcelona and Secretary for Spatial Planning in Catalonia. He has recently co-edited *Cities in the 21st Century* (Routledge, 2016). Contact: oriol.nello@uab.cat

Kirsteen Paton is a Lecturer in Sociology at the Department of Sociology, Social Policy and Criminology of the University of Liverpool (United Kingdom) teaching on the areas of class and urban sociology. She is the author of *Gentrification: A Working-Class Perspective* (Ashgate, 2014) which explores the impacts of gentrification and regeneration policies on a working-class neighbourhood. Her recent research explores regeneration and stigma in Glasgow's East End in relation to the Commonwealth Games and the rise of evictions in relation to welfare reform in the UK. Her forthcoming book *Class and Everyday Life* (Routledge, 2017) explores the everyday lived experiences of class in 21st century neoliberal Britain. Contact: kirsteen.paton@liverpool.ac.uk

María del Socorro Pérez Rincón Fernández is Lecturer at the Departament de Geografia of the Universitat de Barcelona (Spain). She holds a degree on Urban Planning from the Department of Architecture, Universidad Nacional Autónoma de México (UNAM) and a PhD in Geography (*Voces femeninas en barrios en crisis: desorden aparente y realidades paralelas*) from the Universitat de Barcelona (Spain). She has been part of the research group Multiculturalism and Gender of the Universitat de Barcelona (Spain) and of the Inter-university network UB-UNAM, LAHAS Laboratory *Habitat, participation and gender.* Her research interests are urban planning, urban exclusion and urban policies with a gender perspective. Contact: urbperezrincon@gmail.com

Thomas Purcell is Senior Lecturer in Politics and International Relations at Leeds Beckett University (United Kingdom), where he teaches courses in the area of International Political Economy, Marxism and Globalisation. Until recently, he was Sub-Director of the National Centre for Strategies for the Right to Territory (CENEDET) based at the Instituto de Altos Estudios Nacionales (Quito, Ecuador). Prior to this, he held research positions at the Universitat Oberta de Catalunya (Spain), and at the Universities of Manchester and Nottingham (United Kingdom). His research to date has examined the specificities of capitalist development in Ecuador and Venezuela, and also the political economy of Spain and the Eurozone crisis. He is co-author of *The Limits to Capital in Spain* (2014), and has published several book chapters and articles in international journals such as *Third World Quaterly*, *Geoforum*, *Historical Materialism*, the *International Journal of Urban and Regional Research*, *Latin American Perspectives*, and *Science & Society*. Contact: tompurcell90@hotmail.com

João Queirós is a sociologist working in Porto (Portugal). He does research at the Institute of Sociology of the University of Porto and at the Centre for Research and Innovation in Education of the School of Education of the Polytechnic Institute of Porto, where he also teaches. His teaching and research interests cover: social change and class transformation, education, social policies, urban and housing policies, and local and regional development.

In his PhD research, conducted between 2010 and 2014, he studied the social consequences of urban and housing policies as seen in inner city Porto throughout the last half century. Contact: jqueiros@letras.up.pt

Ramon Ribera-Fumaz (MA, PhD Geography, Manchester) is currently the director of the Urban Transformation and Global Change Laboratory at the Internet Interdisciplinary Institute (IN3), Universitat Oberta de Catalunya (Spain). His research focuses on the urban political economy and critical geography of urban transformation and regeneration, including the articulation between urban cultural and economic strategies, smart city politics and the global and national determinants of urban restructuring. He is co-author of *The Limits to Capital in Spain*. He has received the *2010 Progress in Human Geography Essay Prize* to the best article by an early career scholar. Contact: rriberaf@uoc.edu

Onofre Rullan is Professor at the University of the Balearic Islands (Spain). He has participated in several research teams on urban and regional planning. He has collaborated in the publication of over 30 books and over 50 papers. He has participated in nine R+D+I projects financed by the Spanish Ministry of Science and Innovation, the last three of which he has been Principal Investigator. Contact: orullan@uib.es

Fernando Sabaté Bel is a Lecturer at the University of La Laguna (Tenerife, Canary Islands, Spain) and political activist. His research interests centre on historical-cultural geography and the transformation of vernacular rural spaces. He co-authored with Luz Marina García Herrera, the book *Neil Smith. Gentrificación urbana y desarrollo desigual* (Icaria, 2015). Contact: fsabate@ull.edu.es

Tom Slater is Reader in Urban Geography at the University of Edinburgh (United Kingdom). He has research interests in the institutional arrangements producing and reinforcing urban inequalities, and in the ways in which marginalised urban dwellers organise against injustices visited upon them. He has written extensively on gentrification – notably the co-authored books, *Gentrification* (2008) and *The Gentrification Reader* (2010) – displacement from urban space, territorial stigmatisation, welfare reform and social movements. Since 2010, he has delivered lectures in 18 different countries on these issues. His work has been translated into 8 different languages and circulates widely to inform struggles for urban social justice. Neil Smith will always be a huge inspiration to Tom. For more information, see http://www.geos.ed.ac.uk/homes/tslater. Contact: tom.slater@ed.ac.uk

Catharina Thörn is Associate Professor in Cultural Studies at the Department of Cultural Sciences, Gothenburg University (Sweden). She is doing research in housing and urban development with a focus on gentrification. In 2014, she published a book that introduced the concept of gentrification into Swedish and translated classical text by Neil Smith, Tom Slater, Loretta Lees, Eric Clark, Andrej Holm and others. Together with urban movements in Sweden she has published an "anti-gentrification handbook" on how to fight gentrification. Together with Margit Mayer and Håkan Thörn she is also the editor of the book *Urban Uprisings: Challenging Neoliberal Urbanism in Europe* (Palgrave, 2016). Contact: catharina.thorn@kultvet.gu.se

Eryka Y. Torrejón Cardona holds a PhD in Geography from the Universitat Autònoma de Barcelona (Spain). She is Associate Professor of Sociology at the University of Antioquia, Medellín (Colombia). She is member of several international research networks: the Grupo Medio Ambiente y Sociedad (Universidad de Antioquia, Medellín), the Red de Ciudadanía y Medio Ambiente (Universidad Complutense, Madrid, Spain), and the Agroecology Network (University of Cuenca, Ecuador). She is adviser to several Colombian social movements on human rights and land rights: *Colectivos Sociales de la Zona Nororiental de Medellín, Fundación Forjando Futuros and Programa Preparémonos para La Paz-Gobernación de Antioquia*. Her research interests are focused on rural sociology, local development, urban and regional planning, participative methodologies, women's rights, peace processes. Contact: eryka.torrejon@udea.edu.co

Sònia Vives-Miró is a PhD in Geography (University of the Balearic Islands, 2013). She has been visiting scholar at the Center for Place, Culture and Politics (City University of New York) in 2009–10. She is a postdoctoral fellow at the University of Santiago de Compostela (Galicia, Spain) and the University of Lund (Sweden). She is a member of the *Grup d'Investigació sobre Territori i Sostenibilitat* at the University of the Balearic Islands and member of the *Grupo de Investigación de Análise Territorial at the* University of Santiago de Compostela. Her main research interests are urban neoliberalism, financialisation of the built environment in Spain, housing, urban enclosures and urban rent. Contact: so.vives@gmail.com

Antonis Vradis is a geographer and Vice-Chancellor's Fellow at Loughborough University (United Kingdom). He has been conducting research in Athens (Greece) and Rio de Janeiro (Brazil), looking at the effect of major financial restructuring and devaluation upon everyday life in cities and countries of the global semi-periphery. Contact: a.vradis@lboro.ac.uk

Acknowledgements

We are sincerely grateful to all of the following who provided their support so that the International Conference – 'Global Capitalism and Processes of Urban Regeneration. A Tribute to Neil Smith' – could be organised in optimum conditions and that this book could see light of day. Our thanks go out, therefore, to the Fundación Española para la Ciencia y la Tecnología (Spanish Foundation for Science and Technology, research project FCT-14-8526-1); the Ministerio de Economía y Competitividad (Spanish Ministry of Economy and Competitiveness, research project CSO2013-44665-P); the Agència de Gestió d'Ajuts Universitaris i de Recerca (Catalan Agency for the Management of University and Research Grants, AGAUR, research group 2014-SGR-387); the Museu d'Art Contemporani de Barcelona (Barcelona Museum of Contemporary Art, MACBA) who allowed us to use their magnificent auditorium and who did everything possible to ensure the success of the meeting; Betsy Herzog who, selflessly, allowed us to use her well-known photograph of the occupation of Tompkins Square to promote the event; the citizens and associations of the Raval (Associació de Veïns del Casc Antic, Associació Cultural El Raval, El Lokal and Putas Indignadas) for the exemplary nature of their struggles and their unfailing solidarity; Icaria Editorial for continuing to back us over all these years; the Universitat de Barcelona, the Universitat Autònoma de Barcelona and the Universitat Oberta de Catalunya, thanks to whom we are able to dedicate some of our time to projects like this; the scientific and organising committees for their efficiency and constant support; the main speakers, those who presented their papers and all who participated in the conference, who together generated critical thinking of such weight; and our colleagues, students and PhD students who collaborated enthusiastically in the organisation of the event. We are also grateful for Routledge's positive reception of this book proposal and for the favourable reviews received from two anonymous referees, which proved decisive in bringing this project to fruition. Finally, we cannot fail to acknowledge the quality of the texts written by each of the contributors, all of whom participated with enormous interest and dedication to the completion of this volume. Many thanks to David Harvey who promptly agreed to write a foreword to this book, with some personal and warm words on Neil Smith that remind us how much we still need to debate and work on his ideas.

Foreword

David Harvey

I first met Neil in the fall of 1977 when he took up graduate studies in the now defunct Department of Geography and Environmental Engineering at Johns Hopkins University in Baltimore. Joe Doherty, his geography mentor at St Andrews, had persuaded him that Hopkins was the best place to go for a radical/Marxist education. I played no part in bringing Neil to Baltimore since I had been on sabbatical in Paris during the preceding year. But the then Chair of the Department, Reds Wolman, had been on the lookout for good students for me and told me he had signed up a great prospect while I was away. He was not wrong. Neil and I immediately bonded in part because my maternal Scottish grandfather was born just outside of Leith and as a young kid I had on occasion vacationed in Edinburgh and was familiar with the landscape that Neil plainly loved. I had climbed Arthur's Seat in Edinburgh many times.

When Neil arrived, I was in the middle of writing *The Limits to Capital*, while wrestling with various other facets of Marx's thought through the organization of seminars and reading groups. Neil had some background, of course, in this literature and immediately engaged with it creatively and critically. But much more importantly, he had already spent a year as an exchange student in Philadelphia a couple of years before. I have since learned from his diaries that he was at that time far more impressed intellectually with Jane Jacobs in opposition to the modernism of Robert Moses than he was with my book on *Social Justice and the City* which he considered too abstract and a bit of a dead end! But his diaries also tell a tale of his shock and outrage at the racism and ghastly poverty that flourished (if that is the right word) a stone's throw from the University of Pennsylvania where elite students learned their various professional trades in total indifference and seeming blind ignorance at the obvious inequalities and injustices then as now woven into the fabric of urban life in the United States. His sense of shock and outrage at such injustice and racial discrimination never left him and he militated against all forms of elitism and class privilege in academia and beyond for the rest of his life. From then on, also, he insisted that one learned far more from immersing oneself in the city than one did from books and academic seminars. It was probably this that led him to be intensely active for a time in the ISO (International Socialist Organisation) political group that he helped found at Hopkins. He took it hard when he was expelled from the group after some sort of sectarian split for which such groups are famous.

He passionately engaged with both politics and theory. I lost count of the number of picket lines he insisted I join and he was quick to criticize my capacity for bourgeois compromise. He also raced ahead of me in taking up the critique of nature, the production of space and the iniquities of gentrification. It was during this phase that I learned about another dimension to Neil's character. While on the surface he always presented himself as radically intuitive and political, deeply distrustful and scathingly critical of any academic authority, he had a deep respect and love for serious scholarship. He read voluminously and carefully, took copious notes and wrestled intensely with difficult concepts before synthesizing them into communicable ideas that related to the world of daily practice.

I had originally imagined he would write a dissertation on gentrification and urbanization – an arena of work in which his work later became definitive. But his dissertation took me totally by surprise. It was the draft of what became his masterpiece, *Uneven Development: Nature, Capital and the Production of Space*. While, obviously, he had pieces of this work already in place, the bulk of it was written in about four months, at breakneck speed and with an astonishing intensity, during the first months of his tenure in the Geography Department at Columbia. It was and is brilliant stuff.

I suppose I should not have been surprised, because as a participant and organizer in seminars and discussion groups, Neil always showed not only an amazing level of intellectual and political commitment but also a critical capacity that deeply impressed almost everyone who came in contact with him. Neil engaged with many faculty from other disciplines and soon had a reputation across the campus as an intellectual force to be reckoned with. He plainly enjoyed the freedom that came with being in Geography, where his position was secure, while fiercely contesting and arguing for radical positions in the seminars of others and creating not a little political turmoil on campus. And when Neil fiercely contested, you really felt it!

The interdisciplinarity of the Hopkins experience plainly appealed to him. It shaped, I believe, the genius he later exhibited for structuring interdisciplinary dialogue first at the Center for the Critical Analysis of Contemporary Culture at Rutgers and subsequently at the very successful Center for Place, Culture and Politics which he founded at the CUNY Graduate Center. But he always felt rooted in the discipline of Geography and in urban geography in particular. His transformative influence within the discipline was immense both in terms of content (always radical and critical) and style (rooted very much in politics and daily experience). Neil played an important role in the formation of the International Critical Geography Group.

Neil's prolific writings on the production of nature and space, on gentrification and urbanization, and on geopolitics and the history of geography, played a leading role in re-orienting geographical work along critical and

Marxist lines. These writings helped shape the 'spatial turn' in academia, and he therefore acquired a deserved reputation across many disciplines.

It was also while a graduate student that Neil conceived the idea of studying the life and career of Isaiah Bowman, some of whose papers were lodged at Hopkins. With the eventual publication of the award-winning Bowman book many years later, Neil brought this particular project to fruition. The book went through a number of iterations (I thought for a while he was never going to finish it) but when it came out it was greeted as a tour de force, an honest labour of love about a geographer for whom Neil felt very little empathy. The book combined his respect for scholarship with his intuitive sense of what the political should be all about.

But his greatest passion centered on urbanization and in this his work on gentrification was both formative and outstanding. He early on saw that gentrification was a movement of capital and not of people (as it was so often depicted in the literature of the time) and that the space upon which it concentrated was that of what he termed the 'rent gap' (a concept which he loved to remind me I had thoroughly disapproved of at the time he first formulated it!). The class, racist and discriminatory character of this process rekindled all the outrage he had experienced earlier in Philadelphia. He found it difficult to understand how anyone could think that the movement of capital into distressed spaces in the city might be a good thing, that it would come to be seen in certain circles as a cutting edge of urban revitalization. He was horrified at the literature that attached gentrification to the supposedly beneficial activities of some 'creative class'. As Neil so correctly saw it, the revanchist city (with New York's Mayor Giuliani in the driver's seat), coupled with planned and state supported gentrification, would produce a result in which the city might do very well even as more and more of its least privileged residents became far worse off (there are now 60,000 homeless people in New York City).

So it is fitting that this commemorative volume focuses on this aspect of his work. Neil hated the thought of hagiography. I think he would appreciate the diversity of approaches and scenarios here so deftly assembled to illustrate the commonalities and differences of gentrification processes in different places and times. He would have loved to argue and debate with the contributors.

Neil lived his life all along in a passionate, seamless and thoroughly engaged way. He was always gregarious and social, adventurous and deeply committed. His charisma and his warmth and generosity of feeling created a glow around him wherever he went. He loved an argument, inspired people around him. But, as the essays assembled here attest, his spirit lives on and for that we are all eternally grateful.

PART I

Neil Smith

A tribute

1 Neil Smith and gentrification

Abel Albet and Núria Benach

Neil Smith's legacy and his militant commitment

[espais crítics] is a university-based research group that emerged with the aim of disseminating the potential of critical spatial thinking for understanding and transforming the world.[1] More than ten years ago, we decided it was necessary to bring together and spread the thinking of various authors who have analysed inequality and social injustice from a spatial perspective. We proposed translating these ideas into Spanish, publishing them and interpreting them in the light of the context in which they were elaborated, so that this thinking could be put to good use in other contexts. This is how the *Espacios críticos* series was born, with the close support of our publisher, Icaria Editorial. The series already comprises several volumes dedicated to leading authors of this critical spatial perspective, including Edward Soja, Doreen Massey, Richard Peet, Francesco Indovina, Franco Farinelli, Horacio Capel, Neil Brenner, Jean-Pierre Garnier, David Harvey, William Bunge, Maria Dolors Garcia Ramon and Claude Raffestin,[2] and several other volumes are currently being prepared (including Milton Santos, Yi-Fu Tuan, Jordi Borja, Denis Cosgrove, José Manuel Naredo and Colin Ward).

From the outset, it was clear that Neil Smith had to occupy a prominent place in this series: our analyses of urban processes and of the tremendously harsh spatial nature of neoliberal capitalism would not be the same without Neil's inspiring and provocative texts, words and actions. Indeed, without the stimulation of Neil's work, any thinking about how to put our ideas into practice, any self-criticism about our inability to work together more effectively, and any self-belief that our very survival depends on our own doggedness would not be present in the way that it most assuredly is in our work. In our own very specific case, as well as in that of hundreds of other academics and people on the street, without Neil, geography would not be what it is today.

The friendship offered by Neil (in particular, his enormous generosity and candour) smoothed the way for us: Luz Marina García Herrera and Fernando Sabaté Bel, experts on Neil's work, agreeing to take responsibility for preparing the corresponding book. The message that Neil sent in response to our proposal is highly eloquent: 'This is a little stunning. I don't really feel worthy of a book. But, of course, I would be delighted to have something in the *Espacios Críticos* series. Luz Marina, I can't believe you would do this! I very gratefully agree and we can talk details later, can't we? I travel so much that an interview could happen whenever and wherever you like. I have to be in Paris in September, Poland and Bulgaria in October, London in ... Thanks again for your warm appreciation'. Sadly, the trip Neil made to Paris was to be his last; he died in New York on 29 September 2012. His death left us in a state of shock, and the numerous obituaries and events in his memory (above all, those organised during the meeting of the Association of American Geographers held in Los Angeles the following year) showed how well-loved and influential he had become. Luz Marina and Fernando both felt that, despite everything, they owed Neil the book that they had started together, even if this now meant that it would have would have to be written with a mix of grief, fondness and recognition (Figure 1.1).

This book[3] (García Herrera and Sabaté Bel 2015) was presented between 14 and 16 September 2015 at the Museu d'Art Contemporani de Barcelona (MACBA)[4] as part of the International Conference entitled 'Global Capitalism and Processes of Urban Regeneration. A Tribute to Neil Smith'. The event was organised by *[espais crítics]* and was dedicated specifically to disseminating the thinking of Neil Smith and to appreciating his work. Experts were invited to discuss his ideas in a theoretical frame and in terms of their practical application, and the occasion served to bring together those of us who were fortunate to know him well and those who have benefited enormously from his proposals and contributions.

Figure 1.1 Neil Smith (Leith, Scotland, 1954 – New York, United States, 2012).
Source: Tom Slater.

Eric Clark, Deborah Cowen, Luz Marina García Herrera, Maria Dolors Garcia Ramon, Don Mitchell, Fernando Sabaté Bel and Tom Slater were the guest speakers invited at the conference. In addition, around fifty proposals for papers were received from over twenty countries on three continents. Of these, thirty were selected and presented during the event in Barcelona. The long and intense discussions generated among the more than 200 participants bore testimony not only to the quality of the keynote addresses and papers but also to the significance of Neil Smith's ideas across the disciplines. Three days were spent in appraising the value of Neil Smith's work and his influence in fields such as urban gentrification, the rent gap theory, revanchism, appropriation and exploitation of nature, uneven development, the financialisation of the economy, the politics of scale, contemporary politics, political geography and the geopolitics of imperialism, Marxist theory, the limitations of geography and the need for revolution, among others (Figure 1.2).

One of the three days of the conference was held in the streets of the Raval, the neighbourhood in which the MACBA stands, and which is characterised by the harshness of its social marginalisation and the forces of the processes of gentrification to which it is now being exposed.

Several organised routes highlighted the real estate operations that have been transforming this Barcelona neighbourhood since the 1980s. All the routes came together at the *Ágora Juan Andrés Benítez*, a squatted plot of

Figure 1.2 Keynotes, papers and debates during the International Conference 'Global Capitalism and Processes of Urban Regeneration. A Tribute to Neil Smith' that took place at the MACBA (Barcelona, 14–15 September 2015). Ramon Ribera, Anne Clerval and Luz Marina García Herrera appearing in the photo.

Source: Léa Regulier.

Figure 1.3 The Ágora Juan Andrés Benítez (16 September 2015). Social movements and local actors raising their voices against gentrification processes taking place in the Raval neighbourhood of Barcelona as part of the International Conference 'Global Capitalism and Processes of Urban Regeneration. A Tribute to Neil Smith'.

Source: Abel Albet.

land managed by various groups that defend the right of people who live in this part of the city to remain here. At the same time, they are trying to halt the advance of real estate capital and tourism in this area, and to promote the public use of this plot. Conversations with members of these movements focused on their understanding of the dynamics evident in the Raval today, including gentrification, the persecution of prostitution and relations with the immigrants that have settled there. The graffiti in the *Ágora Juan Andrés Benítez* were witness to this building of a common critical understanding of the effects, struggles, resistance and creativity of citizens, in the face of the devastating action of tourism gentrification in Barcelona. During this long session, the *Associació de Veïns del Casc Antic*, the *Associació Cultural El Raval, El Lokal, Putas Indignadas* and a range of other citizen movements raised their voices in protest, with enormous clarity, courage and honesty: Neil would most certainly have enjoyed their ideas and their hospitality (Figure 1.3).

Neil Smith: the importance of his ideas and commitments

This book brings together the keynote presentations and a selection of the rest of the papers presented at the event.

In the foreword, David Harvey reminds us of Neil Smith, the person. Harvey considered Smith one of his most notable followers, and they enjoyed a very close intellectual relationship and friendship.

The first part re-covers that part of the conference dedicated to remembering the work and figure of Neil Smith. In the other parts of the book, some of the texts recognise the theoretical contributions of Neil Smith as an intellectual, who was interested in understanding the strategies capitalism employs to ensure its dominion and reproduction in the neoliberal setting both globally and in the urban environment. Other texts focus more directly on a wide range of cases and specific situations that are often not well known in the English-speaking world. In this respect, the book seeks to recall and recognise the remarkable repercussions and influence that Neil Smith's ideas had, and continue to have, in our universities and in professional, institutional and public life. Although we are reluctant to quantify this influence based on citations and references to his work, it is true that Neil Smith's texts are always, necessarily present in any rigorous study, in any professional debate and in any journalistic analysis of urban gentrification.

The book also seeks to take up a position regarding the true value of theoretical work; in short, such work should have diverse applications in many, often unexpected contexts; the theoretical concepts developed in a specific context should be fully valid and useful beyond the geographical and disciplinary area in which they were initially formulated; and after a reconsideration and modification of the concepts, they should continue to have crucial repercussions. Neil Smith himself provides a perfect example of a scholar interested in 'other contexts' beyond those of both the academic and English-speaking worlds: he held strong political commitments and was involved in many forms of social activism, and he was interested in European, Asian and Latin American experiences, as was clear during his many visits to Spain. Thus, many of the chapters that follow are focused on analysing the relationships between capitalism and gentrification, and stressing the impact of these processes on the daily lives of the inhabitants of European and Latin American cities. The interest lies not only in stressing the enormous impact of Neil Smith's ideas but also in highlighting the multiple applications of these ideas, which, while often not very well known (particularly if they arise outside the English-speaking world), can have great repercussions.

Contents of the book

The chapter by Luz Marina García Herrera and Fernando Sabaté Bel ('A political-geographic project against capitalism: an essay on the work of Neil Smith') examines the continuity of the intellectual career of Neil Smith, his lifelong defence of Marxism as a political and intellectual project and the importance he permanently attributed to space in the dynamics of capitalism. His work is analysed here through a review of the main fields of investigation – discussions of capitalism, the production of nature, the city and imperialism – that produced his most significant contributions including ideas firmly embedded today in the discourse of various social disciplines. Fernando Sabaté Bel is also the author of a supposed apocryphal diary of Neil Smith: Sabaté allows himself the liberty of imagining such diary written by Neil throughout his life, composed by fragments which reflect key moments of his life and significant aspects of his thinking and personality. The 'diary' was conceived after reading Smith's academic work in great detail and enriched with memories of direct dealings with him.

The second part of the book ('On gentrification and the rent gap theory') is devoted to analyse the close relations between gentrification and the rent gap theory, being Neil Smith's most substantial contribution to the field of urban political economy. The rent gap theory helps to explain where, when and to what extent certain neighbourhoods can be transformed by the concerted interests of the capitalist state, private developers and property owners, and what the effects are in terms of gentrification, ground rent dispossession and displacement. The rent gap theory, initially forwarded to only explain gentrification, is increasingly seen to cast light on land use change, capital flows into built environments and uneven development more broadly.

In his chapter 'The state of gentrification has always been extra-economic', Hamish Kallin turns to an ambitious and ill-fated regeneration project in north Edinburgh (Scotland) to foreground the necessity of theorising gentrification not as a standalone, purely economic process, but as part of a wider analysis of capitalism. The state has been central to gentrification scholarship for over thirty years, in particular through the work of Neil Smith, and in the current era of 'state-led gentrification', the use of extra-economic force is increasingly normalised. According to Kallin, gentrification should not be seen as a static, uniform process in 'the West', but just as dynamic, changing and crisis-prone as it is elsewhere.

These ideas are somehow confirmed by the following chapters. Francesco Indovina and Oriol Nel·lo offer a critical overview of gentrification in southern European countries ('Gentrification: disaster, necessity, opportunity? Notes for a critical use of the concept'). From the standpoint that gentrification processes are in this area highly specific and distinctive, the authors consider that the use of imported models of analysis can lead to distorted scientific conclusions and misguided political attitudes. Five issues are essential in order to understand how urban segregation and gentrification are currently operating in these countries: the importance of tourism, the prevailing tenancy regime, the role of transnational political agents, the action of urban social movements and the aims of public policies.

Antonis Vradis ('Toward a theory of gentrination: global capital flows and the reshaping of the global semi-periphery. The cases of Greece and Brazil') puts the question on what happens to the countries of the semi-periphery when global capital flows return inward, when – much in the style of urban gentrification – they hollow out national territories in order to be reused and reintegrated in the global capitalist core. Vradis uses the cases of Greece and Brazil to discuss what a new theory fit for this restructuring could potentially look like and he proposed the idea of gentri-nation. That is, one that utilises Neil Smith's urban gentrification theory in order to try to grasp what is now happening at the global scale, to the populations living at the ever-drifting semi-peripheries of the global neoliberal core.

In 'Making rent gap theory not true', Eric Clark argues that four processes are constitutive of rent gap formation: commodification of space/land/nature, polarizing inequalities, exchange-value-driven decision-making (financialisation) and the ideology of market fundamentalism. Consequently, making rent gap theory not true requires de-commodification of space (social practices of commoning), effective floors and ceilings on income and wealth, use-value-oriented decision-making (deepening democracy) and ideology rooted in solidarity with full recognition of our interdependence.

Closing this part, Ernesto López-Morales ('Rent gap theory is a political resource') states that, despite its criticism during the 1980s and 1990s as being excessively determinist, the rent gap theory has evolved not only to be applicable to cities experiencing neoliberal redevelopment-induced gentrification in both the global south and north, but it has also become politico-economic knowledge useful to urban social organisations and neighbourhood activists. His chapter discusses the basic Marxist grounds of the theory and some of its recent applications.

The third part ('Dispossession and class struggle') focuses on different types of dispossession as examples of present-day class struggle. Don Mitchell examines, in 'From Boise to Budapest: capital circulation, compound capitalist destruction, and the persistence of homelessness', the processes of capital circulation, abstract space production and accumulation – and their contradictions – in the built environment, showing along the way that compound capital destruction is as vital a part of compound capitalist growth. Such destruction necessarily entails destroying built environments and thus exacerbating homelessness and its unsettling effects on those relatively immobile fractions of capital that feel the brunt of such destruction.

After recapitulating Neil Smith's concept of revanchism, Tom Slater scrutinises the emergence of revanchist political economy in Britain, with particular reference to the housing crisis ('Revanchism, ignorance and class struggle in austerity Britain'). In order to explain how revanchism has become so ingrained in British society, his text analyses the production of ignorance via the activation of class and place stigma, where free market think tanks play a crucial role in deflecting attention away from the causes of housing crisis. It is argued that the production of ignorance carves an economic and political path for gentrification on a scale never before seen in the UK, where speculation, rentier capitalist extraction and the global circulation of capital in urban land markets are resulting in staggering fortunes for those expropriating socially created use values.

By drawing on the role bars take in a gentrifying neighbourhood of the capital city of Majorca (Spain), Marc Morell looks into the labour cooperation that makes rent gaps happen and shows it in his chapter 'The class gap in gentrification: a political reading of the rent gap hypothesis'. This approach not only allows us to explore the productive fabric of gentrification, but it also helps to think of the potential class politics rent gaps contain.

In the aftermath of the financial crash, Vickie Cooper and Kirsteen Paton show how state-led gentrification has become even more punitive with hundreds of evictions occurring daily in austerity Britain: 'The new urban frontier of everyday evictions: contemporary state practices of revanchism'. Evictions underpin processes of gentrification where profit is actively generated from devalued property and stigmatised places. The authors show how evictions are actively endorsed by the state through a swathe of policy interventions and welfare reforms that, for the poor, increase the risk of eviction, housing debt and homelessness, while securing accumulative wealth for owners and investors in the private rented sector.

Putting into conversation the bodies of theory on the rent gap and the financialisation of the built environment, the chapter ('Capturing urban rent through evictions: home dispossessions in the historic centre of Palma, Majorca') written by Sònia Vives-Miró and Onofre Rullán seeks to analyse the role played by evictions in the production and appropriation of urban rents. To do this, they analyse the process of devaluation/revaluation and its relationship with the eviction of tenants and mortgage foreclosures in the Historic Centre of Palma (Majorca, Spain). The conclusion leads to argue that evictions in urban spaces with high heritage value can increase the potential rent and thereby serve as a gateway to gentrification.

The fourth part ('Policies and strategies') begins with Laura Calbet i Elias's chapter ('Financialised rent gaps and the public interest in Berlin's housing crisis: Reflections on N. Smith's "generalised gentrification"') deepening Neil Smith's concept of 'generalised gentrification', used to explore the shift in politics and finances in Berlin's housing production following the global financial crisis: a consideration of the financialisation processes is used to identify new reasons for the emergence of rent gaps, and the concept of 'generalised gentrification' allows the author to move beyond the analysis of urban regeneration policies as a strategy for productive growth and to challenge the legitimacy of competitiveness policies that encourage speculative financial investments in land.

Using the case of Inner Paris, Anne Clerval analyses the gentrification processes experienced after the public policies led by the French Socialist Party since 2001, which is quite different from revanchist New York under Rudy Giuliani. Her chapter 'Beyond the "revanchist city": when public policies softly support gentrification in the name of social mix. The case of Inner Paris' appeals to deconstruct 'social mix' as being the keyword to re-launch social housing production, because, in fact, those public policies became a 'soft' way to promote gentrification.

By taking up Neil Smith's original focus on how rent performs a crucial coordinating role in contemporary cities, Greig Charnock, Hug March, Thomas Purcell and Ramon Ribera-Fumaz explore the more recent regeneration processes of Barcelona under the official guise of promoting knowledge-based competitiveness. In 'Urban regeneration, rent and labour: insights from Barcelona's "Knowledge District"', they show how the unevenness of, and limits to, knowledge-based urban transformation has been manifest not only in economic crisis, but also in the sudden and everyday visibility of very different working subjects to that of the 'creative class' of human capital or 'talent' that were supposed to populate the new 'city of knowledge'.

Drawing on some results of a sociological fieldwork-based approach to the consequences of urban and housing policies in inner city Porto (Portugal), in 'A disappearing world: the ever-expanding "frontier of gentrification" through the eyes of Porto's historic centre long-time residents', João Queirós focuses on the effects of gentrification as seen by locals, their lived experience and the ways they envisage the possibility of deterring the seemingly irresistible expansion of the 'frontier of gentrification' that is conquering their place of existence.

Sonia Maria Taddei Ferraz, Luis Gustavo Rosadas Campos, Mayra Duarte Lima, Paula Ramos C. Carvalho de Mendonça and Leticia Lyra Acioly analyse both real and symbolic effects of the relation between the current forms of reproduction of capital from the real estate market and the multiplication of anti-beggar architecture: this architecture is presented as an evident strategy of banishment of beggars and homeless from upscale areas of global cities during the processes of gentrification and urban regeneration, and is presented in their chapter 'Architecture of violence: "anti-beggar architecture" as the "eureka" of urban regeneration'.

Neil Smith argued that gentrification relies on the myth of the 'the urban frontier'. As a concept, a frontier marks both a strategy and physical borders. The frontier is a place for battles, both symbolic and material. According to Smith, the frontier ideology rationalises displacements and exclusion as something natural and inevitable. Yet the struggle it implies can also be seen as an opportunity for action, creativity, consciousness, commitment and resistance. The fifth part ('Activism and resistance') includes different analyses and examples where gentrification has brought forms of social mobilisation and organisation in order to report revanchisms and build alternative discourses and actions.

The chapter by Catharina Thörn and Katarina Despotovic ('The urban frontier: gentrification as ideology and class politics in the remaking of marginal urban space') takes into account the demolition and redevelopment of a former industrial area (Kvillebäcken) in Gothenburg (Sweden) that was used as an important site of social and commercial activities for lowincome individuals and households. Having documented both the planning process – with the intricate relations between the municipality and private real estate owners – as well as the consequences for people who used this site, the authors make visible and transparent, in text and image, the inherent class conflict and class struggle of gentrification and the brutality of the process felt by people on the ground.

In 'Alternative geographies for social action in Medellín', Eryka Y. Torrejón Cardona addresses the situation of the city of Medellín (Colombia), in describing three characteristics of neoliberal cities according to the contributions of Neil Smith: capital recovery, both local and international, through large-scale construction projects; the production and scalar changes in the uneven geographic developments of the city; and the marketing of elements from the local culture (landscapes, identities and images) through the internal and external promotion of the city in order to generate urban income. The conclusions of this chapter examine a constant relationship between the local government and financial capital, a reproduction of urban income based on marketing of the local culture and a distinct and attention-attracting social action that focuses on the effects of scalar restructuring of the city and its global sense of place.

Women in disadvantaged areas that are temporarily forgotten by the capital are silently involved in everyday innovative practices that transgress hegemonic urban policies. Often, however, they take part in creative strategies that focus on the way spaces are used. In her chapter 'Alternative narratives from an invisible city: gentrification, counter-proposals, and women activism', María del Socorro Pérez Rincón Fernández shows the involvement of women fighting against the renovation urban processes that drive gentrification in Barcelona (Spain), while defending their rights in the places they live.

The text by Myrto Dagkouli-Kyriakoglou ('The onslaught against the Greek squatting movement and the value that it produced') argues about the alliance between state tolerance, real estate strategies and fascist activity in Athens (Greece) in order to discredit, attack and eliminate urban squatters because of their role acting as a barrier to capitalistic urban speculation and political repression.

Deborah Cowen and Nemoy Lewis suggest, in 'Revanchism and the racial state: Ferguson as "internal colony"', that Neil Smith's work on gentrification and revanchism are invaluable resources for understanding struggles in places such as Ferguson, precisely because of his emphasis on urban geopolitical economy as not only a space-time of capitalist transformation and anti-capitalist struggle, but one of imperial renovation and anti-imperial insurgence. The authors flag transformations in the domestic face of empire – shifting geographies of Black dispossession through gentrification and the subprime crisis – in order to bring more attention to the relationship between repression and accumulation in the urban homeland of American imperialism.

Notes

1 See http://espaiscritics.org.
2 See www.icariaeditorial.com/coleccion_detallada.php?id=41.
3 As in the other books in the *Espacios Críticos* series, this volume includes an anthology of texts, an analysis of the personal and intellectual journey of Neil Smith and interviews that, in this case, are not with the person in question but with those who knew Neil Smith and his thinking well.
4 See www.macba.cat/en/seminar-neil-smith.

Reference

García Herrera, L. M. and Sabaté Bel, F. (2015). *Neil Smith. Gentrificación urbana y desarrollo desigual*. Barcelona: Icaria Editorial.

2 A political-geographic project against capitalism

An essay on the work of Neil Smith

Luz Marina García Herrera and Fernando Sabaté Bel

As a firm believer in the importance of geography for understanding the dynamics of capitalism, Neil Smith played a key role from the late 1970s onwards in reorienting the discipline towards critical and Marxist approaches, two bodies of thought that, as Harvey notes (2010: vii), were very far apart at the time. The integration of geographical perspective and Marxist theory allowed him to introduce 'spatial sensibility' into the latter and confer a historical materialist dimension on the former (Smith 2010a: 2–3), in a bid to merge social and spatial issues in a single theoretical framework in his studies.

Politics, class, capital and revolution are the essential issues pervading his work. Critical analysis, dialectics, rigour and imagination are the tools he used to address these issues. From his earliest publications, penned between 1979 and 1984, he tackled questions such as the criticism of Marxist structuralism, gentrification and uneven development, the concept of nature in Marx and the figure of Isaiah Bowman, giving advance notice of the directions of the main theoretical contributions on which he was to work in depth for over thirty years until his death.

This chapter is divided into four main sections. The first outlines Smith's theories on the production of Nature, the production of space and uneven development in relation to capital flows. The second considers his theory of rent gap, its explanatory role in the gentrification process and towards the revanchist city, understood as processes of class violence. The third looks at his contributions concerning the close ties between the production of geographical knowledge and its links to American imperialism. The final section examines an equally important facet of the author's personality and scientific practice, namely, the interrelation between his intellectual work and his political activities.

Exploring nature as an accumulation strategy

The concepts of *production of nature*, *production of space* and the ensuing *theory of uneven development* in relation to capital flows were permanent centres of attention for Smith, areas in which his pioneering contributions continue to inspire numerous researchers even today.

An early focus of Neil Smith's research interest was the deconstruction of the concept of nature. Shortly after settling in the United States, he wrote an article with Phil O'Keefe in which they dissected the social and historical construction of the concept of nature, which later provided the basis to understand also the processes of the construction of space. In their article, they noted that, although 'nature' is one of the most commonly invoked concepts in science (natural and social), it has been the subject of surprisingly little critical methodological discussion (Smith and O'Keefe 1980: 30). Rather, the predominant view has been idealised and positivist, in keeping with an age in which industrial development, which required intensive appropriation of nature, was taking off. According to the two authors, the paradox lay in the fact that, at a time when nature was being theorised as something external, the last vestiges of that externality were being practically destroyed: nature was being turned increasingly into a social product (Smith 1984, 2000a). Using Marx's original texts and his views on nature and its relationship to social activity, they introduced the concept of *production of nature*, taking the innovative approach that the mode of capitalist production was not limited to creating and expanding spaces of second nature but also the production of the first nature. Indeed, second nature is not produced outside the first, but rather the latter is produced by, and within, the confines of the former. Thus, the form of all nature has been modified by human activity.

The first question challenged by Smith in his PhD, completed in 1982, was the dualist and static approach to the nature-society relationship. He explored the notion that the positivist and enlightened perspective masks a highly conservative vision that legitimises the domination of nature by capital. Accordingly, he proposed replacing the notion of *domination* with *production of nature*: capital redefines the initial unity between nature and society through the production process. Under capitalist production conditions, the main issue becomes *how* we produce nature and *who* controls said production.

For Smith, the idea of space as a universal receptacle in which objects exist and social processes take place required extensive review. In his view, the unity between society and space is also defined – as in the case of Nature – via the notion of production (*production of space*). The social processes of the transformation and production of nature and space do not follow chaotic models but rather are situated along two major lines in capitalist development: the tendencies towards equalisation and differentiation (Smith and O'Keefe 1980, Smith 1982). By examining the dialectics between the two tendencies, and their verification on concrete scales (urban, global and nation-state), he constructed an *uneven development* theory based on what he suggested calling the *'seesaw' movement of capital*. Uneven development is thus the geographical expression of the contradictions inherent in the logic of the functioning of capital (Smith 1984: xi). Underdevelopment, like development, occurs at each of the different spatial scales. Capital moves from a developed area to an underdeveloped area and back in a seesaw motion in order to avoid the tendency towards a fall in the rate of profit.

As Smith himself was to put it two decades later, 'the central argument was that the specific logics of capital accumulation embody equally specific geographies of economic expansion and that the unevenness of geographical development – between developed and underdeveloped areas on different spatial scales – owes not to geography in the old sense (differences in geographical endowment) but to the inherent logic of economic expansion per se' (Smith 2003: xxii).

Although he initially considered that his theory posed certain limitations at a regional and, above all, global level where the aforementioned seesaw movement, which generated development and underdevelopment, appeared to manifest itself less clearly (Smith 1984: 153), subsequent capitalist dynamics rectified these considerations and made his arguments even more relevant and valuable.

Various authors (Castree 2000: 269, Swyngedouw 2000, Harvey 2010: ix–x) have noted that the theory of uneven geographical development demonstrates that capitalism is an essentially geographical project and that space is central to the survival of capital. For this fundamental reason, any emancipatory programme opposing it must also include an alternative geographical project.

In later years, and as he did for all his fields of research, Smith continued to probe the interrelationship between nature, society and capital. He became increasingly aware that a new dimension of capitalist production was opening up that drastically transformed not just nature itself but also our social relationships within it. In 2007, twenty-five years after the completion of *Uneven Development: Nature, Capital and the Production of Space*, his masterpiece according to Harvey, he published an article under the very meaningful title of 'Nature as Accumulation Strategy' (Smith 2007a) in which he warned that 'we are currently living through a period in which the core socio-economic relationship with nature is being dramatically transformed'. He drew attention to the emergence of a new generation of 'ecological commodities': protected natural areas, wetland mitigation, carbon sequestration and other storage programmes, and CO_2 emission quotas. The conversion or otherwise of these commodities into raw materials for subsequent transformation is irrelevant since their value lies in the fact that *they cannot be consumed productively*. In many cases, they do not even require manufacture, since they are reinserted or remain embedded in socialised nature. Smith warns of the special importance of such phenomena, given that they radically intensify the penetration of capital in nature and become the main strategy for the commodification, marketisation and financialisation of the environment. The production of nature is mutating from a more or less incidental and fragmented reality into *a systemic condition of social existence*, from a local oddity into a global ambition with major spatial, economic and political consequences (Smith 2007a).

Another subject which frequently attracted Smith's attention was a specific aspect of nature, namely certain catastrophic events. In the joint article published with O'Keefe in 1980, he had already made a very early foray into the topic, the theoretical merit of which lay in the linkage he established between the notion of so-called 'natural' disasters and the different conceptions of nature. He also urged focusing attention to the analysis of the *uneven human vulnerability to risks*: 'exposure to nature produced under capitalism is universal, but the consequences of this exposure are far from uniform' (Smith and O'Keefe 1980: 37).

Another of his major contributions in this field arose from the devastation caused in New Orleans by Hurricane Katrina and was significantly titled *There's no such Thing as a Natural Disaster.* Here, Smith argued that in all phases and dimensions of natural disasters considered inevitable, the extent of the disaster is largely the consequence of a social and *political* process. In effect, Katrina demonstrated that the chances of surviving a disaster – not just in the Third World – depend essentially on membership of a given race, ethnic group or social class. The vulnerability of the population of New Orleans and many other parts depended not just on physical geography, but rather, differentiating initial social conditions, brutally exacerbated by the dismantling of the welfare state in this historic era of neoliberal hegemony. They are just as decisive in the long run, if not more: 'There's no such thing as a natural disaster; and the supposed naturalness of the market is the last place to look for a solution to this disastrous havoc' (Smith 2006).

Gentrification, revanchism and class violence

The study of the phenomenon known globally as gentrification is the best-known facet of Neil Smith's research in Spain. The development of the gentrification process in cities attracted his interest from his university days and he realised early on that it was a manifestation at an urban scale of a broader spatial phenomenon intricately linked to capitalism: *uneven development*. His 'rent gap theory' and the evolution of urban dynamics towards what he coined the *revanchist city* – the outcome of class violence processes – marked a watershed both in the study of gentrification and in analyses of the production of urban space.

Neil Smith was a *rara avis* from his earliest days as an academic. Doherty (2013) recalls how, as a student at the University of St. Andrews (Scotland), he wrote two substantial works that were to determine the direction of two of his main research interests. One was his study of the gentrification of Society Hill in Philadelphia, the subject of an undergraduate dissertation in 1977.

A few years later, while studying for his PhD at The Johns Hopkins University, he published his first theoretical approach to the subject. In *Toward a Theory of Gentrification: A Back to the City Movement by Capital not People* (1979a), he challenged the urban models of the Chicago School with his customary irony (Slater 2012: 8). In contrast to the hegemonic explanation based on the preferences of *individually considered* consumers, Smith argued that it was vital to factor in the *collective* action of the agents of capital when seeking to interpret the transformation processes of gentrified districts. He focused his attention on the exploration of Marxist thought and, in particular, a direct reading of Marx in a bid to understand the importance of capital flows, which did not feature in explanations of gentrification at that time. As he himself noted, his main challenge was to insert a sense of 'historic spatiality' into a social theory in which space was conspicuously absent (Smith 2010b: 97–98). The identification of the existence of *potential ground rent* in depressed central neighbourhoods as a mechanism driving gentrification processes allowed him to introduce urban spatial differences in the explanation, linking suburbanisation and gentrification (in other words, the simultaneous processes of *valorisation, devalorisation and revalorisation*) as expressions of uneven geographic development.

This theoretical argument marked a true revolution in an academic world dominated by the theories of urban filtering and revitalisation. As correctly noted by Cowen (2012) and Mitchell (Mitchell and ARC 2013), an examination of the course of the prolonged debates on gentrification clearly shows not just the impact of Smith's works on the scientific community but also how they have strongly influenced the direction and content of research.

Against the backdrop of the consolidation of postmodernist perspectives, 1996 saw the publication of *The New Urban Frontier: Gentrification and the Revanchist City*, considered Smith's key work on the subject, in which he collates, extends and refines his previous research (Doherty 2013: 2–3). In response to previous debates, he used the book to also contextualise the dialectic between economic and sociocultural processes and offer a new perspective, which, far from underplaying the importance of cultural phenomena, seeks to provide consistent material underpinning cultural theories.

From the title onwards, Smith uses American frontier imagery, a metaphor for 'westward expansion', to draw a simile with the approach to gentrification in the English-speaking world: from pioneer terminology to the urban renaissance that transforms decadent centres. An imagery of conventional American culture, which has a far from innocent ideological charge and rationalises differentiation and social exclusion beneath the halo of a natural, inevitable process. In Smith's words, 'the frontier ideology justifies monstrous incivility in the heart of the city' (Smith 1996: 18). Above all, however, the frontier discourse has served to legitimise a process of 'conquest' or, to be more precise, the 'economic reconquest' of already developed spaces – a manifestation of uneven development, therefore – using an internal differentiation strategy, transforming them into enclaves for the affluent (Smith 1996: 22–23).

The book disentangles the logic underpinning disinvestment practices by financial capital, in other words, disinvestment 'as an integral dimension of the uneven development of urban space' (Smith 1996: 299). Without overlooking the historical roots of many specific gentrification processes, such as the case of New York's Lower East Side, Smith draws out the links between local transformations at neighbourhood level and global capital circulation flows. He highlights the centrality of urban development for extended capital circulation processes, which places gentrification at the vanguard of a transformation aimed at 'the class remake of urban centre landscape': a remake that produces 'a redifferentiation of the cultural, social and economic landscape, and to that extent one can see in the very patterns of consumption clear attempts at social differentiation' (Smith 1996: 39).

In his heartfelt tribute, Tom Slater (2012: 7–8) clearly set out the reasons which, in his opinion, led Neil Smith to investigate a theoretical and convincing explanation for gentrification. In effect, Neil understood gentrification as a process of class violence, characterised by conflict and combative resistance on the streets (as in the iconic Tompkins Square Park), but also as a process of equally serious *symbolic violence*, present in the theoretical arguments of the pro-gentrification camp. It was against this violence that he waged his battle of ideas, deconstructing the terminology of the 'pioneers' (which argued that 'no one lives in the areas being pioneered – at least nobody worthy of notice') and the terminology of 'revitalisation', when 'it is often [also true] that very vital working-class communities are culturally devitalized through gentrification as the new middle class scorns the streets in favour of the dining room and bedroom' (Smith 1996: 32–33). From his earliest works, he consistently highlighted the importance of the eviction of the popular classes as a constituent element of the process and denounced the striking absence of such analysis in dominant mainstream approaches (Smith 1982, Smith and Williams 1986, García Herrera, Smith and Mejías 2007, Smith 2008).

Moreover, the book sets out a key argument: the constitution of what Smith coined the *revanchist city*, which he considered a significant trait of American urbanism: a city built on a blend of revenge and reaction where the *reconquest* of the urban centre becomes a strategy for revenge against the working class. In 1986, he had already predicted a tendency towards socio-geographic polarisation and its foreseeable increase and had drawn attention to a 'siege mentality' on the gentrification border (Smith and Williams 1986: 219). It was precisely this revanchist mentality that drove public order discourse, the obsession with security and the resulting fight against crime – together with cuts in social spending – that encouraged and justified police actions in cities. The 'zero tolerance' policy of the New York mayor, Rudolph W. Giuliani, was its clearest exponent and was later touted as the model for numerous cities in the United States and beyond.

The *revanchist city* phenomenon is not accidental, according to Smith, but is rather part of a new regime that displaces the geographical dynamic of uneven development from national and regional economies towards the connection between global and local. Hence his investigations from 1996 onwards focused on the change undergone by gentrification from an anomalous local process to a generalised urban strategy (Slater 2012: 11–12).

In effect, at the start of the 21st century, gentrification, by now an instrument of public policy, is firmly established as an urban strategy on a *world scale*, intertwined with capital and global culture (Smith and DeFilippis 1999, Hackworth 2001, Smith 2001, 2002: 427, Atkinson 2003, Gotham 2005: 1115). Neil Smith was highly critical of urban regeneration policies promoted and funded by public administrations (such as the European Union) and of public interventions geared towards 'social mixing'. He showed that such discourses anaesthetise our understanding of gentrification, particularly if they neglect to point out that the creation of this mix always entails the movement of the middle class to working class neighbourhoods, never the reverse (Smith 2002: 446).

Geography and imperialism

This third section examines the close ties between geographic thought and practice; between geography, politics and economic expansion in the specific case of American globalism.

The history and philosophy of geography received early attention from Smith, who by 1979 had already published *Geography, Science and Post-Positivist Modes of Explanation* (1979b), an article in which he addressed in greater depth the second of his substantial undergraduate works in Scotland. It was also an early indication of his insistence that academic geography can never – and should never – seek to place itself above the changing socioeconomic and political contexts (presenting facts as if such contexts were of no importance whatsoever).

Smith held that geography needed to be critical and incorporate political economy and class analysis. Convinced of the need to rewrite the history of the discipline (Smith 2005a: 265–266), as he had already shown with a controversial article on the closure of the Department of Geography at Harvard in 1948 (Smith 1987: 167), he continually reflected on the internal deficiencies that made geography so vulnerable. He offered his most complete response in his book *American Empire: Roosevelt's Geographer and the Prelude to Globalization*, published in 2003 after a lengthy gestation of over two decades: a book in which the biography of geographer Isaiah Bowman, whose career and influence on American Foreign policy he had investigated since his PhD years in Baltimore, served as the backdrop to uncover the specific nature of American imperialism (Kearns 2012: 955).

American Empire is much less theoretical and abstract than *Uneven Development*, yet, through its reflection on the life of one person, it embodies the practical application of many of the theoretical considerations contained in Smith's first book. Mitchell and ARC (2013: 35) correctly noted that it offers a complex and lucid double narrative. On the one hand, it tells the life story of Bowman, his rise to fame, his contributions to the shaping of the foreign policy of Wilson and Roosevelt (and by extension to the peculiar deployment of US imperial ambitions), and above all his central place in the framing of 'the equivocal role of geography in this emerging American Empire' (Smith 2003: 2). On the other hand, it narrates the history of said Empire with great accuracy.

The examination of Bowman's life and his involvement in key moments of the 20th century allow Smith to show that behind the so-called 'American Century' lies not a weakening of Geography but a radical mutation towards a different geographic project (Smith 2003: xviii). In Smith's words, 'the American Century, understood as a specific historical period, was built with an equally specific but largely unseen geography' (Smith 2003: xvii), a geography for which Bowman was a prime architect. Thus, he critically rewrites the development of American geography as a discipline in the 20th century and shows how its history is intertwined with the expansion of the American Empire throughout the century.

Regarding the particular US form of imperialism, Smith details its historical geography[1] during the different stages of construction (1898–1919 and 1945[2]), discerning two clear phases: an earlier geopolitical colonialism, grounded on territorial expansion, and a later economic imperialism, which did not involve direct political occupation of territories. This enables him to arrive at his central thesis, namely, the *priority of geo-economics over geopolitics*. The American Empire is characterised primarily by the fact that power is exercised in the first instance via the world market and only in the second instance – at specific points in time if necessary – in geopolitical terms.

In summation, even those most critical of *American Empire* recognised that it marked a significant reorientation of our understanding of the nature of American imperialism and of the ambivalent role of geography – as a material process, fact and discipline – within it (Mitchell and ARC, 2013: 36–37).

Neil Smith's chief response to the criticism of *American Empire* emerged less than two years later with the publication of his fourth and last solo book: *The Endgame of Globalization* (Smith 2005b). Even the title alludes to what he viewed as the culmination of American global ambition, expressed in the invasion of Iraq, which Smith expected to end in defeat (*endgame lost*). He argued that, from a long term and international perspective, the war in Iraq marked the continuity of, rather than a break with, the imperial project driven by both liberals and democrats: The imposition of US hegemony in the globalisation process, which is shaped to suit national interests.

Through a combination of geo-economic and geopolitical analysis, Smith exposes the contradictions of a globalist yet nationalist Empire – in other words, a universalist political discourse (defence of democracy, equality, etc.) that vanishes with the application of overtly protectionist economic policies. In reality, behind the mask of the 'War on Terror', launched following

9/11 and the invasion of Afghanistan, lay a war designed to complete the geo-economic globalism of the American Empire.

Imagining an alternative future

This chapter would be (even more) incomplete if it had omitted to mention the close relationship between Neil Smith's academic and political activities, that is, how he translated radical theory into revolutionary political practice. As Deborah Cowen (2012) notes, he devoted a substantial part of his life to this endeavour that can be seen by the many invitations he received from social movements to talk about his work, thus converting his theories and research into tools of understanding to be used in the struggle for change.

Joe Doherty, his first mentor, made this clear in the following words:

> The production of knowledge was for Neil a political act as much as his participation and indeed leadership in community based anti-gentrification movements in Baltimore and New York, in his forays to the picket-lines to express solidarity with workers struggles and in his engagement more recently with the 'Occupy Movement'. Neil saw no division between his academic activities and political activism, they informed and fed off each other.
>
> (Interview with Joe Doherty, in García Herrera and Sabaté Bel 2015: 60)

From a very early age, he became imbued with the political atmosphere of Scottish socialism (Mitchell and ARC 2013: 3). Later, at Johns Hopkins University, he became involved with the International Socialist Organisation (ISO), a Trotskyist Left-wing movement from which he was expelled in 1984, an event he was to recall with no small irony in later years. His most fruitful and important activism, however, was always related to the battle for ideas, to which he attached great importance ('an issue too important to be left aside', as noted by Mitchell and ARC 2013: 2). For Neil, politics was 'about making connections from person to person and place to place' (Smith and Desbiens 1999, Smith 2000b: 338). He continually urged social movements and struggles to focus their gaze towards the global dimension, on the quest for world organisation as a means to tackle globalised capital.

In the latter years of his career he devoted his political and intellectual efforts to recuperating in the collective mindset, the possibility of revolutionary change, an idea he materialised in both his teaching and his research. Between 2007 and 2011 he published various articles in which he explored the possibilities and uncertainties of revolutionary change, rethinking 'the history and future of the revolution' (Interview with Deborah Cowen, in García Herrera and Sabaté Bel 2015: 74). He intended to culminate this line of investigation in what would have been his fifth solo book, a project which he pursued with the same passion as his previous books but remained unfinished.

In 2009 he published *The Revolutionary Imperative* in which he openly advocated the need to recuperate and broaden political imagination as a prior step to conceiving the possibility of revolutionary change. The urgent need to overcome the inability to imagine a different world requires, in his view, an intellectual effort to press, from a grass roots level, for democratic traditions, institutions and practices situated outside or on the margins of hegemonic politics. It also requires the use of democratic imagination to look beyond the immediately visible (Smith 2007b, Aguiló 2014).

Given the current empowerment of the dominant global class and weakness of emancipatory forces, Neil argued that a policy of progressive changes that lacks the ambition to bring about revolutionary change is in itself unrealistic (Smith 2007b: 193). In this regard, his stance coincides with that of Henri Lefebvre who, four decades earlier, expressed the same idea, arguing that his reflection was part of an 'urban reform with a revolutionary bearing'.

Smith also took up Marx's idea concerning the *endogenous* nature of revolutions, in the sense that capitalism constantly creates its own revolutionary conditions. He also considered that there is not just one single possibility of revolution, circumscribed to the protagonism of the industrial proletariat. Rather, he advocated a rethinking of the traditional definition of working class, which need not be based on an immutable historical category but can be widened to incorporate aspects such as gender, race, nationality, or sex (Smith 2007b: 192). In fact, he had already anticipated this idea almost two decades earlier: 'any successful revolutionary movement [...] will be a revolutionary feminism, and a revolutionary antiracism and probably a lot more depending on place and time' (Smith 1991: 416).

The influence of his thinking among social movements demonstrates just how exceptional Neil Smith was. The many articles penned as tributes are testimony to this: Max Rameau, of the Miami-based Take Back the Land group, with which Neil had particularly close ties, notes that the socio-political analyses and campaigns by the group to successfully oppose gentrification (occupation of land and empty or foreclosed homes, forms of defence against evictions) are explicitly based on the 'rent gap' theory coined by Neil Smith (Rameau 2012: 951–952). Don Mitchell (Mitchell and ARC 2013: 40–41) describes how Smith's analysis of Hurricane Katrina was used by urban activists; Margit Mayer (2012: 690) recalls his longstanding links to the Berlin activists and the important role of the Right to the City discussion group, created with David Harvey and Peter Marcuse in the Center for Place, Culture and Politics at CUNY; Jeff Derksen has highlighted his permanent commitment to the Downtown Eastside Neighbourhood activists in Vancouver. Similarly, Smith worked closely with the Harlem Tenants Association, the network of organisations grouped under The Right to the City banner and the Occupy movement. He was also involved with the resistance organised to preserve the community gardens in New York. Equally important were his involvement in Canada with the struggles of Aboriginal populations and his role in suburban activism and university battles (interview with Deborah Cowen in García Herrera and Sabaté Bel 2015: 73–75).

In his commitment to social and community causes, and with the empathetic candidness of his theoretical reflections, Neil Smith consistently defended the importance of *understanding in order to act*. He encouraged all activists to examine the deepest socioeconomic roots of the problems they faced in order to devise an adequate reversal strategy. The need to integrate academic debate and political reflection as two facets of a common endeavour is arguably one of the most important legacies.

Notes

1 Neil Smith stressed that an understanding of the workings of the capitalist economy required an understanding of the spaces that made it possible, achieved through a reading of historical geography (Smith 1986: 88 and 92).

2 First phase: The years leading up to and around World War I, with W. Wilson and his global *Monroe* doctrine. Second phase: The years leading up to and around World War II, with Franklin D. Roosevelt and his New World Order.

References

Aguiló, A. (2014). 'Radicalizar la imaginación democrática'. Available at: www.eldiario.es (accessed 26 January 2014).

Atkinson, R. (2003). 'Introduction: Misunderstood saviour or vengeful wrecker? The many meanings and problems of gentrification', *Urban Studies*, 40(12), pp. 2343–2350.

Castree, N. (2000). 'Classics in human geography revisited. Smith, N. 1984: *Uneven development: nature, capital and the production of space.* Commentary 2', *Progress in Human Geography*, 24(2), pp. 268–271.

Cowen, D. (2012). Contribution in 'Neil Smith: A critical geographer', *Environment & Planning D: Society and Space*, 30, pp. 947–949.

Doherty, J. (2013). 'Neil Smith, 1954–2012: The future is indeed radically open', *Urban Geography*, 34(1), pp. 1–4.

García Herrera, L. M. and Sabaté Bel, F. (2015). *Neil Smith. Gentrificación urbana y desarrollo desigual.* Barcelona: Icaria Editorial.

García Herrera, L. M., Smith, N. and Mejías Vera, M. A. (2007). 'Gentrification, Displacement, and Tourism in Santa Cruz de Tenerife', *Urban Geography*, 28(3), 276–298.

Gotham, K. F. (2005). 'Tourism gentrification: The case of New Orleans' Vieux Carré (French Quarter)', *Urban Studies*, 42(7), pp. 1099–1121.

Hackworth, J. (2001). 'Inner-city real estate investment, gentrification, and economic recession in New York City', *Environment and Planning A*, 33(5), pp. 863–880.

Harvey, D. (2010). 'Foreword', in N. Smith (ed.). *Uneven Development. Nature, Capital and the Production of Space.* 3rd edition, London: Verso. pp. vii–x.

Kearns, G. (2012). Contribution in 'Neil Smith: A critical geographer', *Environment and Planning D: Society and Space*, 30, pp. 955–957.

Mayer, M. (2012). 'Neil Smith: A tribute from Berlin', *City: analysis of urban trends, culture, theory, policy, action*, 16(6), pp. 689–691.

Mitchell, D. and Advanced Research Collaborative, Graduate Center, City University of New York (2013). *Neil Smith, 1954–2012: Radical Geography, Marxist Geographer, Revolutionary Geographer.* Available at: www.maxwell.syr.edu/geo/faculty/On_Neil_Smith/.

Rameau, M. (2012). Contribution in 'Neil Smith: A critical geographer', *Environment & Planning D: Society and Space*, 30, pp. 951–952.

Slater, T. (2012). 'Rose Street and revolution: A tribute to Neil Smith (1954–2012)'. Available at: www.geos.ed.ac.uk/homes/tslater/tributetoNeilSmith.html.

Smith, N. (1979a). 'Toward a theory of gentrification. A back to the city movement by capital not people', *Journal of the American Planning Association*, 45, pp. 538–548.

Smith, N. (1979b). 'Geography, science and post-positivist modes of explanation', *Progress in Human Geography*, 3, pp. 356–383.

Smith, N. (1982). 'Gentrification and uneven development', *Economic Geography*, 58(2), pp. 139–155.

Smith, N. (1984). *Uneven Development. Nature, Capital and the Production of Space.* 2nd edition, Oxford: Basil Blackwell, 1990. 3rd edition, Athens: University of Georgia Press, 2008.

Smith, N. (1986). 'On the necessity of uneven development', *International Journal of Urban and Regional Research*, 10(1), pp. 87–104.

Smith, N. (1987). '"Academic war over the field of geography": The elimination of geography at Harvard', *Annals of the Association of American Geographers*, 77, pp. 155–172.

Smith, N. (1991). 'What's left? A lot's left', *Antipode*, 23(4), pp. 406–418.

Smith, N. (1996). *The New Urban Frontier. Gentrification and the Revanchist City.* London: Routledge.

Smith, N. (2000a). 'Classics in human geography revisited. […] Author's response', *Progress in Human Geography*, 24(2), pp. 271–274.

Smith, N. (2000b). 'Afterword: Who rules this sausage factory?', *Antipode*, 32(3), pp. 330–339.

Smith, N. (2001). 'Nuevo globalismo, nuevo urbanismo', *Documents d'Anàlisi Geogràfica*, 38, pp. 15–32.

Smith, N. (2002). 'New globalism, new urbanism: Gentrification as global urban strategy', *Antipode*, 34(3), pp. 434–457.

Smith, N. (2003). *American Empire. Roosevelt's Geographer and the Prelude to Globalization.* Berkeley: University of California Press.

Smith, N. (2005a). 'Geographers, empires and victims: A response', *Political Geography*, 24, pp. 263–266.

Smith, N. (2005b). *The Endgame of Globalization.* New York: Routledge.

Smith, N. (2006). 'There's no such thing as a natural disaster', *SSRC Perspectives from the Social Sciences.* Available at: http://understandingkatrina.ssrc.org/Smith/.

Smith, N. (2007a). 'Nature as accumulation strategy', *Socialist Register*, 43, pp. 16–36.

Smith, N. (2007b). 'Another revolution is possible: Foucault, ethics and politics', *Environment and Planning D: Society and Space*, 25(2), pp. 191–193.

Smith, N. (2008). 'On the eviction of critical perspectives', *International Journal of Urban and Regional Research*, 32(1), pp. 195–197.

Smith, N. (2009). 'The revolutionary imperative', *Antipode*, 41, pp. 50–65.

Smith, N. (2010a). *Uneven Development: Nature, Capital, and the Production of Space*. 3rd edition, Oxford: Basil Blackwell.

Smith, N. (2010b). 'Box Three', in L. Lees, T. Slater and E. Wyly (eds.). *The Gentrification Reader*. London: Routledge. pp. 97–98.

Smith, N. and DeFilippis, J. (1999). 'The reassertion of economics: 1990s gentrification in the Lower East Side', *International Journal of Urban and Regional Research*, 23, pp. 638–653.

Smith, N. and Desbiens, C. (1999). 'The International Critical Geography Group: Forbidden optimism?', *Environment and Planning D: Society and Space*, 17(4), pp. 379–382.

Smith, N. and O'Keefe, P. (1980). 'Geography, Marx and the concept of nature', *Antipode*, 12(2), pp. 30–39.

Smith, N. and Williams, P. (eds.) (1986). *Gentrification of the City*. London: Allen and Unwin.

Swyngedouw, E. (2000). 'Classics in human geography revisited. [...] Commentary 1', *Progress in Human Geography*, 24(2), pp. 266–268.

3 The apocryphal diary of Neil Smith

Fernando Sabaté Bel

In the opening lines of his well-documented and vivid tribute article published in 2013, the year after the Scottish geographer passed away, Don Mitchell points out that 'Neil Smith hated hagiography'. I do not know if the same holds true of personal diaries, i.e. whether Neil Smith also harboured a hatred for such periodic and systematic narrations of the events of one's own life, designed to be consulted by oneself or by others at a much later date. Smith was a prolific and versatile writer and publisher, regularly engaging with burning topical issues (which almost always served as a good excuse for an in-depth examination of matters of substance, 'long-wave' phenomena). It seems rather unlikely that someone who was already setting out his concerns on a daily basis in influential scientific journals or in widely read media articles should devote time to doing the same in a diary designed to be read later.

Nevertheless, in this contribution, I shall allow myself the liberty of imagining a diary kept by Neil throughout his life and extracting from it various fragments reflecting key moments of his life and typical aspects of his thinking and personality. Please note this is a work of fiction, hence the use of the term 'apocryphal'. After reading Smith's academic work in great detail and reflecting on my many memories (and those of others) from my direct dealings with him, in some remote corner of my brain's emotional and intellectual zones I came across the notes for this, his (presumed) diary. As the editor, I take full responsibility for the content and exonerate the hypothetical author from any possible (and unintended) misrepresentation of his life and work.

Neil Smith's apocryphal diary (selected extracts)

Dalkeith [Scotland], 20th January 1974:

> *How I love the physical geography of my homeland! At St. Andrews I have come to understand the origins of Scotland's glacial and volcanic landforms. I now have the conceptual tools to understand the processes and stark natural contrasts that have fascinated me since childhood: the*

classic 'crag and tail' of Edinburgh Castle and the Royal Mile, the extinct Arthur's Seat volcano, the magical stretch of water known as the Firth of Forth, the Luffness and Gullane sandhills, the basaltic prisms of Bass Rock. Yet, without losing a single ounce of my love for geography of nature, I would also like the University to equip me with effective tools to understand the reasons underlying the divides separating society, the profound and ever-increasing contrasts between people, the wretched countries of the earth that have been abandoned to their fate as if they were beach sediment carried by the world's political system, the steadfast columns of social movements that resist such harm, and how the capitalist system erodes their strength and cohesion.

Dalkeith, 1st March 1974:

Yesterday was the final day I was supposed to hear news about my research scholarship application to the US Department of Education to study the spread of new cereal storage techniques in the Midwest. No reply! Damn bureaucrat did not even take the trouble to acknowledge my application. [Editor's note: 1. Some expressions in this (and other) extracts of the diary have been deleted or toned down, in line with the publisher's instructions. 2. Next to the above is a note in the margin written in different ink (and no doubt later), in which the author seems glad that his first application to study abroad was rejected. The note reads:] *In the end I went to Pennsylvania for the other scholarship: the deep and very visible contradictions of its urban world persuaded me once and for all of the close and inseparable ties between geography and politics. To the bureaucrat who thwarted my original research project, wherever you may be, thank you!*

Philadelphia, 2nd February 1975:

The Philly air was cold this morning even for a hardy Scot like me! Not even the six inches of snow, however, can hide the massive social contradictions in the central Society Hill neighbourhood, where a transformation process is fully under way with increasing involvement by the city and state governments, it would appear. It seems clear to me that, if the theoretical models explained at university were fulfilled to the letter, half of Chicago's emerging city centre should not be there but submerged somewhere under the waters of Lake Michigan.

Even a cursory study of the urban and social history of this area is enough to help realise its symbolic importance for American history and identity: The big influx of labour in the early part of century, the early civil rights battles of black immigrants from the 1950s onwards, the even earlier abandonment of the centre for the suburbs, and Mayor Clark's policies coupled with the information displayed on the advertising boards of large modern public-private construction projects. The editorials and news – all above full-page ads – you find in dailies such as The Philadelphia Inquirer *as well as other indications and factors such as the weakness of the more organised local working class following the decline of the manufacturing industry, in marked contrast to the arrogance and empowerment with which local elites and major financial corporations appear to be acting again. All these factors combine to form a puzzle illustrating the thriving process of urban gentrification, which is closely linked to social class processes and battles (in which the upper class, needless to say, has the upper hand), along the lines described by Glass many years ago in London. This might prove to be an interesting research topic, certainly more exciting than the spread of farm technologies.*

Saint Andrews, 16th March 1976:

This morning I left in Joe's [Doherty] *pigeonhole a draft of the possible paper I discussed with him last week, based on my field observations and archived materials from Philadelphia's Society Hill neighbourhood. I also left him loads of pieces of paper with my notes and comments on my recent readings of the theory of dependence and the dynamics of underdevelopment: Gunder Frank, S. Amin, Baran, Rodney, etc. They are all very stimulating texts which, in my opinion, all suffer from the same original defect: Space is viewed merely as a static and neutral receptacle as regards the logic of the development of capital. All the dialectic capacity mobilised by Frank and Co. when addressing the social and economic factors grinds to a halt or, to be more accurate, vanishes at this point. I think some of the valuable intuitions, if not lucid and direct suggestions, put forward by Marx in key passages of* Das Kapital and Grundrisse *need to be taken seriously (but without sacralising or fossilising old Beardy's ideas). Some who call themselves his followers ~~are clueless~~* [Editor's note: Deleted in the original] *do not use or take to their ultimate consequences Trotsky's fertile approaches concerning the dynamics of 'uneven and combined development'.*

I will need to find time between case studies to look at this topic monographically. Better still, combine and articulate specific empirical cases with a more general theoretical model. Joe will no doubt give me more helpful pointers. I am driving him mad with my requests! But I am sure it will be worth it.

Baltimore, 3rd May 1979:

I have been at Hopkins [The Johns Hopkins University] *for nearly two years now.* It has been *an extraordinary learning and discussion experience with David* [Harvey]. *My occasional contact with some of the people from philosophy such as Donna* [Haraway] *and historians such as Pocock, among other faculties, have been inspirational. The Baltimore ghetto next door is a permanent reminder of the social contradictions, not to mention an incentive to reflect on their causes and dynamics. Other than that, this is all still a little boring. It gets on my nerves to see so many engineers as well as economic and financial analysts who are incapable of rigorously interpreting what is staring them in the face. They could really do with a bit of serious study of critical social theory, beyond the ~~fucking~~ well-worn Chicago School! I catch the bus to Worcester tomorrow to meet up again with my geography colleagues from Clark* [Clark University] *and other universities. We will use the opportunity to plan our contributions to the next conference of the Association of American Geographers together. Let's see if we can stir it up again like last year! Leaving aside some people's inability to take criticism, the battle for ideas in the academic world is too important to be neglected.*

We will also be holding the fourth meeting of the new team at the Antipode journal. Peet rang a few days ago to tell me that he had a serious proposal for me.

Baltimore, 28th January 1980:

I started work this week at the JHU library on the 'secret' Bowman archives. [Editor's note: The reference is to the geographer and, for many years, President of The Johns Hopkins University, Isaiah Bowman; Neil puts 'secret' in inverted commas and offers an explanation straight after:] *I had to submit documents signed by David Harvey stating that I am the assistant and he the director of a research project on the Philosophy of American Geography. It's almost as if Bowman anticipated there would be a radical shift in the Social Sciences some years after his death and he instructed that his files should be accessible only to researchers aged 40 and over. It's just as well David was 44! Looked at with critical rigour, the life of this guy can prove fascinating, not so much because of his actual life but for his conservative theoretical approaches and his excessive – although near-invisible for the public at large – influence on US foreign policy, which reshaped the world in the aftermath of the two world wars. There is enough material for another two PhDs at least. Luckily (?) I have already started mine. Still, it would definitely be worth paying more attention to Bowman in the coming years.*

New York, 11th September 1981:

I chatted about this to Phil [O'Keefe] *at Clark the other day. I am finding the same hard evidence in every one of these case studies: The social processes of the transformation and production of nature and space do not follow chaotic models but are rather situated along two major lines in capitalist development, the tendencies towards equalisation and differentiation. They do exist, although on different scales and at different times. A knot that would be really interesting to untangle is the relationship between these scales. It is not proving easy but I think that I am finally managing to detect a common thread in all this theoretical investigation.*

New York, 18th January 1982:

What a tough week! I'm exhausted but happy. I saw David [Harvey] *again this evening at Penn Station when he was stopping off on his way back from New Haven on one of his regular visits to Yale. We discussed the draft of another of my thesis chapters, which I had given him on Tuesday at the station when he was on his way to Yale. He made full use of his time on the train, as always! Loads of notes in pencil. All his criticism is based on rigorous analysis. However, the comment that most stuck in my memory was one he made a long time ago when he looked over my first outline proposal for the PhD: 'Neil, I thought you were planning to do a case study on urbanisation and gentrification, but this takes you in the direction of a very serious theoretical and critical investigation of the general production of space. I think you realise what you are getting into, your work aims to expand and complete materialist philosophy in one dimension (space), which few of us have worked on until now.' Neither do I forget my rather sarcastic reply: 'I made the long trip to Baltimore for a reason, or rather for a person, Professor.' He went on: 'I have every confidence in your intelligence and creativity, Neil, but I hope that your many political and social commitments don't prevent you from pursuing such a mammoth challenge to the very end.' Thinking back to all that, and on this one occasion only, today I am going to bed early without calling into the bar.*

New York, 19th June 1984:

I have just been informed of the decision taken by the party's disciplinary committee. I have been kicked out of ISO! My clamours for more feminist radicalness and for the focus to be kept on the fight against racial contradictions in our programme and strategic principles were totally defeated on the management committee, losing out to the 'return to the class initiative'. Even on the streets of Harlem, where we have repeatedly pounded by handing out propaganda and organising assemblies; it appeared to be an unconquerable bastion of anti-racist emancipation, the heart of Black America, another gentrification offensive is under way and poor Afro-Americans are constantly being expelled! ~~¡Cocoons!~~ What sheer blindness and excessive dogmatism. [Editors' note: ISO is the International Socialist Organisation, a Trotskyite political party of which Neil Smith was a member for several years after moving to the United States. Here we again see that Smith occasionally revisited his notes, at times much later, to add reflections inspired by hindsight. An undated marginal note states as follows: *This is still the only time in my entire life that I have been kicked out of somewhere rather than me leaving first.*]

Nonetheless, I am grateful to my comrades for the opportunity, for six intense years, to learn to work hard and with many people in the trenches of labour and internationalist conflicts and battles against all kinds of discrimination. Without them I would not have had the chance to discuss and study in detail the thinking of Emma Goldman, S. de Beauvoir, Perkins Gilman, and similarly interesting people, including as C.L.R. James, Angela Davis or Malcolm X.

My contradictions between politics and academia are over. Politics waged in and from academia! Freedom of action and thought, but with full methodological and theoretical rigour. One thing is clear: Academic geography can never – and should never – seek to place itself above the changing socioeconomic and political contexts. Theoretical frameworks such as those of Althusser that often present facts as if such contexts were of no importance whatsoever are of no use to me. Concrete analysis of concrete reality, open-mindedness and a combination of spatial and temporal scales for the ultimate (but permanent) goal: revolution.

New Jersey, 22th September 1987:

I received another letter today from a Geography Department, this time from the Secretary of Johns Hopkins no less. He's new and I don't know him, but he writes to me like a fossil revived from Bowman's days. Ever since the summer when I released an article in which I offered a critical review of the controversial closure of Harvard's Geography Department in 1948, I have been inundated with letters, some of them congratulatory, the majority indignant and critical [Editor's note: The reference here is to '*Academic War Over the Field of Geography*', published in the *Annals of the Association of American Geographers*]. *At the end of the 1940s, all the Harvard affair explanations were overly simplistic, for example that Derwent Whittlesey, one of the main teachers, was a homosexual, not to mention the list of gossip about and idiosyncratic traits of the other protagonists of that scandal. It's like reverting to a historiography in which the course of history is determined by the personal character of kings, princes and bishops! It didn't just happen at Harvard, there were other closures, some just as important, during that period of restructuring general geographic configurations (and of the discipline that studies them).*

Tempers are frayed again. Since Reagan became President and began to follow the education programme enacted by Thatcher a short time earlier in the UK, many geography departments have been closed (Northwestern, Chicago, etc…) using low productivity as an excuse. I was even caught up in the closure of Columbia! (where I was engaged in intense scientific production). Why do away with geography departments, then and now? Can this be understood from the perspective of a purely descriptive and decontextualised history of US geographical thought, based on anecdotes and always on the defensive? It is crucial that we go beyond this impoverished

> *analysis once and for all and replace it with an adequate understanding of the internal shortcomings that make the discipline so vulnerable. At the same time, and as a matter of urgency, a political dimension needs to be factored into the analysis given that it was precisely this weak and acritical geographic knowledge that propitiated the construction of the United States' geo-economic Empire.* [Editor's note: As noted above, all underlining is by the author.]

New Jersey, 6th August 1989:

> *I have just been to a rock concert held at Tompkins Square to commemorate the riots and awful repression in the Park exactly a year ago. The anti-gentrification struggle on the Lower East Side continues but I have no doubt that sooner or later the horrific Rakowitz case will be seized on by the council to turn the city against the neighbourhood and intensify full-blown ideological and economic revenge. I will use part of tomorrow's class in Rutgers to remind the students of the park's role as a scenario and symbol of social and revolutionary struggles in New York since it was built: The struggles of poor immigrants in 1857, the massive conflicts at the height of the economic and financial crises of 1874, the harsh repression of the socialist organisers in 1877, the anti-Vietnam War protest gatherings in the 1960s and 1970s, and so on, up to the present day. Among the many emancipatory women militants who visited the working class Lower East Side was Emma* [Goldman] *the Red, without doubt the person most hated by the right-thinking NY elites. The memory of these struggles is fading and with it disappears the conscience that fighting for grass-roots social change is feasible. If an Encyclopaedia of Revolution is ever written someday, Tompkins Square will surely merit multiple entries. Although what we should really write is a monograph on revolution and its spaces. On the physical and cultural spatiality of revolution.*
>
> *Back to the concert: A number of us who attended had demonstrated our fury and our commitment to the neighbourhood together last year. Cindi Katz, the CUNY professor (I remembered her from my visits to Clark a decade ago), was also there. We said we would call each other.*

Highland Park, New Jersey, 30th May 1993:

> *It has been a Sunday of introspection and personal reflection combined with my domestic hobbies. I'm not really into this Zen stuff but since Cindi and I moved to this house, so near Rutgers University, I devote part of the time saved for travel to enjoying my gardening (horticultural and ornamental), putting up bird tables, making compost with food waste and enjoying the company of the many friends who drop by. These have been some intense and exciting years, and still are: Hectic scientific production, my teaching and providing research advice to the many students who seek*

intellectual (and human) support and guidance, my growing involvement with grassroots social movements, my work with critical artists (Krzysztof Wodiczko and his colleagues never fail to surprise me!), and the formidable cross-disciplinary forum we are consolidating at the Center for the Critical Analysis of Contemporary Culture at Rutgers. That is why occasional leisure and personal fulfilment time, such as what I enjoyed today, does me so much good.

Early this morning, before saying goodbye to Cindi – who is heading off to give a seminar on infancy, social reproduction and spatial restructuring –, I talked to her about the letter I had just received from Geraldine Pratt, who has asked me to co-edit Environment and Planning (Section D of the journal: Space and Society). 'What are you going to say to Gerry?' Cindi asked. 'That I am willing to do it as long as we commit to stopping E&P being the Gucci *of Geography journals. And I will tell her I will write editorials in swift and, where possible, irreverent and provocative responses to concrete problematic issues that arise every so often', I replied. Before a green heron* (Butorides virescens) *flew down to eat sunflower seeds and distracted us from our conversation, I just had time to add: 'Comrade Katz: the battle for ideas has to be waged on all fronts, a very important one being in mainstream journals.'*

Highland Park, New Jersey, 6th June 1993:

Don [Mitchell] *and Susan* [Millar] *came over yesterday, Saturday. The four of us had dinner, taking full advantage of the early summer evening on the front porch. We talked at length about the urgent need to create an international group of socialist geographers. I mentioned the many international invites I have been receiving and the network of left-wing male and female geographers who appear keen to re-emerge at universities and centres for learning almost everywhere. 'We have to help articulate and consolidate this movement', I argued enthusiastically as I downed my second beer, 'But you know the problem? Many of these people see themselves as isolated or as part of small local groups'. At one point Don said laughing: 'Right, let's see how the heck you persuade our colleagues in Hungary, the Baltic Republics and former Yugoslavia, all of them deeply committed to critical geography, to add the name socialist to the network being built. Or those in El Salvador, with all the problems they are having at present.' 'OK', I replied, 'Call it whatever you want but one thing is clear: We have to organise a global response on the same scale as the development of capital in international space. By that I mean how to build the environment to give meaningful organisation to critical policies.'*

Highland Park, New Jersey, 14th April 1996:

I hope to finish correcting the proofs of The New Urban Frontier *this week. I think we have to step up the battle for ideas by focusing on fundamental and carefully-chosen issues. If gentrification exists, which is more decisive, structure or agency? I have spent years battling against approaches that defend consumer preferences as the decisive factor driving gentrification processes. Today I read an article in the* New Yorker *which again argues that 'urban renaissance', that much-used notion, is a reflection of the new tastes and initiatives of a thriving middle class, professionals, designers, creative types, artists, emancipated women, and God-knows how many other groups (not that many more, to tell the truth). Wrong: That does not explain anything fundamental. They are confusing part with whole and turning superficial manifestations into supposedly telluric forces. We need to offer evidence of all kinds showing the hegemonic role of capital and the decisive intervention of the State in the future of urban neighbourhoods in one direction or another (increasingly the same direction: ejecting the poor from city centres and attracting the affluent classes). I hope the book contributes to disseminating the review I have been carrying out of the cultural phenomenon to a much wider audience, recognizing its existence but placing it in its proper context. We shall see how people react to me naming this new urban order revanchist. Hopefully in years to come it will help broaden the consensus regarding the crucial role of class stratification as a fundamental explanation for gentrification.* [Editor's note: Smith's underlining.]

London, 22nd February 1998:

On the train to the airport I recalled once again Marx's old comparison between a private school and a sausage factory: no matter how much the former works with minds and concerns itself with social reproduction, it is essentially no different to the latter, in the sense of being a motor for extended capital circulation and a source of enrichment for a private proprietor.

It is really disheartening to see the transformation of the state and higher-education system by Margaret Thatcher (now pursued with the same, if not more, vigor by Tony Blair's government). My conversations with old and new friends from British universities serve to confirm what is already apparent in the anguished letters written by young lecturers to the Geography journals on whose editorial boards I sit: The job of turning public universities into a commodified space, attractive to investors and shaped to suit business interests has been well and truly completed. Of course, the neopositivist, 'neocritical' and postmodernist scientific production of many colleagues is not helpful in curbing this trend, but rather the opposite. Will they ever learn that a science that is willing to incorporate without critical reflection any supposedly new critical aspect (if that is not a paradox) other than, of course, criticism originating in Marxist thinking, is the perfect culture medium for an acritical insertion into the sphere of capital, for out and out commodification? We need to encourage people to fight something that is in no way a natural process and to help build, what Harvey named, spaces of hope. Spaces where a different form of social organisation and different type of university are possible. [Editor's note: Underlining again by Neil Smith.]

New York, 6th March 2000:

Had a meeting today in CUNY [Editor's note: City University of New York, which Neil joined as a full professor around this time] *with one of my groups of PhD students. I will meet with the others every evening his week. I distributed the materials I had prepared for each student: For Sandra Gonzales I had various books and some postcards I came across in San Diego, which might be of use to her in her research on Hispanic cleaning ladies in NY. For Ariel Zeledon I had a letter signed by the university President to help him apply to the State Department for certain files (which should have been declassified by now) on US support for the Nicaraguan Contra and the effects of air strikes on civilians from the Honduran side of the border. For Julia Molnar I had a collection of posters and photos of the Hungarian revolution of 1956, which I brought from my last trip to Budapest (I still don't know how it emerged relatively unscathed from the Stalinist repression) to help her finish her thesis. Finally for Tom Wilson, who is returning to Jamaica to finish off his work on 'Islands and Globalisation', I had some materials I collected during my visit last year to Tenerife (Canary Islands), which I have not managed to give him until now.*

That fascist Giuliani [mayor of New York] *does not like this 'Proletarian Harvard' paid for by the NY taxpayer one bit. He's doing all he can to denaturalise it: From promoting the arrival of the children of the new rich from the former Soviet Union (at the expense of places for working class New York students) to 'importing' Spanish teachers from Spain(!) in a city where the Latin American Hispanic population is on the way to becoming the majority. It really is institutionalised racism. We need to combat such policies. That is why I always try and persuade students that we have to view the University in a dual perspective: Firstly, as a place for intense and rigorous work and secondly, as a fundamental political battleground; an environment where everyone should be allowed to develop their own ideas and be accountable for the consequences (even if the effects of ideas can often prove difficult to control). However, in times of crisis of emancipatory ideals, the battle for ideas is even more crucial. Let's leave pessimism for better days.*

New York, 6th October 2000:

I went for another walk through the Lower East Side today. I stopped to talk to Sonia Álvarez, a social activist who arrived from Puerto Rico in 1945 and whose community garden was recently ripped up by the council to make way for more luxury buildings. She told me that you don't see many Puerto Ricans on the avenue on Saturday nights anymore, just white Europeans.

I need to book my ticket for my forthcoming trip to Barcelona, another invite, this time for a course at the Museum of Contemporary Art in the heart of the Raval district. I will use the visit to have a look at the surrounding area and see the definitive effects of the new Rambla which, from what I hear, has cut like a knife through the old Chinese Quarter. Time will tell but, unless I am badly mistaken, another full-blown gentrification process is under way (just as in Vienna, Budapest, Paris, Sydney, Buenos Aires… everywhere). 21st century Haussmann reforms are implemented in some places with all the refined sophistication of the social democrat barons (the Barcelona author, Vázquez Montalbán, has described social democratic sophistication brilliantly… have any of his other novels been translated into English?).

Edinburgh, 9th August 2005:

It was great to meet up again over the past few days with Joe Doherty and some former colleagues from that unforgettable class at university, including Charly Withers. A pity Malcolm Forbes couldn't make it.

Our discussions really left an impression on me. Some friends – not Joe – occasionally recriminate me for the excessive ferocity of some of my theoretical and political responses. To tell the truth, there is good reason these days (as there is almost always has been throughout history) to write, speak out and be indignant about the powerful, the liars, and those who encourage suffering for others. The material foundations of the system of domination are very weak. What we also need to do now is weaken the foundations of its discourse, its unchallenged presence in our minds.

Before crossing the Atlantic I finished a very critical article (although well thought through in my head, it was partly the product of my insides) in which I recall Marx's appeal concerning the need for implacable criticism of everything that exists.

Toronto, 29th July 2006:

I had a wonderful day today with an early start (something I have not had the chance to do for ages) bird watching and listening to the twittering of hundreds of birds in the marshes of the Mountberg Conservation Area, not far from Toronto.

Having awakened from my 'quasi-mystical immersion', on my return I remembered some of my earlier reflections and began to think about the radical transformation being undergone by our most basic socioeconomic relationship with nature. This corner of Canada hasn't been affected yet but I am sure that before too long it will be put forward for commodification under the pretext of 'wetland mitigation', just like the new 'ecological' *commodities such as carbon sequestration and storage programmes, CO_2 emission quotas, quota trading in fisheries and other endangered species which are taking hold virtually everywhere. It may be that natural landscapes like Mountberg do not get transformed physically, but that's not the point because, in practice, similarly immaculate places are used to perfection as an exchange value and reinserted under a new form into socialised nature. Increasingly this new 'green' capitalism is penetrating nature through 'ecological' commodification, marketization and financialisation. Nature is increasingly used as a blatant strategy for the accumulation of capital.*

New York, 25th April 2007:

I have just answered an e-mail from Deborah Cowen. She sent me a long quote from the philosopher Kierkegaard along with her own insightful comments. It was on the subject of our recent conversations about intellectual engagement and the notion of revolution. To tell the truth I was not familiar with the Danish philosopher's writings, imbued with religious thoughts, on the phenomenon and consequences of the stagnation of revolutionary thinking when situations of social regression become prolonged over time. Very interesting!

The production of knowledge is a political act. Political practice should be rooted in specific places, albeit retaining its wide-ranging and relational character. All these years and to the best of my ability I have tried to engage extensively with international activity by being present in many places, debates and struggles for space. I try and be consistent with my deeply-rooted belief in the need to stand up to capitalism through a global struggle that situates anti-capitalist organisations at the same level at which capital operates itself. This is why I consider it crucial for social battles and movements to adopt a global dimension and seek to build a global organisation, which is the only effective means to combat globalised capitalism.

It is vital that we place the issue of revolution back on the agenda. To do so it is indispensable that we recuperate and broaden political imagination as a first step to even contemplating the mere possibility of revolutionary change. We have to undo the ideological road promoted by Thatcher, Blair, Schroeder and Mitterrand, which they managed to persuade the majority to follow. I am going to focus on that task, both in my teaching (maybe an annual seminar on the subject?) and as a priority area for future investigations and articles.

New York, 30th October 2011:

I have not written anything in this old and battered diary for some time. I don't really know why, but today I wanted to write something before letting my dear colleague Blanca Ramírez know of my decision regarding reducing my participation in and taking some distance from the ICGG [Editor's note: International Critical Geography Group, whose main driving figures included Neil Smith]. *I recall with fondness the 1st International Conference on Critical Geography organised by the newly-created group in August 1997 in Vancouver, as well as all the enthusiasm of the young geographers and of the veterans too. The following conferences were good too although I have some doubts concerning how the group has evolved and I have decided to reduce my involvement in its activities. In exchange, I will feel freer and have more time to forge closer ties with grassroots social movements, which I feel is the main shortcoming of the group and the personal engagement I hope to broaden at present.*

At any rate (and this is what I will tell my dear Mexican colleague), the very existence of the ICG is in itself a success for various reasons: It has kept alive an important core of politics against the hegemony of neoliberal thinking and at times has even gone on the offensive (which is what we need so badly). In addition, it has managed to attract a new generation of young academics and left-wing activists – to give them an always questionable identity label –. There are not that many of them but in terms of quality they are particularly important. Lastly – and perhaps most importantly – it has created the opportunity to make stable and extensive international contacts, contacts which for many are unlikely to have come about were it not for the group.

I never forget the founding slogan of the group: 'A world to win'. We must never allow the spirit expressed by this slogan to become lost.

New York, 17th May 2012:

Today is another one of those days when I find it hard to concentrate to be able to write. Could it be the medication? I hope to muster up some enthusiasm by writing here in the diary.

I have just replied to the French geography students to reconfirm my collaboration. My doctor has advised me to cut down on travel but in the end I thought, sod it! I'm going to go to Paris in September. I have never let young people down and I am not going to start now. I want to take advantage of the trip to interview Yves Lacoste and look into some periods of his life and of the evolution of his geographical thinking on which I am still unclear.

I also need to write to Luz Marina [García Herrera]. *It is incredible, in Barcelona they want to publish a book on my work, part of a collection on critical geographers. I'm flattered but I hope at least that they will be sufficiently critical of me too. I have always hated hagiography.*

PART II

On gentrification and the rent gap theory

4 The state of gentrification has always been extra-economic

Hamish Kallin

We are stood at the end of an unfinished road. Here the wind is strong and unrelenting, whistling in off the Firth of Forth, and piles of architectural detritus frame a project left spectacularly unfinished. This is a landscape scarred by two successive crises: the first being that of deindustrialisation, which hit huge swathes of urban Scotland in the latter half of the 20th century and rendered this harbour quite suddenly devoid of ships; and the second being that of finance capital, which has left a new form of dereliction in its wake. As McClanahan (2014: 209) notes, 'it currently feels as if time, here, is *on hold*.' (Figure 4.1).

This is Granton Harbour, where Edinburgh (the capital city of Scotland) hits the water. At the turn of the millennium, an ambitious 'regeneration' project promised to transform this 'cut-off, polluted, contaminated and forgotten part of [the city]' (Llewelyn-Davies 2000: 2) into 'one of the finest development opportunities in Europe' (WEL 2002: 1). The plans fluctuated, but variably included an artificial island, luxury flats, a marina, a cinema, a world trade centre, museums, shops, luxury hotels, and so on. The blueprint proffered a familiar form of doublespeak: the lofty ideals of 'uniqueness', 'luxury', and 'placemaking' were to be conjured up through identikit buildings in the same style found across the UK's myriad sites of speculative development. And almost none of this was built. A few 'luxury' blocks stand isolated amidst open plots. The future of the rest remains open to speculation (literally).

'But is this gentrification?' asks one of my students. The sun is fading, and it is time to go home. A lull in the wind allows me to answer. Is this what gentrification looks like in your heads? No. Can you talk of a closed rent gap, of revalorised land? No. But is gentrification theory *useful* to grasp the politics of what is happening here, of what was supposed to, and what might still happen here? Yes.

A dose of agnosticism

There are few who have contributed more to the study of gentrification than Neil Smith. It was his insistence on placing the concept within a wider theory

Figure 4.1 Granton Harbour (Edinburgh, Scotland).
Source: Photo by Boyd Goff (2015). Used with permission.

of capitalism that allowed it to enter the lexicon of urban studies beyond the select cities in which it had hitherto been observed. If the specific aesthetics and class politics of house-by-house renovations in post-war north London (so famously described by Ruth Glass as 'gentrification') were observable in a handful of similar cities, the deeper mechanics of what was happening and why were far more broadly recognisable. Through his continuous emphasis on the state's involvement in this process, Smith was also instrumental in conceptualising the evolving forms of gentrification as part of 'a systemic change in the way that the state relates to capital' (Hackworth and Smith 2001: 475). Struggles over urban space were one specific arena in a wider shift towards neoliberalism.

One particular criticism of Smith's work, delivered courtesy of Asher Ghertner (2014: 1568), seeks to challenge the way that gentrification 'has come to operate as an umbrella term for a general outcome of class transformation, regardless of its widely divergent underlying causes and implications'. For Ghertner, the concept remains too closely wedded to a 'Euro-American template'. He cites four reasons in particular why this undermines its usefulness, both analytically and politically, in other contexts: (1) The theory presumes *re*investment into previously *dis*invested areas; (2) presumes that this takes place within an established system of private property; (3) presumes that this results in the land being put to a 'higher and better' use; (4)

and, crucially, remains 'agnostic on the question of extra-economic force' (Ghertner 2014: 1555).

Clearly, 'gentrification' does not suffice to explain every kind of urban development, and, when deployed lazily, it can serve to foreclose the impact of other more pressing forces. The emphasis on *re*investment in the theory does necessitate careful thinking about whether or not it can apply in contexts where the land is undergoing a first round of commodification or enclosure. To answer that is not my concern here, and, more broadly, I make no claims as to what is happening or not happening in Indian cities. Yet the 'gentrification' that Ghertner rejects is that theory reduced to its most simplistic, unhelpful, and depoliticised in *any* context. I want to argue that the state of gentrification has *always* been extra-economic. Challenging the supposed 'universalism' of Western theory is a vital political challenge, but if claims to *difference* are grounded in the notion that extra-economic force is alien to gentrification in 'the West', then these are weak claims to difference. Bernt (2016) has already highlighted that the dichotomy Ghertner presents is intriguing, for it implies that the property markets in Europe and North America function in a way that capitalist fantasies have always insisted they will and must: freely, neutrally, and without violence (for all its myriad flaws, Proudhon's *What is Property?* reminds us that this has never been possible). Fading Hayekian fantasies aside, we wait in vain for liberalism to become 'true'.

I want to take this further by demonstrating both that Ghertner's (2014) representation of 'gentrification theory' is useless to explain the Waterfront Edinburgh project and, at the same time, that this was a project that internalised gentrification at its heart. In a volume dedicated to the legacy of Neil Smith, it seems particularly fitting to reassert that gentrification theory (in his hands) was *never* reducible to 'nothing more than a rising rent environment and associated forms of market-induced displacement' (Ghertner 2015: 552). In place of such an implausible definition, I will reassert the necessity of holding onto a more holistic understanding of the process as one of capital accumulation, prone to the variegations of unevenness and crisis-tendency that this implies.[1] My argument may seem a little contradictory, for gentrification theory in this context cannot ignore the extra-economic forces at work, but a dose of agnosticism might be rather useful. Going in search of a neat delineation between 'economic force' and 'extra-economic force' feels like an increasingly lost cause.

Re-stating the politics of rent gaps

The signs that shout DEVELOPMENT OPPORTUNITY sport two logos, one of something called Waterfront Edinburgh Limited (WEL) and another of something called the EDI Group. The first is owned by the second, and the second is owned by the City of Edinburgh Council (Figure 4.2).

Figure 4.2 Granton Harbour (Edinburgh, Scotland).
Source: Photo by Boyd Goff (2015). Used with permission.

Technically, WEL was set up as a 'private company', whose sole purpose was to make profit. This is a company, however, whose assets (in terms of land and funding) came from the state and whose board of directors was made up entirely of elected representatives. It was, to all intents and purposes, a profit-seeking arm of the local authority. As one senior council employee put it to me,

> Granton had the biggest concentration of social and employment issues in the city [but] the site was absolutely fantastic, you know, you stand in the middle of Granton on a sunny morning looking out across the [water] and you think, *wow*, you know, the sea could really do something here. So we felt that, one, the problems were there, but, two, the opportunities were there as well.

The link between 'the problems' and 'the opportunities' is umbilical. It is precisely the deprivation and dereliction in Granton, its position on the fringe of a city that likes to bask in its bourgeois credentials – in short, the disinvestment over decades, symbolically, economically and politically – that made it an attractive development opportunity. In other words, the rent gap (Smith 1979a) was (and remains) crystal clear here. Adjacent to some of the poorest districts in the city, this was a project steeped in pretensions of grandeur, elitism, and luxury.

Figure 4.3 Granton Harbour (Edinburgh, Scotland).
Source: Photo by Boyd Goff (2015). Used with permission.

Fifteen years after the project was started, however, this vast site looks nothing like the glimmering spectacle its original planners imagined. 'The core strategy', wrote WEL (2002: np) with such certainty, was 'to increase land values to the levels where the aspirations of the Master Plan become viable.' In other words, the goal was rising land values; the mechanism for achieving the goal was rising land values. The hyper-valorisation and commodification of the land was both the chicken and the egg in this scenario, quite clearly showing us how the gentrification process was hard-wired into a governmental strategy set to meet the systemic demands of capital. Those demands, of course, are never stable (Smith 1982). If it was the circulation of the land as fictitious capital that made this project plausible, it was the crisis of fictitious capital that destroyed it. For if you invert the golden goose of land valuation, the plan falls flat (Figure 4.3).

What makes this gentrification, then? Perhaps Maloutas (2012) was right to call this kind of tagging an act of 'conceptual stretching'? This was a project that relied on attracting a more affluent class of residents. It was designed explicitly with that in mind, tied to Edinburgh's then buoyant housing market. In the first five years of this millennium, the cumulative value of the city's residential properties jumped up by some 82% (Mathews and Satsangi 2007: 502), and Waterfront Edinburgh only made sense as an addendum to that rise. For those ostensibly in control of the project (the architects, planners, councillors, company board members and so on with whom my research was conducted) this was not so much a choice as a perceived necessity. How else was anything to be built? Yet, to then say, as Maloutas does, that the process of gentrification risks becoming a mere synonym for 'urban regeneration' (and should therefore not be invoked so readily) is to acquiesce to a model of 'regeneration' that can conceive of nothing else. The history of town planning attests to this being a conceptual surrender to the status quo, whilst the chronic need for affordable housing reaffirms the necessity of overturning that 'common sense'.

In the immediate fallout from the financial crisis, building completely stopped. 'It's funny the way things have changed,' notes Sarah, a planning officer with the local authority, reflecting on the years that followed. When negotiations aiming to impose a quota for 'affordable housing' started, the planning team were fully prepared for a struggle. Post-crisis, 'that never happened, and [...] we ended up having far more affordable housing, because we got to a stage [where] affordable housing operators were the only people who were investing'. Quite clearly, to take someone down to Granton now and say 'look: gentrification!' would be insufficient. Yet the failure of a project should not serve to obfuscate its aims, particularly when those aims are fundamental to the explanation of why the project failed. This development was the victim of a crisis caused in no small way by the very kind of 'real estate' speculation it was seeking to capitalise on. There is obviously a good deal of scale-hopping in that assertion – the crisis was not of WEL's making – and it is only with Smith's (2002) understanding of gentrification as an increasingly, but never totally, global strategy that we can begin to understand how this project was tied into the making of *and* the fallout from a global crisis of capital.

The failure of the project highlights the state's centrality to it. There was no market-led revalorisation at work here and, as it turned out, the local authority could not induce one. If viewed as a purely economistic, predictive trajectory (which it was never intended to be), the rent gap model clearly 'fails' here. It is important, therefore, to re*state* the politics at the heart of the model. The power of the rent gap in this instance was the perception and insistence that there was one (reinforced by no shortage of marketing materials, networking, planning documents, and so on), rather than the 'actual' existence of one. How was this framing articulated? Here the state's role is pervasive.

This is not a perversion of the theory of gentrification, and neither is it an observation that awaited the 'third wave' (as Ghertner erroneously suggests). In Neil Smith's undergraduate dissertation, subtitled 'state involvement in Society Hill', the state and financial institutions are considered alongside each other, as they have to be in a capitalist society. When the bulk of this argument made it into print (Smith 1979b), this could barely have been clearer. 'The state', Smith tells us (Smith 1979b: 166), 'played several roles, acting as an economic, political and ideological agent for the project'. A decade later, Zukin (1989: 17) noted that 'by this late point in capitalist economies probably no real estate market develops without state intervention', and if we were at a 'late point' then, we are at a considerably *later* point now. The volume of work that reinforces the centrality of state power in the process of gentrification vindicates this as an observation that has become empirically more true over time (Glynn 2008, Paton 2010, Gray and Mooney 2011, Hodkinson 2012, Kallin and Slater 2014). An investigation into the way in which gentrification lay at the heart of the Waterfront Edinburgh project necessitates an investigation into state power and the multiple ways in which gentrification is internalised as a governmental strategy.

The idea of 'extra-economic force' derives from Marx's (1990) discussion of 'so-called primitive accumulation'. This takes the form of a *separation* between the producer and the means of production, which, as De Angelis (2004: 67) explains, 'appears as a crystal-clear relation of expropriation, a relation that has not yet taken the fetishistic character assumed by capital's normalisation, or the "ordinary run of things".' In Scotland, the shadow of one particular episode in this process of appropriation is still keenly felt – in poetry, memory, activism, and policy – for it was in the Highlands that Marx found his most gripping example of dispossession from the land. Here the landlords 'wittingly and willingly embarked on a policy of substituting a commercial rent economy for the familial economy of traditional Highland society' (Hunter 1976: 33), and a concerted drive of depopulation followed. The 'so-called' is sardonic: Marx was seeking to detail the untold brutality of this process and the degree to which the respectable veneer of private property was built on violence and theft.

Within more general Marxist thought, debate ensued as to whether or not this was a stage in the history of capital accumulation (one left behind, but necessary to move forwards), or a continuous aspect of its systematic character (Glassman 2006). Harvey (2002), seeking to re-deploy the notion of ongoing 'primitive accumulation' without the historicist baggage, uses the term *accumulation by dispossession* to describe the myriad ways in which new avenues for capital circulation are opened up through, among others, privatisation, commodification, creative destruction, enclosure, and so on. This restless process is symptomatic of a capitalism that is lurching from crisis to crisis looking for new places to invest in and profit from. As Ghertner (2014) correctly identifies, the state looms large in these accounts, as it does in the wealth of scholarship on neoliberalisation more generally. However, the exact distinction between 'economic' and 'extra-economic' force remains elusive.

By way of clarification, Ghertner turns to De Angelis (2004), who argues that the *separation* of the producer from the means of production is a permanent facet of capital relations. Within this, the 'primitive' form is characterised by the use of 'extra-economic force', defined, echoing Marx, as a departure from the 'normalised silent compulsion of economic relations' (De Angelis 2014: 70), and underwritten by force (most often, though not necessarily, the force of the state). Crucially, this is not an assertion tied to any particular sphere of the globe (the 'South', the 'East', and so on), but to all of space where capital relations are dominant. Ghertner (2014) is using a concept grounded specifically in a *general, ongoing* aspect of capital relations to legitimate the sense of an unequivocal 'difference' within that.

Likewise, Harvey (2006) makes clear that accumulation by dispossession is a fundamental aspect of the way contemporary capitalism operates. In the UK, it simply makes no sense to try and explain 'gentrification' *without* paying attention to those wider political-economic shifts (the near-annihilation of the public housing stock, the vilification of that which remains, the privatisation of urban land, the punitive changes to social welfare, an 'affordable housing policy' that pushes up house prices, and so on). This is not to say that 'gentrification' as a term can somehow capture all this, but to reduce the idea solely to rent increases or expect it to stand alone is pedantic and unhelpful. The project in Granton was both accumulation by dispossession *and* gentrification. Through privatising the land and throwing it into circulation as fictitious capital, the project literally seizes the space and commodifies it. Through this, in turn, gentrification was designed to occur.

The silent compulsion of debt

To draw this to a close, I want to briefly touch on the question of debt, and how the division between the economic and the extra-economic becomes extremely difficult to pin down. The generalisation of indebtedness acts as a systemic compulsion to gentrify. Through the financialisation of housing, the more unaffordable houses are, the more profit can be squeezed out of those desperately trying to live in them. The link between house prices and whether or not anyone can afford those prices has been fundamentally and very profitably broken (Rolnik 2013). Horsley (2015: 5) identifies this as one aspect of 'an economic system geared toward the singular purpose of lending as much money as humanly and mathematically possible'. On the one hand, this creates an endless pool of potential 'gentrifiers', with very real ramifications for the processes of class formation. Whilst I do not want to resuscitate the tired production vs. consumption debates within gentrification scholarship, the turn towards generalised indebtedness increasingly implicates the second as a form of the first.

At the same time, however, *the state* is increasingly in debt on a variety of scales. City of Edinburgh Council's outstanding debt now stands at around £1.6 billion and counting (Unite 2015). Behind the straitjacket of neoliberal policy making lies a mountain of debt. All policy decisions have to be made with a view of paying off the interest and keeping the debt levels as low as possible. The rationale grows to irrational proportions, and we are now at a point where 'tackling the debt' means abolishing it, ignoring it, or performing an act of abandon that must be at once radical but deeply pragmatic, for as one councillor pointed out with an air of resignation, 'we're never going to pay it'. Why does this matter? Because, as one of his colleagues pointed out, 'one of the things about councils is they're often land rich and cash poor.' Their main asset, the land, becomes a seductive target for a process of self-imposed asset stripping, akin to the passengers of a sinking boat throwing their possessions overboard.

The near-permanent indebtedness of the state in this context is not exceptional. As Streeck (2014) notes, it is now pertinent to talk of *the debt state*. In this light, accumulation by dispossession becomes an increasingly *normal* process of staying afloat. The privatisation of land and its circulation as fictitious capital is reframed as responsible governance. To not strive for gentrification would not make sense, and it is precisely for this reason that the whole rationality needs overturning. 'False choice urbanism' (Slater 2014) is all we are offered. The state in this instance needs to be theorised as a *capitalist state*, not something separable or detached from the 'ordinary run of things'. In this sense, extra-economic force defines the state's role in gentrification, to create new opportunities for capital investment where 'the market' itself is unable to do so (Weber 2002).

The effects of this 'blurring' are quite real. WEL represent that elusive grey area between something we call 'the state' and something we think of as 'the market'. To quote Smith (1979b: 166) once more, '[t]his happy confusion of ruling class interests was no mere imperfection of an otherwise perfect plan', it was fundamental to it. What is willingly lost is any notion of effective accountability, any pretence of enacting democratic values. The profit is outsourced but the financial risk is socialised. The site down at Granton now bears the scars of a project that failed. Yet through this 'failure', the state has intervened to keep the pendulum of uneven development swinging. The empty plots may look like a project abandoned, but the violence of demolition has done its work. The rent gap remains open.

Note

1 Whilst I do not think too much should be vested in a description of particular neighbourhood change fifty years ago, it is worth noting that even in Ruth Glass' (very brief) description of 'gentrification', which Ghertner (2014: 1556) cites, there is the obvious point that 'this is an inevitable development, in view of the demographic, economic and political pressures to which London, and especially Central London, has been subjected' (Glass 2010: 7). How this can be interpreted as a singular fixation on rent increases as the sole *causal* factor is beyond me.

References

Bernt, M. (2016). 'Very particular, or rather universal? Gentrification through the lenses of Ghertner and López-Morales', *City*, 20(4), pp. 637–644. doi:10.1080/13604813.2016.1143682.

De Angelis, M. (2004). ‘Separating the doing and the deed: Capital and the continuous character of enclosures’, *Historical Materialism*, 12, pp. 56–87.
Ghertner, A. (2014). ‘India’s urban revolution: Geographies of displacement beyond gentrification’, *Environment and Planning A*, 46, pp. 1554–1571.
Ghertner, A. (2015). ‘Why gentrification theory fails “in much of the world”’, *City*, 19, pp. 552–563.
Glass, R. (2010). ‘London: Aspects of change’ in L. Lees, T. Slater and E. Wyly (eds.). *The Gentrification Reader*. Abingdon: Routledge, pp. 7–9.
Glassman, J. (2006). ‘Primitive accumulation, accumulation by dispossession, accumulation by “extra-economic” means’, *Progress in Human Geography*, 30, pp. 608–625.
Glynn, S. (2008). ‘Soft-selling gentrification?’, *Urban Research & Practice*, 1, pp. 164–180.
Gray, N. and Mooney, G. (2011). ‘Glasgow’s new urban frontier: “Civilising” the population of Glasgow East’, *City*, 15, pp. 4–24.
Hackworth, J. and Smith, N. (2001). ‘The changing state of gentrification’, *Tijdschrift voor Economische en Sociale Geografie*, 92, pp. 464–477.
Harvey, D. (2002). *The New Imperialism*. Oxford: Oxford University Press.
Harvey, D. (2006). ‘Neo-liberalism as creative destruction’, *Geografiska Annaler. Series B: Human Geography*, 88, pp. 145–158.
Hodkinson, S. (2012). ‘The new urban enclosures’, *City*, 16, pp. 500–518.
Horsley, M. (2015). *The Dark Side of Prosperity: Late Capitalism’s Culture of Indebtedness*. Farnham: Ashgate.
Hunter, J. (1976). *The Making of the Crofter’s Community*. Edinburgh: John Donald Publisher’s Ltd.
Kallin, H. and Slater, T. (2014). ‘Activating territorial stigma: gentrifying marginality on Edinburgh’s “other” fringe’, *Environment and Planning A*, 46, pp. 1351–1368.
Llewelyn-Davies (2000). *Masterplan for Waterfront Granton: Draft Master Plan*. London: Llewelyn-Davies.
Maloutas, T. (2012). ‘Contextual diversity in gentrification research’, *Critical Sociology*, 38(1), pp. 33–48.
Marx, K. (1990). *Capital, Volume I*. London: Penguin Books.
Mathews, P. and Satsangi, M. (2007). ‘Planners, developers and power: A critical discourse analysis of the redevelopment of Leith Docks’, *Planning Practice & Research*, 22, pp. 495–511.
McClanahan, A. (2014). ‘Archaeologies of collapse: New conceptions of ruination in Northern Britain’, *Visual Culture in Britain*, 15, pp. 198–213.
Paton, K. (2010). ‘Creating the neoliberal city and citizen: The use of gentrification as urban policy in Glasgow’ in N. Davidson, P. McCafferty and D. Miller (eds.). *Neoliberal Scotland: Class and Society in a Stateless Nation*. Newcastle: Cambridge Scholars Publishing, pp. 203–224.
Rolnik, R. (2013). ‘Late neoliberalism: The financialization of homeownership and housing rights’, *International Journal of Urban and Regional Research*, 37, pp. 1058–1066.
Slater, T. (2014). ‘Unravelling false choice urbanism’, *City*, 18, pp. 517–524.
Smith, N. (1979a). ‘Toward a theory of gentrification: A back to the city movement by capital, not people’, *Journal of the American Planning Association*, 45, pp. 538–548.

Smith, N. (1979b). 'Gentrification and capital: Practice and ideology in Society Hill', *Antipode*, 11, pp. 24–35.

Smith, N. (1982). 'Gentrification and uneven development', *Economic Geography*, 58, pp. 139–155.

Smith, N. (2002). 'New urbanism, new globalism: Gentrification as global urban strategy', *Antipode*, 34, pp. 427–450.

Streeck, W. (2014). *Buying Time: The Delayed Crisis of Democratic Capitalism*. London: Verso.

Unite (2015). 'Scottish local government debt', *Unite City of Edinburgh Council Branch*. Available at: www.unitececbranch.org/scottish-local-government-debt.html [accessed 3 May 2016].

Weber, R. (2002). 'Extracting value from the city: Neoliberalism and urban redevelopment' in N. Brenner and N. Theodore (eds.). *Spaces of Neoliberalism*. Oxford: Blackwell, pp. 172–193.

WEL (2002). *A Capital Opportunity*. Edinburgh: Waterfront Edinburgh Limited.

Zukin, S. (1989). *Loft Living: Culture and Capital in Urban Change*. New Brunswick: Rutgers University Press.

5 Gentrification

Disaster, necessity, opportunity? Notes for a critical use of the concept

Francesco Indovina and Oriol Nel·lo

Introduction

Although research into this phenomenon is nothing new, the notion of gentrification has created a great stir in academia in recent years, largely due to the influence of radical British and American geographers, most notably Neil Smith. It can be stated unequivocally that denunciation of the processes of gentrification has already had a positive impact, in both academic and socio-political terms.[1]

The concept of gentrification has been effectively used to label and denounce forced replacement of the population in urban areas, whether through public policies or the mechanisms of the land and housing market or through a combination of the two. The concept has also proven instrumental in explaining how this replacement often exacerbates social inequalities and urban segregation. In the study of gentrification, scholars have shown that the removal of low-income populations from their places of residence usually has damaging effects, notably increased transport costs, loss of community networks and, frequently, the political dismantling of subaltern groups. In contrast, it has also described how other urban agents – such as property owners and new residents – can, in principle, obtain substantial benefits from the same process, in terms of income, accrual of assets and welfare.

In our opinion, however, the application of the term 'gentrification' to a wide range of urban social processes is distorting its scientific meaning, while also contributing to questionable political conclusions. The idea of gentrification as a 'generalised' process (Smith 2006) has led to it being used rather schematically and brought to bear on a whole host of urban processes of very different social significance. From the perspective of public policies, urban policy gentrification has been classified as, alternately, a disaster, a necessity and an opportunity. These simplistic visions have triggered corresponding reactions, ranging from powerlessness to resignation or applause. Those who regard gentrification as a calamity often feel powerless to quash it, despite all their indignation; those who accept it as a necessary evil greet it with resignation; and those who see it as an opportunity, especially for private profit, applaud it and strive to encourage it.

Our contribution herein seeks to offer a critical view of gentrification in the countries of southern Europe, with special emphasis on the examples of two highly significant urban realities: Barcelona and Venice. This critical proposal is based on three initial assertions or assumptions. First, as we see it, gentrification is not always either a necessity or an inevitable disaster, even in the absence of a complete transformation of the capitalist economic system. Second, we shall try to show that the urban transformation processes currently underway do indeed offer opportunities to improve the collective well-being, provided these processes can be controlled through public governance and collective action. The fact that such opportunities are often squandered or overlooked does not in any way mean that they should not be explored and, as far as possible, pursued. Finally, we shall maintain that any proper understanding of the evolution of social structures in present-day urban processes demands an analysis both of common trends and of the specificities that explain the development and implantation of these trends in a particular urban setting.

This chapter comprises three parts. The first describes the main features of urban segregation in our cities and notes the apparent trends in gentrification in our times. The second part shows how the processes of urban transformation currently in progress in many cities in southern Europe present specific and distinctive characteristics, which oblige us to approach with caution the concepts and analytical schemes formulated on the basis of research into other realities. We finish by pointing out how the uncritical and decontextualised use of the concept of gentrification can lead to political conclusions that are mistaken, to such an extent that they can prove counterproductive for the design and application of urban policies for social equity.

Social inequality, urban segregation and gentrification

From the conceptual viewpoint, any analysis of urban gentrification and discussion of measures aimed at confronting it must be put into a wider context: the study of social inequality and its manifestation over the space via urban segregation. Gentrification can be defined as the displacement, through pressure from the property market or planning policies, of sectors of the population residing in specific urban areas, in order to replace them with a population with greater spending power. On this basis, gentrification must obviously be considered an expression of urban segregation, which in its turn, as we well know, is both a cause and consequence of social inequality.[2]

Urban segregation is a phenomenon whereby different social groups tend to separate from each other within an urban space according to their economic power, geographical origins, cultures or other social factors. In European cities, the main vector of segregation is income, which operates by means of a mechanism that has been extensively studied and explained: in an urban context distinguished by the existence of substantial differences in prices and urban quality, the ability to choose a residence is a function of personal and family income (Harvey 1973). In principle, this obliges social groups with the least resources to gather together in places where prices are lower and urban quality – in terms of built environment, public space, housing, services and accessibility – is lower. In contrast, groups with greater resources are able to choose, and end up congregating in places where, in exchange for higher land and housing prices, they can enjoy a high level of urban quality, as well as the advantages derived from living amongst their peers (Secchi 2013).

The literature concurs that in the cities of southern Europe, there has been a trend towards increased social segregation in recent years. Although segregation is obviously a structural phenomenon intrinsic to the process of capitalist urbanisation, the situation created by the economic crisis, from 2008 onwards, has tended to aggravate its effects. This is, to a certain extent, a logical consequence of the deepening of social inequalities, which impinge on a substantial part of the urban population of those countries experiencing material difficulties and deprivations.[3] All this inevitably suggests that a growing proportion of the population is faced with serious problems in gaining access to housing and therefore tends to concentrate in areas with lower prices, as well as greater deficiencies and scarcity of services, with a subsequent impact on the image of these neighbourhoods, which often contributes to further decline in prices.

As an example, Barcelona's history has been distinguished over the last two centuries by notable inequalities and social conflicts, as a result of its emergence and consolidation as one of southern Europe's main industrial centres. In the closing decades of the 20th century, however – coinciding with a rapid transformation of the city's economic base and significant metropolitan expansion – social equalities in the Barcelona area were curtailed. This was largely due to the mechanisms of the Spanish welfare state, implemented since the late 1970s, and it was accompanied by an alleviation of the effects of urban segregation. This relative reduction in segregation was visible both on a metropolitan scale – in the differences between the average incomes of the metropolitan rings and the city itself – and within the municipalities that form the urban area (Trullén 2014, Subirats 2015). Since the start of the economic crisis, however, there has been a striking increase in inequality, accompanied by a marked rise in segregation and a social polarisation of the metropolitan area (Nel·lo and Donat 2014). Accordingly, the gap between the incomes of different neighbourhoods and municipalities is widening and an ever-higher number of census tracts in the metropolitan area are presenting extreme values with respect to income distribution. It should be pointed out that this situation is due not only to the concentration in specific neighbourhoods of social groups with lower incomes but also to a trend among wealthier social groups to separate themselves from the rest of the population (Nel·lo and Blanco 2015, Nel·lo 2016).

The situation in Venice, while sharing the main features of the common process, also displays its own specific characteristics. In fact, due to its geographical peculiarities, the city has become a 'saturated' urban space with limited possibilities for new construction. Since the end of World War II, this phenomenon has combined with the changing lifestyle of its inhabitants and the rising cost of living in the traditional city (everyday expenses, property maintenance, technological updates of homes, transport, etc.) to trigger massive emigration towards the mainland. More particularly, migrants have moved in to areas on the outer edges of Veneto, creating specific types of low-density settlements (Indovina 2009). This phenomenon is directly connected with the intensive development of tourism in the city centre, which has sparked notable interventions to transform property and alter its use, along with a clear process of gentrification (DAEST-IUAV 1984–1990, Bianchini 2006, Bragato et al. 2008). This is particularly apparent in the upsurge in offers of accommodation, both at the top end of the market – through modifications to the historic heritage of palaces, especially on the Grand Canal – and in bed and breakfasts, which have often brought together housing units that had been previously separated. There has also been a boom in luxury shopping, to such an extent that Venice now boasts more gift shops than Milan. All this has set into motion the mechanisms typical of the gentrification processes, which, in principle, have been unaffected by the economic crisis (on the contrary, the flow of tourists is growing continually, partly due to the very deepening of inequalities taking place on a world scale).

Barcelona and Venice are dissimilar in some ways, due not only to their different sizes but also to their economic specialisation, their place in the network of world cities and the policies that have been applied to them. Nevertheless, as we have seen, they share common elements as regards the evolution of the urban segregation process – elements which, in our opinion, are also found in many other European Mediterranean cities, and are primarily distinguished by:

a The structural nature of the urban segregation phenomenon, which clearly predates the economic crisis.
b Intensification of the separation of the social groups within the urban area.
c Potential territorial polarisation resulting from an increase in the areas presenting extreme indicators of segregation.
d The trend towards segregation on the part of both the most underprivileged and the wealthiest groups.
e The metropolitan scale of the phenomenon: segregation takes place not only between different neighbourhoods in a single municipality but also between whole municipalities within a metropolitan area.
f The significance of touristic uses of urban spaces as a triggering factor in spatial segregation.
g The coherence and territorial cohesion of segregation, at both the top and bottom ends, which in both cases present clear territorial continuities (Nel·lo 2016).

These are the distinguishing features of what Éric Maurin (2004) called the advance towards the *société de l'entre-soi*. In this context, when considered specifically as the replacement in a particular area of a population with one income level by another with a higher income level, how do gentrification processes evolve?

We offer three hypotheses. The first is linked to a possible *loss of part of the 'autochthonous' social base of the gentrification processes*. The latter, as we know, have an economic basis in land speculation and construction, but it also needs a social basis: a dynamic, emerging middle class set on improving its housing conditions and multiplying its assets. In previous economic phases, this portion of the population enjoyed an ever-increasing income, drove consumption and tried to improve the quality of its homes. The gentrification processes offered these sectors, on the one hand, better housing (heavily supported by tax benefits and policies that provided greater access to credit) and, on the other, renovated urban areas, in which they could identify themselves and find themselves reflected, as well as finding the forms of consumption and social relationships to which they aspired. These social groups have clearly been affected by the crisis and have lost a large part of their capacity to 'disrupt' the property market.

The second hypothesis focuses on the *trend in wealthier social groups towards separatism*, which, as we have seen, has now become one of the main vectors of segregation. From a spatial viewpoint, this does not seem to be creating isolated islands of gentrification but rather seems to be spreading like a spot of oil. The rent gap, the opportunity arising from the contrast between the rent currently paid for land and what could be paid for it, seems to be found more these days at the boundaries of, and adjacent to, districts that have already been consolidated as residential areas for wealthy social groups than in other neighbourhoods. In many of these peripheral areas, especially those with a concentration of lower-income residents, it is not only wrong to talk of gentrification, it is also an impertinence. The processes of population exchange found here are the exact opposite: groups with an income level below the mean for the neighbourhood come in, while those with a higher income leave.

Finally, *the investment capacity that could give rise to gentrification processes seems, at the moment, to have been severely undermined*. This is true for both private enterprise, which faces enormous difficulties in obtaining credit and is confronted by an accumulation of heavy losses, and for the public sector, where the capacity to invest in infrastructure and urban renewal has plunged dramatically on account of its reduced income, as well as an increasingly stringent application of measures designed to reduce public debt. As we shall see, the conjunction of these three factors – lowered expectations of middle-income groups, a trend towards separation among the wealthy and limitations on public and private investment – implies that the engines of the gentrification processes in the cities of Mediterranean Europe are nowadays highly specific and distinctive.

For a critical and rigorous use of the concept of gentrification

In this context, we consider it essential to avoid any mechanical transposition of imported models or interpretations when analysing the processes of urban segregation and, more particularly, gentrification. In our opinion, there are five issues essential to any understanding of how urban segregation and gentrification are currently operating in southern European cities: the importance of tourism, the prevailing tenancy regime, the role of transnational economic agents, the action undertaken by citizen movements and the intervention of public authorities. We shall examine these five issues below, albeit, of necessity, in somewhat schematic terms.

First, the significance of *tourism* must be highlighted with respect to the phenomenon of urban transformation and, more particularly, possible segregation processes. Focusing on the two realities that we know best – Venice and Barcelona – we can infer that, in many cases, the prevailing population replacement processes are being triggered not so much by activities aimed at permanent residence in the residential market but, above all, by the impact of tourism. The well consolidated development of the property market in Venice and the rapid transformation of some neighbourhoods in recent years in Barcelona (e.g. Barceloneta, el Born) are excellent examples of tourism's capacity to disrupt a housing market. It is true that both cities represent extreme cases of the impact of tourism, but it must be recognised that the new forms of residential tourism, such as B&Bs and Airbnb, have an extraordinary ability to insinuate themselves into the property market and expand precipitously.[4] In the context of a highly fragmented property market, property owners originally living themselves in the affected neighbourhoods are often driving the processes of transformation, as they believe that the change in the use of their home provides them with a resource to support themselves. Thus, although the existence of intermediary companies and the progressive concentration of the ownership of tourist apartments and policies on public space all undoubtedly play an important role in accelerating the process, gentrification through *touristification* does not fit into models that attribute population displacement solely to actions taken by big finance and property investors. For this very reason, this type of gentrification is extremely difficult to manage, both technically and politically.

Second, any understanding of the scope and risks of gentrification processes requires a close analysis of the *housing tenancy regime* prevailing in each city. Gentrification processes do not present the same patterns or the same possibilities of thriving in a single context heavily dominated by an ownership regime (as is the case in many Spanish and Italian cities) as they do in another context with a greater rental regime predominance (as is found in many American cities).[5] Nor will these processes act in the same way in neighbourhoods distinguished by a massive presence of rented public housing (as in many French and Dutch cities). In each of these contexts, policies providing facilities and public space will have different effects on the property market: in cases where rented public housing is predominant, the impact on prices will be negligible or non-existent; in cases with a majority of private home owners, such measures will increase the assets of residents, while in a context characterised by a preponderance of private rental they will probably trigger processes of population replacement. One of the main shortcomings of many studies of urban gentrification in our countries has been the actual importation of the concept, and the arguments that have accompanied it, without examining the crucial issue of tenancy regimes.

Third, it must be borne in mind that, on account of the economic crisis and the restricted access to credit, the capacity of local agents in the cities of southern Europe to influence the processes of urban transformation has been weakened. In the light of this sluggishness, the study of gentrification processes must pay particular attention to the *impact of transnational capital flows*. It must be appreciated that in some southern European countries, lack of public and private investment is bringing down land prices in large areas within cities, as well as leading to deterioration.[6] These circumstances can be exploited by local and, above all, transnational capital to gain privileged positions in the land and housing markets, and this may have important implications for the future. The segregation and gentrification of our cities in 2025 is being forged in today's land markets, and the intervention of transnational actors is having a growing impact on these developments.

Fourth, we have to consider that the agents who transform a city are not confined to public authorities or land owners and capital holders. As we well know, every city has its *citizens movements*, organised to a greater or lesser degree, which fight to improve the living conditions of their neighbourhoods and maintain and expand their collective heritage (common assets, public assets, social practices). The actions of these groups have a real effect on living conditions, whether by directly supplying services or through winning improvements that are later implemented by the authorities.[7] In the context of a free land market, such improvements inevitably impinge on property prices and could propitiate displacement processes or barriers to the incorporation of new residents with a lower income. Should we therefore condemn as gentrifiers the residents of the Carmel neighbourhood in Barcelona, who, under Franco's dictatorship, used their own resources to build a sewage system? Are we to imagine that the committees 'Save Venice' or 'Against the Waves' in the canals of the city, or the citizens committed to return to the Arsenal to the city, are all movements that promote gentrification? Certainly, in the absence of appropriate policy government their actions could have unexpected effects, but this would not be the fault of the movements but due to the inanity of politics.

Finally, an accurate analysis of gentrification and its implications demands a *nuanced critical analysis of the nature and actions of the public authorities*. It is the case that, generally speaking, property laws in almost all countries favour land ownership and private development, in keeping with the misguided belief that 'when construction goes well, the economy thrives' (a statement emphatically proved false by recent developments in the economies of the United States, Ireland, Spain and other countries). In this situation, it is certainly possible that urban renewal initiatives give rise to gentrification processes, with all the deleterious effects outlined above: for example, the debate that is so prevalent in Italy and Spain about 'urban voids' (factories, military barracks, warehouses, etc.) and the 'great opportunities' that these provide usually omits the fact that for some people these 'vacuums' are 'full of rent'. It is also true, however, that public administration is inevitably involved, in the midst of contradictory social interests. In this context, areas of the administration (particularly, but not exclusively, councils) can sometimes be taken over by political forces disposed to improving the city in favour of more disadvantaged groups. The resulting policies can prove relatively successful, by both improving living conditions and reducing inequalities in income (from land and earnings).[8] The radical condemnation of any urban improvements – via programmes for rehabilitation, transport, sanitation or facilities – as sources of gentrification would lead to immobility on the part of the public authorities and render them unable to instigate any urban policy at all, because this would inexorably affect land and housing prices. If we push this reasoning to its limits, any urban intervention (by public authorities or social organisations) designed to improve the living conditions of subaltern groups through the physical transformation of the built space would be inadvisable for fear of unintended consequences – at least until the private ownership of land, or the whole capitalist system, is eliminated. This is not only scientifically unsound but also misguided in political terms. Full espousal of this viewpoint would lead, for example, to criticizing any urban improvement measure implemented by the political forces emerging from the social movements that have taken over the councils of some of Spain's major cities (Madrid, Barcelona, Cádiz, Valencia, Coruña, Santiago de Compostela, etc.) since May 2015.

Can society be changed without changing cities?

Precisely because it is not clear when we are going to emerge from this crisis – or rather, because it is not clear if we are going to emerge from this crisis in the current economic system – commitment to the city is vital. As a major social construct and a focal point for inequalities, but also a breeding ground for alternative policies, the city must be the central element of any project to improve living conditions. The first requisite for the creation of projects with this aim is a correct understanding of urban dynamics, as conceptual transpositions and ignorance of the realities on the ground can result in ill-advised and even counterproductive policies.

We have tried to explain how the processes of urban segregation are currently of foremost importance in the cities of southern Europe, and how one of their most striking expressions is the phenomenon of gentrification. In the present circumstances, this tends to display specific forms and behaviour.

In effect, the decline in the living standards of the middle-income social layers, the tendency of the more prosperous sectors to separate themselves and the loss of investment capacity (both public and private) mean that gentrification processes are mostly being triggered by other agents. In cities like Barcelona and Venice, these are primarily tourism and the actions of international financial capital. It must be realised, however, that their effects are largely mediated by three main factors: the structure of the real estate market, the role of urban movements and the intervention of the public authorities.

It can therefore be inferred that urban policies and mobilisation by citizens can be vitally important to cope with the processes of segregation and gentrification – and this is indeed the case. Consequently, such processes can be held back or prevented, to a certain extent, by the public governance of urban transformations. This requires that the transformative impulse is not exclusively entrusted to economic market forces or, even less so, to speculation. A city is a collective asset, resulting from the contribution of many different agents, and for this very reason its transformation must be inspired by a collective vision and governed by public policy. The latter must define where, how and when transformations must take place, and for what purposes, by creating new opportunities and encouraging any decisions (including private ones) that are in line with established objectives, while impeding any others that contravene those objectives.

The notion of a 'collective vision' should, however, be more than just a generic motto. It has to be imbued with *content*, in the sense, for example, of making clear that, generally speaking, criticism should not be directed at processes that renew parts of a city but rather at their objectives, to ensure that they are not intended for private profit rather than improvement of the collective well-being. Similarly, the collective vision must be imbued with *form*, because in urban politics, forms take on a double significance: on the one hand, the *forms of decision-making and administration* reflect both the democratic quality of the processes and the joint creation of policies by the authorities and the citizens, while, on the other hand, the *physical forms of the city* strongly influence the social processes that unfurl within it.

By way of conclusion: on the one hand, we must base the analysis of the causes and effects of the urban segregation processes in our cities on rigorous empirical data, without in any way relinquishing the appropriate critical postulates; on the other hand, we must strive to conceive and drive citizen movements and public policies capable of guiding and leading urban transformation. Only in this way we will be able to avoid both schematics that reduce gentrification to a disaster, a necessity or an opportunity, and their political counterparts of powerlessness, resignation and applause.

The prospect of a profound shift in the prevailing property relationships – particularly the private ownership of urban land – is not only desirable but also possible. In the meantime, however, what is required is an unremitting demand for, and application of, urban policies geared towards improving the lives of the most under-privileged population, because, among other considerations, such policies form some of the steps that need to be taken towards more far-reaching transformations. Accordingly, concerns over gentrification must not bring progressive governments and urban movements to a standstill – in fact, on the contrary, such concerns must encourage them to increase their efforts to fight for public governance of urban transformations.

It is impossible to transform society as a whole merely by changing the city, but changing the city is an indispensable prerequisite for progress towards a more democratic and equitable society.

Notes

1 For an overview of the conceptualisation of the term and the debate around it in the United Kingdom and the United States of America, see Leeset al. (2010). For the dynamics of gentrification on various continents, see Lees et al. (2016). As regards Neil Smith's contribution to the debate, see García-Herrera and Sabaté (2015).

2 There are various sources of data on increases in social inequalities in the countries and cities of southern Europe. For wealth distribution on a global scale, see the reports of the Credit Suisse Research Institute: *Global Wealth Report* (2014) and its main source, Credit Suisse Research Institute: *Global Wealth Databook*, (2014) Still on an international scale, but with a particular focus on income distribution, the OECD provides useful data: *Income inequality update. Rising inequality: youth and poor fall further behind. Insights from the OECD Income Distribution Database* (2014). For the evolution of inequalities in Spain, see Fundación FOESSA, *VII Informe sobre la exclusión y desarrollo social en España* (2014). For a comparison of the European situation with the evolution of the situation in the United States of America, see Someiller and Price (2015). For the study and consequences of the evolution of the inequalities of the present historical period, see three major works: Stiglitz (2012), Piketty (2013) and Atkinson (2015).

3 With respect to the impact of the crisis on southern European cities, see, for the Spanish case, Albertos and Sánchez (2014) and Méndez (2014); for Italian cities, Fregolent and Savino (2014); for Greek cities, Brekke et al. (2014); and, finally, for Portugal, Seixas (2013).

4 With respect to the impact of tourism on our cities, in its traditional forms and its newer expressions, see, for Barcelona, Arias (2015) and, for Venice, Indovina (1988, 1997 and 2003) and Scaramuzzi (2004 and 2009).

5 In most countries in southern Europe, around three quarters of families live in an ownership regime (Italy, 73.1%; Greece, 74%; Portugal, 74.9%; Spain, 78.8%), while in the United Kingdom and the United States of America this proportion falls short of two thirds (UK, 64.8%, USA, 64.5%) (Data *Eurostat* and *USA Bureau of Census* for 2014).

6 According to the Spanish Ministry of Public Works, for example, the mean price of urban land fell by 42.5% between 2007 and 2013 (from 280.6 €/m^2 to 161.5 €/m^2). See Vaquer (2015) in this respect.

7 With respect to the urban and territorial movements in Veneto and Barcelona, see Fregolent (2014) and Nel·lo (2015a and 2015b).

8 In fact, this transformative action ties in, to a large extent, with the best tradition of European city planning, which in many cities succeeded, in the words of

Bernardo Secchi, in making '[...] the material conditions in which rich and poor choose to live, or are induced and constrained to live, less distant from each other than their respective incomes and assets' (Secchi 2013: 38). With respect to the specificity of the European metropolis – the result, to a great extent, of the strength of social movements and the public policies they have achieved – see also Indovina (2014).

References

Albertos, J. M. and Sánchez, J. L. (eds.) (2014). *Geografía de la crisis económica en España*. Valencia: Universitat de València.

Arias, A. (2015). 'Tu a Boston y yo a Airbnb. Un análisis crítico de Barcelona', *La trama urbana*. Available at: https://latramaurbana.net/ [accessed 14 July 2016].

Atkinson, A. (2015). *Inequality. What can be done?* Cambridge: Harvard University Press.

Bianchini, R. (2006). 'Venezia nel 2030: una città vuota niente abitanti ma solo turisti', *La Repubblica*, 25th August.

Bragato, S. (et al.) (2008). *Dal disagio abitativo al mercato privato dell'abitazione: i nuovi termini della questione casa-immigrati*. Venezia: Veneto Lavoro.

Brekke, J., Dalakoglou, D., Filippidis, C. and Vradis, A. (2014). *Crisis-scapes: Athens and Beyond*. Available at: Crisis-scapes.net.

DAEST-IUAV (1984–1990). *Osservatorio casa a Venezia*. Venezia: DAEST-IUAV.

Fregolent, L. (ed.) (2014). *Conflitti e territorio*. Milan: Franco Angeli.

Fregolent, L. and Savino, M. (eds.) (2014). *Città e politiche in tempi di crisi*. Milan: Franco Angeli.

García-Herrera, L. M. and Sabaté, F. (2015). *Neil Smith. Gentrificación urbana y desarrollo desigual*. Barcelona: Icaria.

Harvey, D. (1973). *Social Justice and the City*. London: Edward Arnold.

Indovina, F. (1988). 'Turisti, pendolari, residenti', *Co.S.E.S Informazioni*, 32–33, pp. 27–36.

Indovina, F. (1997). 'Venezia, opera d'arte deperibile', *L'Illustrazione Italiana*, 94, pp. 20–29.

Indovina, F. (2003). 'La salvaguardia di Venezia. Problema e opportunità', *AreaVasta*, 6–7, pp. 84–94.

Indovina, F. (2009). *Dalla città diffusa all'arcipelago metropolitano*. Milan: Franco Angeli.

Indovina, F. (2014). *La metropoli europea. Una prospettiva*. Milan: Franco Angeli.

Lees, L., Slater, T. and Wyly, E. (2010). *The Gentrification Reader*. New York: Routledge.

Lees, L., Shin, H. B. and López-Morales, E. (2016). *Planetary Gentrification*. Cambridge: Polity.

Maurin, É. (2004). *Le ghetto français. Enquête sur le séparatisme social*. Paris: Seuil.

Méndez, R. (dir.) (2014). *Atlas de la crisis. Impactos socioeconómicos y territorios vulnerables en España*. Valencia: Tirant lo Blanch.

Nel·lo, O. (2015a). *La ciudad en movimiento. Crisis social y respuesta ciudadana*. Madrid: Díaz & Pons.

Nel·lo, O. (2015b). 'Movimientos urbanos y defensa del patrimonio colectivo en la región metropolitana de Barcelona', *Ciudad y Territorio. Estudios Territoriales*, XLVII(184), pp. 309–325.

Nel·lo, O. (2016). 'Desigualdad social y segregación urbana. Una reflexión a partir del caso de Barcelona', in A. Orellana (ed.) *Urbanización planetaria y la (re) construcción de la ciudad: Teorías y prácticas*. Santiago de Chile: RIL Editores, pp. 287–318.

Nel·lo, O. and Blanco, I. (2015). *La segregació urbana a la regió metropolitana de Barcelona*. Barcelona: Pla Estratègic Metropolità de Barcelona.

Nel·lo, O. and Donat, C. (2014). 'Los efectos territoriales de la crisis en la región metropolitana de Barcelona', in J. M. Albertos and J. L. Sánchez (eds.). *Geografía de la crisis económica en España*. Valencia: Universitat de València, pp. 565–608.

Piketty, T. (2013). *Le Capital au XXIe siècle*. Paris: Seuil.

Scaramuzzi, I. (2004). 'Venezia 1951–2001: un film da montare', *L'Architettura. Cronache e Storia*, 590.

Scaramuzzi, I. (2009). 'La città dove è sempre Carnevale', *Dialoghi Internazionali Città nel Mondo*. Milan: Bruno Mondadori.

Secchi, B. (2013). *La città dei ricchi e la città dei poveri*. Bari: Laterza.

Seixas, J. (2013). *A cidade na encrucilhada: repensar a cidade e a sua politica*. Lisbon: Afrontamento.

Smith, N. (2006). 'Gentrification generalized: From local anomaly to urban "regeneration" as global urban strategy', in M. S. Fisher and G. Downey (eds.). *Frontiers of Capital. Ethnographic Reflections on the New Economy*. Durham: Duke University Press, pp. 191–208.

Someiller, E. and Price, M. (2015). *The Increasingly Inequal United States of America. Income Inequality by State, 1917–2012*. Washington: Economic Analysis Research Network.

Stiglitz, J. E. (2012). *The Price of Inequality: How Today's Divided Society Endangers Our Future*. New York: W.W. Norton and Company.

Subirats, M. (2015). *Barcelona: de la necessitat a la llibertat. Les classes socials al tombant del segle XXI*. Barcelona: l'Avenç.

Trullén, J. (2014). 'Crisi econòmica, creixement de les desigualtats i transformacions socials', in IERMB, *Informe general. Enquesta de Condicions de Vida i Hàbits de la Població de Catalunya, 2011*.

Vaquer, M. (2015). 'Alcune chiavi (di lettura) dell'equidistribuzione dei benefici di oneri nel diritto urbanístico spagnolo', in D. D'Orsogna (ed.). *Perequazione urbanistica. Materiali per la comparazione giuridica*. Turin: Giappichelli, pp. 93–121.

6 Toward a theory of gentrination

Global capital flows and the reshaping of the global semi-periphery. The cases of Greece and Brazil[1]

Antonis Vradis

In John Berger's 1972 novel, *G.*, the life of the story's main protagonist is conveyed to the reader almost entirely by proxy, through the narrations and stories of those who would find themselves close to him at one time or other. In recent times – and in our non-fictitious but ever-so dreamlike world – the fiscal crisis that stemmed from the capitalist core (and which has hit particularly hard on the semi-periphery since) was narrated largely in the absence of its protagonists, their plight articulated through a narrative and narrators robustly fixed in capital's pre-existing decision-making locations: a plight that by now concerns near-complete national territories and their populations. Greece in particular has acquired a kind of a negative cause célèbre status: *be careful or you may end up like the Greeks* seems to be the mainstream mantra, a story drawn and then articulated through and out of the lens of the core, pointing at the troubled semi-periphery. The story of this principal protagonist of the fiscal crisis is conveyed almost entirely by proxy in the prolonged wave of the current capitalist crisis, G is for Greece.

From the moment George Papandreou (Greek PM at the time) announced his government was to enter into a loan agreement with 'troika' (IMF/EU/ECB) in May 2010, the level and pace of change that has befallen virtually all life aspects in the country (from the economic to the social, from the outright political to the cultural) has only just fallen short of a society at war. From the rapid dismantling of whatever social provisions existed earlier on to severe wage cuts, galloping unemployment, vast privatisation schemes, the abandoning of state infrastructure and even a sharp increase in suicide rates, the situation in Greece has rapidly come to resemble a battle-zone. To be more precise, a battle-zone exactly *after* the battle has succeeded; a state and a territory in tatters. This is the long moment of post-catastrophe and post-argument that people seem to be living through now, now that the memoranda are long-integrated in the daily reality of the country, and now that ever-so ironically, the same force that was decrying the carnage of the memoranda has stepped in as an unorthodox saviour, an administrator of the wreckage left in their midst.

What caused this carnage in the first place? It was not a war (not in a battle sense, at least) that hit the Greek territory. Now, six years into what commenced as a fiscal crisis, it seems as if we are still grasping for words that will adequately describe this violent readjustment of capital. Yet it has long become evident that we will have to look beyond the fiscal sphere if we are to find a vocabulary allowing us to comprehend what is truly going on. This crisis is no longer about an economy gone awry, if it ever was so. What we stand witness to instead is little less than a seismic transformation of the entire core-periphery relationship of the now rampant neoliberal model. Yes, this has been argued before, most recently by Fraser (2013) who in her otherwise excellent analysis claimed that Greece had by now been reduced to a 'protectorate'. The need to develop a theory of 'gentrination' comes precisely in response to a need to look beyond a simply linear (even if transgressive) reading of human history: The colonial times are behind us, and, just like with all its other aspects, history will never repeat itself.

Something else is going on, and sure enough, there is a major change under way. It is now evident and transparent that power is shifting away from local elites, and offshore, it is heading towards an ever-centralised mode of governance. Yet we need take further complexity into account. For the first time in capitalism's neoliberal era, these 'protectorates', as per Fraser, are to be found *inside* what had previously comprised a part of the Global West's core (the European Union and the Global West as a whole, in the cases of Greece and Brazil, respectively) yet a part of its semi-periphery in practice. This situation is then historically unprecedented, because it includes an unevenness of power that resembles that between the heartland and the colony, only this time, with a staggering geographical proximity between the two.

In addition, and not at all unrelated to their geographical position, the process of subordination and radical restructuring of the economic, social and political function of neoliberalism's newly found proximal protectorates for the largest part is bloodless. A restructuring without the level of conflict one would have possibly expected to see, one that would match the level of restructuring under way. Under the pretext of austerity, the semi-periphery is subjected to no less than a state- and nation-building exercise. Much akin to western neoliberalism's quests for creative destruction in distant territories post-9/11, this dual exercise involves the doing away of the economic, social and political structures of national territory, and its near-simultaneous replacement with other structures, more favourable to the project of neoliberal expansion.

Carnage to rebuild it all: nation to state, to continent

Even for the most well-informed and well-intentioned observer, the process of the ostensibly intentional self-destruction of much of the Greek state structure surely makes little sense. Take a single but telling example: For a government to order the closure of its own state-run media (ERT) is in all likelihood historically unprecedented and, of course, entirely unexpected even until a few months ago. With Brazil, the story is even more jaw-dropping in many senses, as an extensive bribing scandal has swiftly turned into a case study for dismantling democratic institutions and installing an emergency government, much akin to those installed in the countries of the troubled Southern Europe during the most recent crisis. Why would authorities here be tearing apart the structures that supposedly bind them together?

Two notions may help us in solving this apparent mystery: state-building and nation-building. The two, often used intermittently, largely came back in favour following the US-led operations in Afghanistan and Iraq during the post-9/11 era. Here lies one of those cases where the absence of words (even the single prefix: 're-') speaks louder than the words themselves. However critical it may have been to the nation-state building processes, bibliography on both has established the use of the verb 'build' instead of 'rebuild'. The severe destruction and annihilation that will, by definition, precede the reconstruction process thus meticulously tucked under the historical shallowness of the verb of choice. To build something implies nothing was there before; such is the pernicious power of discourse, and then, there is the not-so-subtle difference lying between the two terms. Whilst state-building refers to the [re]construction of a state's institutions (its infrastructure and functions), nation-building instead describes the process of trampling over the social, political and cultural diversity that may have been previously existent within a national territory in favour of the devastatingly levelling homogeneous force of national unity: Think Syriza's appeal to some kind of 'common good' as it was selling off crucial national infrastructure, or the 'unity' of Temer's cabinet (un)reflected in its all white, all male, all privileged composition.

The operation playing out in the Greek territory in the years that have followed the first memorandum, in May 2010, comprises nothing less than a long-lasting prelude to such a dual nation and state-building process. It is, in this way, notable and also completely unprecedented. It is notable, as is the case with the majority of countries undergoing the IMF treatment, as the Greek experiment has been seen as the complete tearing apart and then the rebuilding of a state without it ever going through the physical destruction of war. Still, the level of levelling and reconstruction is such that these 'restructuring' and 'readjustments' exercises ought to be included in the nation-state building literature. In the case of Greece, it was unprecedented; this dual process plays out in a territory that was previously, until *extremely* recently, firmly inside the capitalist semi-periphery. One only has to look back a few years, when this type of pre-planned dismantling and rebuilding of an entire state apparatus would have appeared completely implausible for an EU member-state.

This change, however, is not exclusively tangible, far from it. First and foremost, austerity has been justified and defended at the level of discourse, as a supposed necessity and in the name of national 'salvation'. Suddenly, Temer's reassuringly empty words more closely echo Tsipra's 'salvation by any cost'. The entire onslaught against social foundations, against those conceived to be society's 'weak' or 'marginal', is built in its entirety upon this power and under the veil of discourse.

Austerity, disinvestment, devaluation

The fact that the long-term goal of the austerity program was to revitalise the Greek economy through devaluation and disinvestment is no secret. In the case of Brazil, the unprecedented collapse of the economy is seen by many as an equally lucrative opportunity to do the same. 'Bring labour costs down to an adequate level', goes the line of reasoning, 'and capital will be lured to step in'. Reduce the flow of funding towards an already faltering public infrastructure (including anything from hospitals to motorways, railways to schools and electricity to water) and justification to privatise these becomes all that much easier. Even though there are promises to the ever-squeezed population of a positively linear path to investment lying ahead, the conceptual drawing underlying the austerity programme would actually resemble a more circular one. It needs to fall a great deal before rising again; even when this happens, the revitalised economy will most likely be finding itself in a rather transformed state. Austerity, in other words, creates a gap between the actual and potential value of assets in Greek and Brazilian territories alike. This idea, concerning the existence of a gap that explains and attracts the inflow of capital into a certain geographical territory, is, of course, far from novel; it was thirty-five years earlier that Neil Smith introduced it in his seminal rent gap theory. Smith defined the rent gap as 'the disparity between the potential ground rent level and the actual ground rent capitalized under the present land use' (1979). Now, why would one ever want to draw lines of comparison between a theory of gentrification on the one hand and plans of austerity, disinvestment and devaluation, on the other? Simply because the parallels are far more striking than what meets the eye.

Although the speed and scale of these disinvestment programmes are staggering, it is not conceptually novel: Privatisation is, after all, at the core of the neoliberal model. What has been taking place intertwined with such disinvestment is a wide form of devaluation. The former commences and stays grounded in the world of finance, but the latter, even if also setting out from the financial world, has a far greater outreach. If disinvestment is the financial cause, devaluation is the effect. Financial yes, but also social, cultural, and political. It is in this way that the two match the state-building and nation-building pair. State-building concerns infrastructure, and it is disinvestment that lowers state infrastructure almost to the point of annihilation, creating the gap necessary for private capital to step in. Nation-building, as shown, concerns a wider set of social, cultural and political relationships. It is in these wider relationships that devaluation takes place. Here, then, lies a crucial difference between devaluation in the case of the gentrification of urban neighbourhoods, on the one hand, and when it plays out at a national-scale, austerity-led, on the other hand. Neil Smith's purposefully tight definition of the rent gap concerned a single and exclusive value, rent level. At the national scale, devaluation caused by conscious disinvestment spans across a much wider range. It commences from faltering property values and depreciation of the labour force and strikes through the cultural and the political sphere. The overall shrinkage and abasement of political discourse is a prime example of devaluation, and one that presupposes and explains the rapid rise of the monstrosity that is the Far-Right in Greece and the white-led, neoliberal and right-wing agenda of 'market reform' in Brazil.

In essence – much akin to Smith's rent gap – disinvestment and devaluation create a vast profit gap laying across an entire national territory, instead of a segment of the urban setting. If austerity causes a prime, first step of disinvestment, it is only logical to expect this to then become part of process similar to gentrification, only this time at a national level.

Enter gentrination

Gentrification does not only presuppose a rent gap for capital to pour itself into. Much more crucially, it presupposes a single urban entity, within a segment of which it will take place. Ever since Ruth Glass coined the term in 1964[2] an entire school of thought has developed around both the causes as well as the effects of gentrification upon this urban segment (whether a street, housing block, neighbourhood or wider area) as well as the urban entity (a city). Across an entire spectrum of otherwise conflicting views[3] concerning what drives gentrification, a consensus emerges in that it comprises the process of seeing often low-income, derelict or even decaying parts of a city being taken over and rehabilitated by members of a class that is higher than those occupying them previously. There is an additional agreement in the understanding that gentrification occurs in urban areas where there is an adequate profit margin to allow for it. In an economy and society dictated and mediated by proximity and distance, the areas most likely to fall prey to gentrification are those close enough to the urban core, in other words, those offering a lucrative potential turn-over. Yet in most cases, these are not close enough to belong to the urban core itself: this, after all, would have been recuperated by capital earlier on. The ideal gentrification target, so to speak, is the urban semi-periphery: a segment of the urban that will allow either for the outright expansion of the core or for the outsourcing and facilitation of certain functions necessary for its unobstructed operation.

Let us shift our viewpoint and geographical scale for a moment. Would it not be possible, in face of the discourse about European unification, to think of the European and then even the global territory in a manner equivalent and similar to how we think about the urban territory? Would it not also be possible, in the discourse of globalisation, to think of a single state (even if a vast one, such as Brazil) as the equivalent of a neighbourhood in an urban territory? Would it not then finally be possible to conceive the restructuring of a sovereign state within the EU, and a vast sovereign state within the global system order, similar to the restructuring of an urban segment?

This conception involves a discursive leap both in terms of the role played by this segment within the whole – whether urban, European or global – and in terms of the rules governing the segment itself (whether a neighbourhood or a nation state). It is, yet, entirely feasible to draw the parallel that the role of a gentrified urban segment may be acted upon, at a continent-wide level, by a continental or global segment (i.e. one or more member-states within the EU, or Brazil within the global economy). I will term this process 'gentrination' and I will argue that it is exactly what has been playing out in post-austerity Greece, and in modern day Brazil, as it goes through its longest and most severe moment of crisis in over a century.

Gentrination is the scalar equivalent of urban gentrification at the national territory level. Similar to gentrification, it is a gradual process. First, the physical infrastructure as well as the social, political and cultural foundations of the nation state experience devaluation through the disinvestment of austerity. Second, having produced an adequate profit-gap much akin to gentrification's rent gap, the host national territory sees the influx of private capital and the simultaneous radical overhaul of its foundations, as well as key social and cultural foundations. Third comes the final and most crucial step, wherein undesirable voices are pushed out of public discourse, while undesirable populations are either forced out of public space into detention or expelled from the national territory altogether. Meanwhile, this national territory is restructured in such a way so as to facilitate the supra-national core. In the Brazilian case, the country's enormous export sector (sugar, coffee, soy, ethanol) was never meant to benefit the local populations, but a weak economy would make for a steady and uninterrupted cheap supply of raw materials to the global capitalist core. In the Greek example, its national territory is designated to act as a deterrent of human movement and a facilitator for the flow of capital and goods. On the one hand, it is then set to facilitate the inflow of goods into the EEA and their forwarding onto Northern Europe; the Chinese state investment in both the port of Piraeus and the country's rail infrastructure play witness to this. On the other hand, the national territory now acts as a gate, a deterrent and hindrance for human inflows: The building of a 'security fence' along its Turkish border, the mobilisation of the supra-national policing force FRONTEX and NATO along the Greek border as well as the erection of the now infamous hotspots across the country, all point at this newly found key function of the Greek territory following its gentrination.

Gentrination therefore has it both, the plight of displacement for large populations and the restructuring and functioning anew of the territory in which it takes place. So what would be the potential use for a national territory in the capitalist semi-periphery to go through the process? Surely, one would think, this entire process would not be under way simply in order to tighten the border, or to facilitate trade?

This is the moment to widen our scope, to look at a potential new role for Greece in Europe and Brazil in the Global West as a whole. To do so we need to bring in another term from the geographical tradition, albeit with a twist: the 'spatial fix'. In its current form, David Harvey (1985) tells us that spatial fix is the theory that capitalism can tackle its crises of over-accumulation through investment in the urban environment. Urbanisation, in other words, has so far been capitalism's one way out of its recurrent crises, but a second way out has been through territorial expansion (Harvey 2003). Here, Harvey sees capitalism's expansionist political and military projects (specifically the post-9/11 US 'War on Terror') as a form of a spatial fix that is now taking place on an international scale.

This is not, therefore, the first time that such a direct line between capitalist and urban expansion is drawn. Sure enough, the parallels in the evolution patterns of capitalism and cities are striking. In both cases, the expansion comes first. In capitalism's case, moving outward from the global western core to the colonised lands has ensured its perpetuation through geographical spread; colonisation was the archetypal spatial fix. As for cities, the endless growth of the American suburbia in the 1950s and 1960s adhered to a very similar principle. Of course, in the urban case, a wave in the opposite direction would emerge and gentrification was once again pioneered in US cities throughout the 1980s. Soon enough, the affluent middle classes would make their way back into the urban cores all across the Global West.

Our current global moment marks a crucial (re)turn: the return, through gentrination, of capitalism's rampant, destructive force now edging *close* to the core. Capitalism's quest to create a large enough profit margin is now played out at the Global West's semi-periphery. This is crucial because it evidently rebukes the claim that the periphery is subjected to a type of neo-colonisation process by the core. Even if the balance of power within the European Union structure does strike an alarmingly similar chord to that between a protector state and its protectorate[4] relationship, to see Greece as a colony, or Brazil as merely the 'crooked', failed state and either as some form of 'neo-colony' only creates the illusion of distance. Gentrination, to the contrary, highlights precisely the element of proximity of these territories to the core. On the one hand, this proximity is necessary for the gentrinied territories to perform their assigned functions as above. Conversely, the proximity shows a potential element of weakness for capital, having largely reached its capacity in pushing its crises away, it signals an attempt at creative destruction: a spatial fix that hits strikingly and alarmingly close to the core.

Caught in an array of personal dramas, hounded and haunted as he made his way across the European continent, John Berger's character, G, appeared oblivious to the rapid shifts that were taking place around him. Against his foresight, his will and his expectations, he was himself eventually caught up in the cataclysmic events of WWI. Living through our undoubtedly turbulent times, we must not lose sight, and we need to grasp and to meticulously record the full extent of the violent social restructuring that is now taking place within the territories primarily hit by the global financial crisis, although, at the same time, we must open our eyes to grasp the result of this restructuring upon the entire European and the global capitalist structure. This is not a storm that will be weathered; it is a change of paradigm as a concept gentrination may be able to make a modest contribution in helping us understand the spatial configuration of the new paradigm ahead of us.

Notes

1 With special thanks to Christos Filippidis for a highly illuminating discussion on nation- and state-building.

2 A concise overview of the otherwise abundant and diverse gentrification literature can also be found in Slater (2011).

3 One of the key conflicts among theorists of gentrification include production vs consumption-driven models of gentrification. See Slater (2011).

4 Which is, as shown earlier, how Nancy Fraser described the Greek state in an otherwise succinct analysis of the shifting-core relationship: 'Greece is reduced to a protectorate, Spain, Portugal and Ireland are ruled from Brussels…' (Fraser 2013: 126).

References

Berger, G. (1972). *G.* London: Weidenfeld & Nicolson.

Fraser, N. (2013). 'A triple movement?', *New Left Review*, 81, pp. 119–132.

Harvey, D. (1985). *The Urbanization of Capital: Studies in the History and Theory of Capitalist Urbanization.* Baltimore: The Johns Hopkins University Press.

Harvey, D. (2003). *The New Imperialism.* Oxford: Oxford University Press.

Slater, T. (2011). 'Gentrification of the city', in G. Bridge and S. Watson (eds.). *The New Blackwell Companion to the City.* Oxford: Wiley-Blackwell, pp. 571–585.

Smith, N. (1979). 'Toward a theory of gentrification: A back to the city movement by capital not people', *Journal of the American Planning Association*, 45, pp. 538–548.

7 Making rent gap theory *not* true

Eric Clark

Introduction

Neil Smith presented rent gap theory in the context of a critique of consumer sovereignty and the naturalisation of markets. As ideological lodestar, market fundamentalism legitimates far-reaching processes of commodification and privatisation, opening up spaces for financialisation. Financialisation forges social relations conducive to the penetration of finance into the production, exchange and consumption of (built) environments. It enhances financial control over the governance of human niche construction by expanding spheres for financial 'investments', whereby unearned rentier revenues increasingly flow to a resurgent rentier class. Exchange value becomes master; use value becomes slave (Harvey 2014). Consequently, changes in built environments become increasingly determined by where rent gaps can be created and appropriated – commonly involving displacement, loss of livelihoods and human suffering – rather than as outcomes of conflict-laden democratic processes. My point of departure in this chapter is the question: How might we go about making rent gap theory *not* true? I start, however, on a personal note.[1]

Although we met only a dozen or so times, Neil Smith influenced my life immensely. As a student of urban planning at Stockholm University, I would seek inspiration in the library of the Department of Human Geography at Norrtullsgatan 2, a quiet place with old brown bookcases and a helpful librarian. Escaping a wintry day and a bachelor's thesis, a special supplement on neighbourhood revitalisation in the latest issue of the *Journal of the American Planning Association* drew me into a long read and a head-spin. Here was Chester Hartman arguing that displacement was a major problem, Peter Marcuse analysing redlining, and most notably Neil Smith's (1979) now classic theory of gentrification. Here was a strikingly clear critique of the myth of consumer sovereignty and a powerful explanation of forces underlying urban change. I knew then that I had to learn more about urban land rent, but I could not have imagined that some fifteen years later I would be nicknamed Eric 'Rent Gap' Clark.[2]

In keeping with Neil's preference for 'hard-nosed engagements with ideas' over the 'uncritical and celebratory' fashions of disciplinary hagiography

(Mitchell 2013: 1), I summarize here by way of introduction my hard-nosed engagements with rent gap theory before addressing the question at hand. In the 1980s, it was important to show that rent gap theory was true, faced as it was by rejection from leading scholars (e.g. Ley 1986). Dusty numbers from the Malmö archives empirically verified the theory (Clark 1985, 1987, 1988), but also revealed an oversight in Smith's original formulation: that capitalised land rents rise with speculation on future rental revenues, once finance capital and landed developer interests condemn an area to devaluation and redevelopment. A more adequate representation of rent gaps should consequently reflect this tendency for the gap to close during the years prior to redevelopment.[3]

A second hard-nosed engagement focused on the petrified and unhelpful bifurcation into capital-oriented, supply- and production-side explanations of gentrification, and culture-oriented, demand- and consumption-side explanations (Clark 1992, 1994). In arguing for the relevance of Bohrian complementarity to gentrification theory, I emphasised that the prevalent understanding of rent gap theory as production oriented and rent gaps as somehow unrelated with culture, demand and consumption is misguided.[4] Ever forging critique into new strengths for radical theory, Neil saw the 'vigorous reintegration of cultural analyses of gentrification with a less fashionable but nevertheless vital political economy' as a 'much richer project than simply reconciling production-side and consumption-side arguments' (Smith 1995: 126). Indeed, Bohrian complementarity insists on resisting simplistic reconciliatory syntheses.

A third engagement with rent gap theory focused on its relationship with Harvey's elaborations on Marxian theory of interplay between differential rents (Clark 2004). Harvey saw rent gap theory as unsophisticated – 'much too simple and definitely obvious' (Smith 2010: 97) – compared with his masterly treatment and development of Marx's rent theory in *The Limits to Capital* (1982). Smith made much of differentials and 'differentiation of ground rent levels between different places' (1982: 145) as he advanced rent gap theory, but in his analysis of 'the rhythm and periodicity of … movements of capital' (1982: 149) in the uneven development of urban space, he didn't explicitly juxtapose rent gaps to the interplay of differential rents I and II. Somewhere, I thought, these relations needed to be sketched. I attempted to do so, drawing on an empirical case I already knew well: Malmö. The interplay between differential rents 'emphasizes a *synchronic comparison across space of differences in capital investment*, especially in terms of normal and above normal', while the concept of rent gap 'emphasizes a *syntopic comparison across time, of differences in actual and potential land rent* which correspond to different types and volumes of capital investment' (Clark 2004: 155). With different emphases, they reveal the relational spacetime ('the hyphen disappears', Harvey 2009: 137) dynamics of distinct yet imbricate rhythms of capital circulation, flowing through while affixed to land.[5]

It may have made sense in the 1980s to show that a 'simple and definitely obvious' theory is true, but it definitely does not make sense now. More important is to engage in showing how we might go about making rent gap theory *not* true. If we wish the 'theory of the urban land market to become *not* true', we need to 'eliminate those mechanisms which serve to generate the theory' (Harvey 1973: 137). These mechanisms are, above all (extending points made in Clark 2005):

- Social relations of private property (commodification of space/nature)
- High degrees of inequality
- Exchange-value driven decision-making (financialisation)
- Myths of market fundamentalism, frontier, consumer sovereignty and related myths.

Consequently, what can make rent gap theory *not* true are:

- New and reinvigorated traditional forms of shared ownership (de-commodification, commoning)
- High degrees of equality (institutionalised floors and ceilings on income and wealth)
- Use-value-driven decision-making (deepening of democracy)
- Egalitarian myths of interdependence, solidarity, complementarity between autonomy and community, and related myths and metaphors.

(De)commodification

Private property in land constitutes the very foundation of rent gaps, as it allows for extraction of capitalised land rents, speculative bidding on future rents and the discernment of potential rents under 'higher and better' land uses. Private property rights 'confer on the owner near-monopoly control over land and improvements, monopoly control over the uses to which a certain space is put' (Smith 1979: 541). These include the right to unearned rentier incomes, a form of free-riding on the work of others (Sayer 2015). Property rights in land come in many forms, the complexity of which 'is quite staggering' (Harvey 1982: 276).

Making sense of this staggering complexity must, as Blomley suggests, centre on 'the legal construction of both place and mobility' (1994: 225). Place becomes broken into isolated spatial commodities for which land markets are formed: 'perhaps the weirdest of all the undertakings of our ancestors' (Polanyi 2001: 187). The expansion of private property, 'by restricting access, can deprive others of a place to live, even of the right to life', thereby raising the question of 'whether people can be said to have a right to, literally, a place in the world' (Smith 1994: 41–42). With forced displacement among the most widespread human rights violations in the world (COHRE 2009), it seems 'as if there is a systematic plan to expel low-income and unwanted populations from the face of the earth' (Harvey 2010a: 245). This mobility is forced. It has nothing to do with the right to mobility and everything to do with violating the right to stay put, the right to place (Clark 2011).

To make rent gap theory *not* true, the ongoing commodification of land, nature and space must be ceased and turned around. Commodified lands need to be brought back into the commons. The right to place must override the right to extract rent. With a much longer history, commons are even more diverse and complex than private property rights in land. Moreover, these are far from mutually exclusive, absolute, either/or categories. Commons are continuously being co-created and 'just as continuously being enclosed and appropriated', primarily through extraction of rents (Harvey 2012: 77). There is a rich and neglected social and legal history of commons (Bollier 2014) that offers insight and encouragement in seeking ways to institutionalize new social practices of commoning and to reinvigorate traditional forms of shared ownership. This won't happen overnight, however, and in the current conditions of many cities steeped in revanchist urbanism, if 'we are truly to embrace the city as the new frontier today, then the first and most patriotic act in pioneering, if historical accuracy is to be observed, will be squatting' (Smith 1996: 232).

(In)equalities

That the growing literature on inequality attracts so many new readers and analysts is not only because there is so much of it to study but also because it makes such a great difference in so many ways, from a broad array of social and health problems, to trust, democracy and willingness to assume social and environmental responsibility (Wilkinson and Pickett 2009). The most common measure of inequality is the Gini index of income (a single number reflecting distribution across a population). This can, but seldom does, include income of capital, which is ubiquitously more skewed. Vastly more skewed still is distribution of wealth. Oxfam calculated that in 2013 the 85 richest people on the planet possessed as much wealth as the poorer half of the global population. Updated in 2014 with new data, the headline read, 'The 67 people as wealthy as the world's poorest 3.5 billion' (Moreno 2014). Try to imagine: 0.000000957% of the global population possesses as much as 50%. This level of concentration is hard to fathom, let alone its consequences.

Inequality is, however, 'not just about the size of wallets. It is a sociocultural order, which (for most of us) reduces our capabilities to function as human beings, our health, our self-respect, our sense of self, as well as our resources to act and participate in this world' (Therborn 2013: 1). The costs of *The Killing Fields of Inequality* are immense: human suffering, unrealised flourishing, disabled democracy, impaired trust, loss of solidarity and security, and a whole raft of social, psychological and physical health problems. Relatedly, inequalities also underlie rent gaps, especially the dynamic strength of rent gaps as mechanisms driving urban change. With extreme concentrations of income and wealth comes the power at the one end to make and take rent gaps, be they in inner-city neighborhoods or the re-scaled rent gaps of large land grabs in the global periphery. At the other end, those weakened by inequalities are more vulnerable to displacement, as rent gaps are created and captured: Their homes and livelihoods cannot compete in the spatial market with 'higher and better' uses geared to expand rentier revenues.

To make rent gap theory *not* true, inequalities need to be radically diminished. As with commons, we have a long egalitarian history of keeping inequalities in check (Wilkinson 2001). There are also recent and ongoing processes of equalisation that provide inspiration and reassurance that struggles for greater equality can achieve enhanced conditions of life not only for the poor, but for the better off as well (Therborn 2013). In eliminating the killing fields of inequality, rent gap theory also becomes *not* true. To be more specific, policies such as universal provision of basic income, coupled with regulations on maximum income (Daly 1996) set in relation to basic income and maximum inequality (e.g. 20 to 1, see Dorling 2014), and the return of highly-concentrated wealth to the commons (see above), would go far in rendering rent gap theory irrelevant.

Financialisation vs democracy

The 'increasing tendency to treat the land as a pure financial asset' underlies 'the form and the mechanics of the transition to the purely capitalistic form of property in land' (Harvey 1982: 347). The same can be said today about music, words, ideas, organisms and ourselves, as 'intellectual property rights', bio-prospecting and branding open up new spheres for financial 'earnings' through speculative 'investment'. Once treated as pure financial assets with expectations on financial yield, these are also reduced to just another 'special branch of the circulation of interest-bearing capital' (Harvey 1982: 347). Financialisation involves the subordination of use values to exchange values, in sphere after sphere, thereby expanding the volumes of 'investment opportunities' for ever more concentrated centres of financial wealth.

Establishment of private property rights in land, in its broadest sense including bodies of water and elements of land commonly called natural resources, creates a foundation for the commodification of environments by judicially and administratively rendering specific parts tradable on markets, where their exchange value can guide decisions on *investment*. The profit- and rent-seeking behaviour of finance capital and landed-developer interests drives the formation of market relations through the privatisation and commodification of built and natural environments, extending the process wherever property relations retain the characteristics of commons that hinder the free flow of financial *investment*. Environments are securitised and, treated as pure financial assets, enter the orbit of finance capital as potential sites for *investment*, or disinvestment, depending on their expected yield to shareholders.

Financialisation is a process whereby privatisation, commodification and securitisation of the environment allow for the penetration of financial control and decision-making into the fabric of societies and (built) environments. It has involved 'the phenomenal expansion of financial assets relative to real activity (by three times over the last 30 years)' and 'the absolute and relative expansion of speculative as opposed to or at the expense of real *investment*' (Fine 2013: 6, emphasis added). Ever in search of new fields to securitize and *invest* in, the financial sector actively engages in the creation of conditions allowing nature 'to circulate as financial capital' (Prudham 2007: 259), entailing enclosures of resource commons and the displacement of people, their livelihoods, knowledge and practices.

I emphasize *investment* because, as Sayer convincingly argues, it is 'the most dangerously ambiguous word in our economic vocabulary' (2015: 34). Masking the difference between wealth extraction and wealth creation, it camouflages the former as the latter. Sayer distinguishes object-oriented definitions that focus on what is invested in (enabling production of new use values in goods, services and skills) from 'investor'-oriented definitions that focus on 'the financial gains of the 'investor' from any kind of spending, lending, saving, purchase of financial assets or speculation – regardless of whether they contribute to any objective investment, or anything socially useful' (Sayer 2015: 34–35). The slippage between these usages is a source of mystification, concealing the subordination of use value to exchange value, while obscuring the moral difference 'between contributing to the creation of something useful and just getting a return, no matter what' (Sayer 2015: 36). Sayer associates the rise of exchange-value-oriented 'investment' relative to use-value-oriented investment to 'the emergence of 'financialised' capitalism, which prioritises making money out of money, instead of the tricky business of organising people to produce goods and services. It's truly extraordinary that we treat these different things as one and the same without even noticing' (Sayer 2015: 36).

Finance capital claims to 'see the world as full of potential', indeed, to 'see potential everywhere' (HSBC billboards). Financialisation reaches into everyday life as we increasingly consider our homes, our education and even ourselves, as financial assets we 'invest' in for the sake of financial returns (Martin 2002). It reaches into education systems, healthcare, infrastructure of various kinds, urban planning as well as political life. Its hallmark is the displacement of use-value-oriented decision-making by exchange-value-driven decision-making. Financialisation is dialectically entwined with the previously considered aspects, exploiting while intensifying inequalities in economic and political power, and driving the privatisation of commons in order to expand the sphere of property as investment opportunities open to speculation on changes in exchange values.

Making rent gap theory not true requires political, economic and judicial reforms geared to foster use-value oriented decision-making processes in all spheres of investment, i.e. the deepening of democracy. This in turn requires effective constraints on exchange-value-oriented decision-making ruling over investments in various systems of provision, especially housing and the built environment more widely. Neil perceptively concluded his 'revanchist city' book by quoting Peter Marcuse's defence of degentrification: 'The opposite of gentrification should not be decay and abandonment … but the democratization of housing' (Marcuse 1991, quoted in Smith 1996: 231).

Market fundamentalism vs egalitarian ethos

The myth of market fundamentalism is changing everything, from workplaces, communities, education, health care and public institutions, to our relationships with our environment and understandings of self (Michaels 2011, Verhaeghe 2014). One form of social economic integration, market exchange, is mythologised at the expense of other forms: state redistribution and community reciprocity (Polanyi 2001). Market fundamentalism 'conveys the quasi-religious certainty expressed by contemporary advocates of market self-regulation', bearing an 'affinity with religious fundamentalisms that rely on revelation or a claim to truth independent of the kind of empirical verification that is expected in social sciences' (Block and Somers 2014: 3).

For Neil, the struggle for ideas was central to the making of the future (Mitchell 2015). Early on, he saw gentrification as 'increasingly constructed through the vocabulary of the frontier myth' (Smith 1996: 13). More recently, he insisted that 'we are in a moment when the future is radically open' (Smith 2015: 964), and that forging alternatives requires that we free ourselves of 'the economic conveyor belt of capitalist common sense which lurches the social body from crisis to crisis' (Smith 2011: 265). Just as the roll-out of variegated neoliberalisation since the 1970s was largely orchestrated and guided through ideational path dependencies (Blyth 2002, Harvey 2005) – one moment in a longer history of imposing visions to reshape the world (Smith 2005) – the creation of alternatives cannot otherwise than build to some extent on reinvigorated traditions with very different integrative myths and metaphors. It is not only so 'that without metaphors, scientific inquiry would go nowhere' (Harvey 2010b: 198). Social movements and societal change cannot either move without the power of better ideas, metaphors and foundational myths through which to perform the creation of alternatives.

If the myth of market fundamentalism is a co-evolutionary partner with privatisation, polarisation and financialisation, working together to make rent gap theory true, then making rent gap theory not true must involve engaging alternative myths and metaphors. Against the rugged pioneer individual on the market frontier, the 'self-made man', we must emphasize our fundamental interdependence, and the interdependence of individual and collective. Bauman (1994: 43) stresses the importance of 'awareness of the intimate connection (not contradiction!) between autonomous, morally self-sustained and self-governed (often therefore unwieldy and awkward) citizens and a fully-fledged, self-reflective and self-correcting political community. They can only come together; neither is thinkable without the other'. Indeed, 'the two fundamental traits of our enduring egalitarian ethos, valuing sharing and autonomy, are connected at their roots' (Clark and Clark 2009: 316; cf. Clark and Clark 2012). Over the long stretch of human history, we have displayed 'patterns of behavior that systematically prevented overreaching individuals from achieving dominance' (Shryock and Smail 2011: 255). We have a rich history from which to forge more beneficial myths and metaphors conducive to making rent gap theory *not* true.

Conclusion

These are the forces that make rent gap theory true. From a dialectical co-evolutionary perspective, none of the four spheres 'prevails over the others, even as there exists within each the possibility for autonomous development': we should not 'see one of the spheres as determinant' (Harvey 2010a, 128: 132). In seeking 'openings for the construction of viable political-economic alternatives' (Harvey 2016: 322), the line of analysis I have sketched suggests that in order to make rent gap theory *not* true, our political economies need to be reconstructed such that we: de-commodify land, and work together to cultivate and institutionalize social practices of commoning; institutionalize ceilings on inequalities by legislating floors and ceilings on incomes and wealth; move decision-making from shareholders, boardrooms and the trading floors of stock exchanges to democratic bodies, placing use-values in focus; and replace myths and metaphors of market fundamentalism with recognition of our interdependence, how we mutually constitute one another, how we are dependent on and owe solidarity to others.

Notes

1 This research benefited from funding of the European Union Seventh Framework Programme (FP7/2007-2013) under grant agreement no. 266800 FESSUD (fessud.eu). I am especially grateful to Abel Albet and Núria Benach for organizing such an intensely engaging conference.
2 Sharon Zukin kindly called me this at a workshop on urban transformation held in Stockholm in June 1995, where Anders Gullberg and I presented a study on the making and taking of rent gaps in the massive post-war renewal of Stockholm's central business district (Clark and Gullberg 1997).
3 Neil, 'tremendously warm' and keen 'to nurture and encourage emerging scholars' (Slater 2012), generously acknowledged this work as a 'superb, critical study' (Smith 1992: 111), a 'landmark study' (Smith 1996: 72). By the way, this emerging scholar was a mere two years younger than Neil.
4 Neil appreciated this critical engagement as 'pathbreaking' (Smith 1995: 125).
5 I wrongly thought bringing Harvey's and Smith's analytically distinct rent dynamics into dialogue would be of interest, but the paper was desk-rejected twice, and has not drawn interest among land rent scholars.

References

Bauman, Z. (1994). *Alone Again: Ethics after Certainty*. London: Demos.
Block, F. and Somers, M. R. (2014). *The Power of Market Fundamentalism: Karl Polanyi's Critique*. Cambridge: Harvard University Press.
Blomley, N. K. (1994). *Law, Space, and the Geographies of Power*. New York: Guilford Press.
Blyth, M. (2002). *Great Transformations: Economic Ideas and Institutional Change in the Twentieth Century*. Cambridge: Cambridge University Press.
Bollier, D. (2014). *Think Like a Commoner: A Short Introduction to the Life of the Commons*. Gabriola Island, BC: New Society.
Clark, E. (1985). 'Kvarteret Stralsund i Malmö 1832–1984', *Nordisk Samhällsgeografisk Tidskrift*, 2, pp. 48–56.
Clark, E. (1987). *The Rent Gap and Urban Change: Case Studies in Malmö 1860–1985*. Lund: Lund University Press.
Clark, E. (1988). 'The rent gap and transformation of the built environment: Case studies in Malmö 1860–1985', *Geografiska Annaler Series B: Human Geography*, 70, pp. 241–254.
Clark, E. (1992). 'On blindness, centrepieces and complementarity in gentrification theory', *Transactions of the Institute of British Geographers*, 17, pp. 358–362.
Clark, E. (1994). 'Towards a Copenhagen interpretation of gentrification', *Urban Studies*, 31, pp. 1033–1042.
Clark, E. (2004). 'Rent rhythm in the flamenco of urban change', in T. Mels (ed.). *Reanimating Places: A Geography of Rhythms*. Aldershot: Ashgate, pp. 147–160.
Clark, E. (2005). 'The order and simplicity of gentrification – A political challenge', in R. Atkinson and G. Bridge (eds.). *Gentrification in a Global Context: The New Urban Colonialism*. London: Routledge, pp. 256–264.
Clark, E. (2011). 'Dispossession, displacement and human security', *Industrial Transformation, Urbanization and Human Security*, Proceedings of IHDP (International Human Dimensions Programme) International Conference, Taipei: Academia Sinica.
Clark, E. and Clark, T. (2009). 'Isolating connections, connecting isolations', *Geografiska Annaler Series B: Human Geography*, 91, pp. 311–323.
Clark, T. and Clark, E. (2012). 'Participation in evolution and sustainability', *Transactions of the Institute of British Geographers*, 37, pp. 563–577.
Clark, E. and Gullberg, A. (1997). 'Power struggles in the making and taking of rent gaps: The transformation of Stockholm city', in O. Källtorp et al. (eds.). *Cities in Transformation – Transformation in Cities: Social and Symbolic Change of Urban Space*. Aldershot: Avebury, pp. 248–265.
COHRE (2009). *Global Forced Evictions Survey – 2007–2008*. Geneva: The Centre on Housing Rights and Evictions (COHRE).
Daly, H. E. (1996). *Beyond Growth: The Economics of Sustainable Development*. Boston: Beacon Press.
Dorling, D. (2014). *Inequality and the 1%*. London: Verso.
Fine, B. (2013). 'Towards a material culture of financialisation', *FESSUD Working Paper Series*, 15.
Harvey, D. (1973). *Social Justice and the City*. London: Edward Arnold.
Harvey, D. (1982). *The Limits to Capital*. Oxford: Blackwell.

Harvey, D. (2005). *A Brief History of Neoliberalism.* Oxford: Oxford University Press.
Harvey, D. (2009). *Cosmopolitanism and the Geographies of Freedom.* New York: Columbia University Press.
Harvey, D. (2010a). *The Enigma of Capital and the Crises of Capitalism.* London: Profile.
Harvey, D. (2010b). *A Companion to Marx's* Capital. London: Verso.
Harvey, D. (2012). *Rebel Cities: From the Right to the City to the Urban Revolution.* London: Profile.
Harvey, D. (2014). *Seventeen Contradictions and the End of Capitalism.* London: Profile.
Harvey, D. (2016). *The Ways of the World.* London: Profile.
Ley, D. (1986). 'Alternative explanations for innercity gentrification: A Canadian assessment', *Annals of the Association of American Geographers*, 76, pp. 521–535.
Martin, R. (2002). *Financialisation of Everyday Life.* Philadelphia: Temple University Press.
Michaels, F. S. (2011). *Monoculture: How One Story is Changing Everything.* Canada: Red Clover Press.
Mitchell, D. (2013). 'Neil Smith hated hagiography – A long obituary'. Available at: www.maxwell.syr.edu/uploadedFiles/faculty/geo/Neil%20Smith%20Hated%20Hagiography%20-%20A%20Long%20Obituary.pdf.
Mitchell, D. (2015). 'The future is radically open. Introduction', *ACME: An International E-Journal for Critical Geographies*, 14, pp. 954–956.
Moreno, K. (2014). 'The 67 people as wealthy as the World's poorest 3.5 Billion', *Forbes*, March 25.
Polanyi, K. (2001 [1944]). *The Great Transformation: The Political and Economic Origins of our Time.* Boston: Beacon Press.
Prudham, S. (2007). 'Sustaining sustained yield: Class, politics, and post-war forest regulation in British Columbia', *Environment and Planning D: Society and Space*, 25, pp. 258–283.
Sayer, A. (2015). *Why We Can't afford the Rich.* Bristol: Policy Press.
Shryock, A. and Smail, D. L. (2011). *Deep History: The Architecture of Past and Present.* Berkeley: University of California Press.
Slater, T. (2012). 'Rose Street and revolution: A tribute to Neil Smith'. Available at: www.geos.ed.ac.uk/homes/tslater/tributetoNeilSmith.html.
Smith, D. (1994). *Geography and Social Justice.* Oxford: Blackwell.
Smith, N. (1979). 'Toward a theory of gentrification: A back to the city movement by capital not people', *Journal of the American Planning Association*, 45, pp. 538–548.
Smith, N. (1982). 'Gentrification and uneven development', *Economic Geography*, 58, pp. 139–155.
Smith, N. (1992). 'Blind man's buff, or Hamnett's philosophical individualism in search of gentrification', *Transactions of the Institute of British Geographers*, 17, pp. 110–115.
Smith, N. (1995). 'Gentrifying theory', *Scottish Geographical Magazine*, 111, pp. 124–126.
Smith, N. (1996). *The New Urban Frontier: Gentrification and the Revanchist City.* London: Routledge.
Smith, N. (2005). *The Endgame of Globalization.* New York: Routledge.

Smith, N. (2010). 'Box 3: Toward a theory…', in L. Lees, T. Slater and E. Wyly (eds.). *The Gentrification Reader*. London: Routledge, pp. 97–98.

Smith, N. (2011). 'Uneven development redux', *New Political Economy*, 16, pp. 261–265.

Smith, N. (2015). 'The future is radically open', *ACME: An International E-Journal for Critical Geographies*, 14, pp. 957–964.

Therborn, G. (2013). *The Killing Fields of Inequality*. Cambridge: Polity Press.

Verhaeghe, P. (2014). *What About Me? The Struggle for Identity in a Market-Based Society*. London: Scribe.

Wilkinson, R. (2001). *Mind the Gap: Hierarchies, Health and Human Evolution*. New Haven: Yale University Press.

Wilkinson, R. and Pickett, K. (2009). *The Spirit Level*. London: Allen Lane.

8 Rent gap theory is a political resource

Ernesto López-Morales

In 2009, I wrote Neil Smith an email. I wanted to know his opinion on a paper I had presented at a conference in Chicago earlier that year that I now wanted to publish. He kindly replied that he liked the paper very much. He also advised me to publish in the *International Journal of Urban and Regional Research* (IJURR), which I subsequently did. His words were both generous and encouraging.[1] However, he was unsure about the concept of Capitalised Ground Rent 2 (CGR2) that I had invented by drawing on his conceptualisation of Potential Ground Rent (PGR) (López-Morales 2011). He argued that, theoretically, they were the same and that there was no need to invent something new. I agreed with him in that they are conceptually the same *most of time*. Everything depends on the amount redevelopers or property owners are able to monopolistically capture.

At that time, I was trying to shed light on a point Smith had addressed earlier. This was the fact that, in every process of redevelopment, the PGR is actually transformed into Capitalised Ground Rent (CGR) at a certain maximum level dictated by zoning and construction policies. This is in regard to the rationale choice redevelopers have to make in order to balance the advantages of internalising the enormous geographical opportunities supplied by the state (transport infrastructure, centrality, etc.) versus the marginal costs of additional units that redevelopers are able to assume. In Chile, the CGR2 can be so high that it makes it impossible for most owner-residents and tenants to cash in. This has also been documented in East Asia, especially China and South Korea (see Lees, Shin and López-Morales 2015, 2016, and Shin 2009 for a more specific narrative).

In recent well-documented cases of rent gap formation and exploitation, the state has been identified as a very present agent in the formation of CGR2 by, on the one hand, increasing centrality through infrastructure provision and, on the other hand, zoning and increasing construction codes in certain places, making the floor area ratio so high that it indirectly 'creates' higher profits for redevelopers. This is a basic factor of gentrification in countries like Chile, Mexico, South Korea and China. In some cases, the PGR created by zoning is limitless. However, from the perspective of the capitalist class – investors and redevelopers – the real limit to PGR is the marginal revenues redevelopers experience when they come to internalize the cost of the additional units (namely extra m^2, flats, etc.) produced. This state-led fictitious land valuation is pure rent gap creation, however, not by devaluation of the capitalised ground rent as it was seen by Smith in 1979, nor by Clark in 1987 in Malmö, but by the extreme valuation of the potential ground rent and its accumulation as Capitalised Ground Rent 2. A little earlier, Hackworth (2007) had also seen something like this happening in the USA which he identified as a 'post-recessionary' form of gentrification. I have developed the rent gap theory further in more recent papers (López-Morales 2013, 2015a, 2015b, 2016, Lees, Shin and López-Morales 2016), and the following points shall draw from those works.

In Chile, urban central neighbourhoods face intensive trends of privately led redevelopment. As a result, the influx of middle-income households into traditionally deprived inner areas has also been massive. In these areas, almost all of the CGR2 is captured by large-scale real estate redevelopers and investors.[2] The CGR2 is a politico-economic device for reproducing real estate capital and is a by-product of the capitalist class imposition of socio-spatial changes in the city.

The ideas of CGR and PGR are present in Neil Smith's earliest works (the most famous ones being in 1979 and 1982). Later, this idea was further developed by Clark (1987, 1995), Hammel (1999) and Hackworth (2007), among several others. I think that these approaches are now very useful for addressing the increasing conflicts of housing unaffordability, class-led displacement and polarised power relations that are currently seen in several Latin American and East Asian cities and neighbourhoods facing gentrification-induced redevelopment. The similarities between all of these cases are the very present role of the state in the provision of the necessary conditions for the PGR to increase, often along with the existence of corporatized, large-scale redevelopers (Lees, Shin and López-Morales 2016).

I believe that rent gap theory is powerful and useful enough to counteract the radical post-colonial call to mentally 'unlearn' what we know about the workings of capitalism. According to some radical post-colonial scholars, rent gap theory would correspond to a type of metanarrative that is incapable of observing the variegated and unpredictable spatialities and the multiple and heterotopic array of socio-cultural narratives that currently emerge from cities in the global south (see Robinson 2011 for instance). In my opinion, the problem is that the radical, so-called post-colonial perspective is sometimes also a call for the depoliticisation of urban theory, by overshadowing the everyday mechanisms that capitalism uses to dictate how and where social reproduction happens in the city. This is something that we must not overlook (López-Morales 2015b). Rent gap theory was also heavily criticised during the 1980s and 1990s for its supposed determinism and rejection of individual and social agency as factors that determine neighbourhood redevelopment (Duncan and Ley 1982, Hamnett 1991), a critique that was somewhat short-sighted given the longevity of the theory and how useful it has become when interpreting current neoliberal urban systems in different parts of the world.

Rent gap theory is both a powerful theoretical and political device. It rests on the Marxist conceptualisations of the use and exchange value of land, the class-monopoly nature of land value (Harvey 1974, 2006) and the differential and absolute forms of rent (Marx 1995). The potential ground rent captured by the capitalist class, namely CGR2, has a class-monopolistic nature at the moment of its capture by redevelopers. In Chile, the lowered ground rent that petty land owners and original owner-residents are able to cash in at the moment they sell to private redevelopers amidst redevelopment frenzies in their neighbourhoods, is the residual value that is left to the appropriation of current landowners or tenants after most of the ground rent is absorbed by capitalists. CGR is a residual value from the monopoly nature of PGR.

Research on ground rent formation and capture can show the complex 'relational' ways in which redevelopment is produced and experienced by residents in certain neighbourhoods, and an increasing number of users beyond the 'traditional' public sector, property developers and residents. That is to say, there are also local and transnational investment funds, a vast array of estate agents, policy makers, speculators, some corrupt bribers and state officials, and lobbyists who manage to cash in portions of that gap, and sometimes contradict each other.[3] Today in Latin American cities, one can see an increasingly complex chain of multi-agent rent gap formation/absorption, and sometimes disputes among economic interests (see Gaffney 2015 for an interesting account of the situation in Rio de Janeiro). Key questions are: How many parties and who exactly captures the rent gap? What social and political effects are generated by this capture on different scales in the city and across society in general?

Bourdieu's 2005 book *The Social Structures of the Economy* pointed out that urban economics does not refer to some kind of equilibrium between supply and demand by perfectly rational agents, but more often is a product of the irrationality behind the unequal power exerted by the holders of capital. Sociological changes cannot be dissociated from the economic foundations of the process of urban change. Earlier, however, Neil Smith had been far more influential in these discussions. His materialistic and highly-politicised theory helps examine the socially complex nature of the process of formation, capture and accumulation of land value as understood as the use and exchange value of land. In my opinion, the riches of Smith's work are, first, that he pushed the very important (social) use value of urban land to the forefront of the discussion and, second, that he showed the geographical nature of certain economic agents' behaviour. This question implies the idea that the logic and practices of urban neoliberalism are far from being 'parallel' to social reproduction. Instead, neoliberalism embeds and transforms how spatial production is perceived and the way in which social and economic actors respond and participate (either as promoters, victims and/or opposers) within the neoliberal logics of urban production and reproduction, and neighbourhood change. Neil Smith founded a new radical geo-economic thinking.

We need a materialistic approach to the politics of urban neoliberalism

Neil Smith's perspective was always carefully scientific but his epistemology was always politically engaged with grassroots subaltern experiences and rights. His earliest work was the creation of a liberal theory of land value formation to demonstrate that land demand is driven essentially by a fraction of the capitalist class who not only have the power to decide where and when redevelopment occurs, but also to impose when it is valuable to speculate, using property and/or neighbourhood disinvestment as a means to generate enormous rent gaps that imply private profit in a future cycle of investment into redevelopment.

It is essentially a Schumpeterian idea of deliberate deterioration of the urban fabric. Smith clearly talked about 'creative destruction' and the need to control its deep social consequences in the cities of capitalism (Smith 2002). Much earlier, Schumpeter (1976) had synthesised capitalism as a 'process of creative destruction' that revolutionises its economic structures from 'within', incessantly creating new niches of investment by destroying the old ones (a very impressive work by Rachel Weber 2002 also makes this point even clearer). This means destruction and replacement of fixed capital (or consumption funds, to be more accurate in Marxian terms; an accuracy Weber fails to have), means of production, and physical and social infrastructures, according to the pace of new technological advances. This is done in order to keep high levels of competitiveness among suppliers in the market. Rent gap absorption is precisely one of the key forces that sustain the capitalist city as a machine of production and competition. Additionally, gentrification, according to Neil Smith, is a means of opening up spaces for new rounds of profit and accumulation, especially when the scale of gentrification involves comprehensive state policies for city transformation and the involvement of large-scale redevelopers and financiers.

In his 1996 book *Gentrification and the Urban Frontier*, Smith added the fundamental point that the ruling classes in urban societies not only lead the process of urban redevelopment as they 'discover', 'conquer' and 'colonise' key areas of the city, but they also exert significant power in order to discipline and control the residing society, especially by reshaping the way public spaces are lived in and perceived. This idea of so-called *revanchism* has been seldom applied to Latin American contexts with the exception of Swanson (2007) in Ecuador, who saw how indigenous beggars were expelled from the improving downtown areas as part of a generalised 'whitening' of the city. Whitening is also used by some neighbourhood activists in Mexico City to denote the many sides of revanchist gentrification over central areas, being (1) Ethnic whitening, (2) whitening of the buildings, both as result of the influx of affluent middle classes and their tastes, and (3) money laundering via property redevelopment. Yet the idea of revanchism has been paramount for critical urban studies, as it opens up a very political debate on what seems to be consubstantial with highly ideologised capitalist urban development following Haussmann's authoritarian reconfiguration of Paris in the 19th century. This is namely the heavy involvement of the state in the structure of geographical and economic urban change (Merrifield 2013).

Schumpeter had agreed with Marx that capitalism would ultimately come to be definitively exhausted; capitalism's apparently eternal waves of crises and post-crisis recoveries, however, have driven capitalists to continuously switch from the primary circuit to investing in the secondary circuit, in forms of fixed capital and, more specifically, the built environment and urban space (Harvey 1973, 2006). An example is the subprime mortgage financial crisis that first exploded in 2008 in the United States, the United Kingdom and parts of Western Europe, with devastating consequences for low-income, elderly and ethnic minority populations. There it was capital that, for more than a decade, had regularly switched from the secondary circuit into the speculative financial markets of mortgage-backed securities. Then, following the crisis, this money returned from the financial sphere to the secondary circuit to be converted into real estate capital again via the 'hyper-commodification' of urban land as CGR2. This is an important part of what gentrification essentially is. Urban ground rent lies at the foundation as it is the mechanism where investment is actually amplified by the highest and best use of land. Many urban redevelopment policies in cities around the world are actually applied in already inhabited spaces from where capitalism is extracting value without giving anything in exchange, similar to the extractive mining or fishing sectors in the Global South.

Marx and neoliberal urban land markets

The rent gap theory considerably helps with the measurement of gentrification-led social displacement and exclusion (López-Morales 2011, 2015a, 2016 and Shin 2009, both drawing on Slater 2006 and earlier works by Marcuse 1985), as the private capture of ground rent is always of a class-monopoly nature. In the words of Marx, ground rent only comes to exist when commodified and appropriated by a parasitic landowning bourgeoisie (Marx 1995). Notwithstanding, its value is explained by classical theorists as produced by physical characteristics like centrality. In Marx's categories, rent starts with the social transformation of nature. In fact, differential rents can be produced by differences in land productivity. The first form is Differential (ground) Rent I, explained by Marx as the surplus generated by the social work performed on a piece of land with better agricultural productivity. Second, due to the intensity of the capital invested in that land and the capacity of the capitalist class to technically transform the natural properties of land (or space), Differential Rent II emerges. Both types of rent are to be accumulated at the end of the day by a landlord, or the land-owning class (Borisov, Zhamin and Makárova 1965, Marx 1995). In the city, however, the natural characteristics of land are less important, while the social agency produced around it by the needs and interests that the city agglomerates are far more crucial. For instance, with rent levels established by human-produced scarcity, the power exerted by landowners or redevelopers can be decisive in the determination of a level of rent, since it allows them to hold back this resource, creating artificial scarcity, until they receive a positive return above all the differential rents.

Smith (1979) addressed and reinvigorated what Marx had said about ground rent, that rather than being an efficient rationing device for organising land and its related resources for meeting the productive needs of a society, it is produced by a social group, acting as a class and aiming at incrementing its price without producing anything for it. The city is a space where power relations matter. The urban ground rent is shaped by the interests of the land owners acting together as a class in order to raise the rate of return when there is not enough scarcity. This is a particular form of ground rent, seldom stressed by classical or neoclassical authors, and that usually largely exceeds differential land rents. Urban rent as 'class-monopoly rent' is based on the interests of the owning bourgeoisie (who in the case of Chile, gradually had come to own not only immense amounts of central land, but also the means of urban redevelopment. For a historical narrative on Chile's powerful Chamber of Builders, see Kornbluth 2013). By the 1990s, the corporate real estate sector was intervening in Chile's political and institutional arrangements, and influenced the urban land market, seeking on the one hand to obtain a better rate of return, and on the other to generate a surplus, as they owned a scarce commodity, land – the well-located city space and/or buildings – rented or sold on the market at a much higher price than its real production cost. They also owned the means of urban production, namely highly-sophisticated high-rise construction capacity.

The highly-influential Ecuadorian urbanist Fernando Carrión (2005) and the Chilean Inter-American Development Bank officer, Eduardo Rojas (2004) had convincingly advocated Latin American states to create the best conditions for market profit and reduce uncertainties as the only way to assure privately-led redevelopment of inner areas. Carrión's and Rojas' theories drew on a sort of 'there is no alternative' (other than privately-led redevelopment) type of urban policy. Largely heard by the Chilean state, their ideas reflected the kind of rationale the ruling elite needed to exercise in order to reduce uncertainty in land-use competition, and they were achieved through governmental regulation (planning controls, provision of infrastructure, subsidies, redlining, or even sometimes political corruption), providing the speculators or developers with reasonable expectations over the long run. Chile's state institutions and regulations, a 'superstructure' made by the ruling class, maintain this existing organisation intact, while allowing growth and accumulation but avoiding cyclical crises and controlling social discontent. The process of redevelopment of urban inner areas in Chile rests, essentially, upon the nature of liberalism, and its ideological, economic and political principle of private freedom, regulated by the exertion of class power (Jessop 2002).

In urban economies like Chile's, where the rate of owner-residence is high (around 80%), certain building regulations, other than upzoning certain areas, have played a role for the devaluation of certain spaces. For instance, guidelines that limit small-budget construction, namely the little-known policy exerted by the Santiago Centro municipality until 2006 that severely restricted small-building construction but simultaneously allowed high-rise private reconstruction by allowing the higher Floor Area Ratios, provided certain basic conditions (this case shows a sort of 'institutional' redlining, as seen by López-Morales, Meza and Gasic 2014). The latter means certain geographical areas are aimed at certain income groups, while social access to those areas is constrained insofar as land and property prices increase. The impact on housing prices results in large sectors of society 'finding barriers to operate in the market insofar as their "need for housing" is not paralleled by a payment capacity that guarantees [sufficient] levels of accumulation to the producers of this commodity' (Jaramillo 1981: 21; my translation of the original Spanish).

This particular form of class-monopoly rent in Santiago is a reified mechanism; it responds to the need for creating money without producing value. In fact, the parasitic class that monopolises the ground rent absorbs value created by the state via zoning regulations and the provision of infrastructure. This process also responds to the rules of the financial market, where the primary circuit of capital is radically moved into the secondary circuit – speculative and land-rent based – always under the condition that the urban realm offers better rates of return than the financial or commercial markets. This is also a plausible explanation for the process of massive reinvestment and renewal in formerly dilapidated areas in many modern global cities, and a powerful scientific political instrument to see their injustices created (Lees, Shin and López-Morales 2016: Chapter 3). I believe this is also one of Neil Smith's most important politico-economic contributions.

As Eric Clark (1995: 1497) stresses, the tension between actual and potential ground rents is far from being 'clinically clean of ties to power in social contexts nor of ties to the imagery of agents'. In fact, it is the opposite. For the appropriation of the rent gap, the state on the one hand, and private owners and investors on the other, have specific roles. The former creates the economic, legal, and administrative framework; the latter two respond to their private interests over land rent accumulation. However, the artificial creation of rent gap by diminishing the quality of life of an environment is inherent to the previous phases of every process of gentrification (Smith 1996). Obviously, this devaluation has plenty of social connotations, the most important one being the displacement pressure rent gap that private accumulation generates. Other effects are the diminishing quality of life and increased levels of stress often experienced by those users of a gentrifying neighbourhood that manage to stay put, and who do not face just friendly upper-income newcomers but the influx of real capitalists whose aim it is to rip off the value of their neighbourhood by creating strange environments that come from the hyper-exploitation of the land.

Conclusion

Some critical researchers in the Global South visualise gentrification as depending on state policies and huge economic investment in redevelopment. In this context, the classic 'icon' of the middle-class gentrifier is weak if compared to the great land purchasing power of developers. Lees, Shin and López-Morales (2016) claim that, as a result, even the terms 'state-led' gentrification and 'gentrifier' ought to be reconsidered in order to comprise a different kind of relationship between the agent and the producers. As shown in a significant number of cases, it is actually impossible not to see the power exerted by bankers, redevelopers and state officials both in the devaluation and revaluation process of urban change. Contrary to many sophisticated post-structural or post-colonial theorisations, rent gap is a simple idea that anti-gentrification activists can read, understand and debate, and then apply to their everyday realities to help nurture their struggle to challenge the dominant systems of urban segregation. This is the main reason of my work as an academic and activist in Chile and Mexico.

The concept of *laissez faire* does not exist in neoliberalism, but instead a market protected by continuous flows of tight redlining state regulations, housing subsidies and market-friendly land flexible upzoning. Both neoliberal and authoritarian capitalist states are systems of rapid capitalisation via private accumulation of land rent, thus resulting in privately-led urban market expansion. As Lees, Shin and López-Morales (2015) claim, cases show how urban space has become not only a condition for production (e.g. land that hosts manufacturing) but the element for production itself, subject to exploitation and hence accumulated without intervention of the primary circuits of capital. In short, with capital switching into secondary circuits, the ground rent and the enormous opportunities for its absorption and accumulation by the capitalist class via the rent gap, instead of profit extracted from labour exploitation, is exacerbated as a profitable commodity *per se*.

Still very few people have attempted to link gentrification economics with displacement and exclusion. Some 'effects' manifested in cities of the Global South (not clearly seen in the 1980s and 1990s when rent gap theorisations were at their height in the Global North), however, make it possible to argue that the conflicted nature of the land-economic side of gentrification is one of most crucial aspects of the injustices of capitalism in its current phase for many cities in the world. Smith's theoretical foundations are still vividly nurturing critical reflections on these areas.

Notes

1 I later met Neil Smith again during the 2010 AAG Conference in Washington DC. It gave me the opportunity to learn from his Trotskian perspectives on urban geography and understand how to create social power from radically variegated movements and socio-environmental agendas. In 2012 I invited him to visit Chile, as I wanted local people to know more about his legacy and be able to meet him in person. He accepted, but sadly passed away the following year.

2 Historically, the construction sector in Chile has been so powerful that since the 1970s, these redevelopers and investors have obtained enormous state subsidies aimed at increasing residential demand, with enormous numbers of housing purchase subsidies issued by the state in order to keep this system working. For instance, between 2000 and 2009, an equivalent of 25% of Chile's 2016 estimated number of households received a total of 1,296,683 housing subsidies aimed at the purchase of new units. Around 40% of these have not been met by actual purchases and still remain 'on hold' (Observatorio Habitacional 2016, www.ine.cl/canales/chile_estadistico/mercado_del_trabajo/encuestas_suplementarias/pdf/ingresoscompleta.pdf).

3 For instance, private real estate advice to mid-income property owners has become a very lucrative niche in Santiago over the past five years or so.

References

Borisov, O. S., Zhamin, V. A. and Makárova, M. F. (1965). in A. Vidal (ed.). *Diccionario de Economía Política*. Available at: www.eumed.net/cursecon/dic/bzm/index.htm.

Bourdieu, P. (2005). *The Social Structures of the Economy*. Cambridge: Polity Press.

Carrión, F. (2005). 'El centro histórico como proyecto y objeto de deseo', *EURE*, 31(93), pp. 89–100.

Clark, E. (1987). *The Rent Gap and Urban Change: Case Studies in Malmö 1860–1985*. Lund: Lund University Press.

Clark, E. (1995). 'The rent gap re-examined', *Urban Studies*, 32(9), pp. 1489–1503.

Duncan, J. and Ley, D. (1982). 'Structural Marxism and human geography: A critical assessment', *Annals of the Association of American Geographers*, 72(1), pp. 30–59.

Gaffney, C. (2015). 'Gentrifications in pre-Olympic Rio de Janeiro', [Special Issue 'Latin American gentrifications' edited by E. López-Morales, H.B. Hyun and L. Lees], *Urban Geography*, 37(8), pp. 1132–1153. doi:10.1080/02723638.2015.1096115.

Hackworth, J. (2007). *The Neoliberal City: Governance, Ideology, and Development in American Urbanism*. Cornell: Cornell University Press.

Hammel, D. J. (1999). 'Re-establishing the rent gap: An alternative view of capitalised land rent', *Urban Studies*, 36(8), pp. 1283–1293.

Hamnett, C. (1991). 'The blind men and the elephant: The explanation of gentrification', *Transactions of the Institute of British Geographers*, 16(2), pp. 173–189.

Harvey, D. (1973) *Social Justice and the City*. Baltimore: The Johns Hopkins University Press.

Harvey, D. (1974). 'Class, monopoly rent, finance capital and the urban revolution', *Regional Studies*, 8(3–4), pp. 239–255.

Harvey, D. (2006). *The Limits to Capital*. New York: Verso.

Jaramillo, S. (1981). *Producción de vivienda y capitalismo dependiente: el caso de Bogotá*. Bogotá: Uniandes.

Jessop, B. (2002). 'Liberalism, neoliberalism, and urban governance: A state-theoretical perspective', *Antipode,* 34(3), pp. 452–472.
Kornbluth, D. (2013). *El papel de los gobiernos y la Cámara Chilena de la Construcción en la configuración de las políticas de vivienda para la ciudad de Santiago de Chile: Continuidades y disrupciones.* Master Thesis. Instituto de Historia, Facultad de Filosofía y Educación, Pontificia Universidad Católica de Valparaiso.
Lees, L., Shin, H. B. and López-Morales, E. (2015). *Global Gentrifications: Uneven Development and Displacement.* Bristol: Policy Press.
Lees, L., Shin, H. B., and López-Morales, E. (2016). *Planetary Gentrification.* Cambridge: Polity Press.
López-Morales, E. (2011). 'Gentrification by ground rent dispossession: The shadows cast by large scale urban renewal in Santiago de Chile', *International Journal of Urban and Regional Research*, 35(2), pp. 330–357.
López-Morales, E. (2013). 'Gentrificación en Chile: aportes conceptuales y evidencia para una discusión necesaria', *Revista de Geografía Norte Grande*, 56, pp. 31–52.
López-Morales, E. (2015a). 'Gentrification in Santiago, Chile: a property-led process of dispossession and exclusion', [Special Issue 'Latin American gentrifications' edited by E. López-Morales, H. B. Hyun and L. Lees], *Urban Geography*, 37(8), pp. 1109–1131. doi:10.1080/02723638.2016.1149311.
López-Morales, E. (2015b). 'Gentrification in the global South', *City*, 19(4), pp. 557–566.
López-Morales, E. (2016). 'Assessing exclusionary displacement through rent gap analysis in the urban redevelopment of inner Santiago, Chile', *Housing Studies*, 31(5), pp. 540–559.
López-Morales, E., Meza, D. and Gasic, I. (2014). 'Neoliberalismo, regulación ad-hoc de suelo y gentrificación: el historial de la renovación urbana del sector Santa Isabel, Santiago', *Norte Grande*, 58, pp. 161–177.
Marcuse, P. (1985). 'Gentrification, abandonment and displacement: Connections, causes and policy responses in New York City', *Journal of Urban and Contemporary Law*, 28, pp. 195–240.
Marx, K. (1995). *Capital – Volume Three: The process of capitalist production as a whole*, in Marxists Internet Archive (ed.). Available at: www.marxists.org/archive/marx/works/1894-c3 /.
Merrifield, A. (2013). 'The urban question under planetary urbanization', *International Journal of Urban and Regional Research*, 37(3), pp. 909–922.
Observatorio Habitacional (2016). *Ministerio de Vivienda y urbanismo.* Available at: www.observatoriohabitacional.cl.
Robinson, J. (2011). 'Cities in a world of cities: the comparative gesture', *International Journal of Urban and Regional Research*, 35(1), pp. 1–23.
Rojas, E. (2004). *Volver al Centro. La recuperación de áreas urbanas centrales.* Washington, DC: IDB.
Schumpeter, J. A. (1976). *Capitalism, Socialism and Democracy.* London: Routledge.
Shin, H. B. (2009). 'Property-based redevelopment and gentrification: The case of Seoul, South Korea', *Geoforum*, 40(5), pp. 906–917.
Slater, T. (2006). 'The eviction of critical perspectives from gentrification research', *International Journal of Urban and Regional Research*, 30(4), pp. 737–757.
Smith, N. (1979). 'Toward a theory of gentrification: A back to the city movement by capital not people', *Journal of the American Planning Association*, 45, pp. 538–548.

Smith, N. (1996). *The New Urban Frontier: Gentrification and the Revanchist City.* New York: Routledge.

Smith, N. (2002). 'New globalism, new urbanism: Gentrification as global urban strategy', *Antipode*, 34(3), pp. 427–450.

Swanson, K. (2007). 'Revanchist urbanism heads South: The regulation of indigenous beggars and street vendors in Ecuador', *Antipode,* 39(4), pp. 708–728.

Weber, R. (2002). 'Extracting value from the city: Neoliberalism and urban redevelopment', *Antipode,* 34(3), pp. 519–540.

PART III

Dispossession and class struggle

9 From Boise to Budapest

Capital circulation, compound capitalist destruction and the persistence of homelessness

Don Mitchell

It is the considered legal opinion of the United States Department of Justice that laws targeting homeless people in the small western American city of Boise, Idaho – and, by extension, similar laws across the United States – are *cruel*. In August 2015, the Department of Justice announced its intention to join a lawsuit in Boise on the side of the homeless plaintiffs. The plaintiffs charged the city with 'cruel and unusual punishment' – that is, with a violation of the Eighth Amendment in the Bill of Rights – because the city cites and sometimes arrests homeless people who 'camp' in public spaces when there are no available shelter beds.[1]

In United States cities, 'camping' has a particular meaning. Homelessness has been a crisis in the United States for thirty-five years now, and in all those years, cities have continually experimented with laws to regulate homeless people or to banish them altogether from the streets, frequently running into legal problems as they have done so. For example, early efforts to criminalize sleeping in public foundered on uneven enforcement: Dozing office workers on their lunch breaks or folks having a post-picnic siesta were not charged with sleeping in public while visibly homeless people often were. Such class legislation is frequently invalidated, and so cities have replaced anti-sleeping ordinances with anti-camping ordinances. Camping is sleeping with any 'appurtenances' such as a blanket or sleeping bag, a tent or other shelter. In many cities, this includes anything that could be used for camping purposes, such as a warm jacket, a cardboard box (used as a mattress or shelter) or as in the case of Boulder, Colorado, a tree or bush that might provide at least a bit of shelter from the rain and snow (as well, perhaps, as privacy) (Mitchell, Attoh and Staeheli 2015). Anti-camping in American cities means anti-sleeping if you do not have approved shelter.

In a press release, the Department of Justice said '*it should be uncontroversial* that punishing conduct that is a universal and unavoidable consequence of being human violates the Eighth Amendment… Sleeping is a life-sustaining activity – i.e. it must occur at some time and in some place. If a person literally has nowhere else to go, then enforcement of the anti-camping ordinance against that person criminalizes her for being homeless'.[2] *It should be uncontroversial*, but it's not. By the end of the month, for example, former New York Mayor Rudolph Giuliani weighed in, complaining about what he perceived as a declining quality of life in his Upper East Side neighbourhood. 'Do you know when people lived on the streets and didn't use bathrooms inside?' he asked in an interview on local television, 'It's called the Dark Ages' (Bellefante 2015). New York tabloids simultaneously went on a rampage against the presence of homeless people on the city's streets. New York's homeless population has continued to grow, as it did all throughout Giuliani's 'quality of life' campaigns in the 1990s, and as it continues to do across the country. As the National Law Center on Homelessness and Poverty annually documents, laws against homeless people, and people who aid them by handing out meals in public, continue to proliferate. Overriding it all seems to be a simple, stark desire to just be rid of homeless people. As the mayor of San Francisco remarked in mid-August, all the homeless people in his city 'were going to have to leave' in time for the Super Bowl in February. He did not say how this was to be accomplished, but the same article noted that in San Francisco whether it was worth it to 'even bother trying' to find shelter for homeless people was a 'contentious debate' (Nir 2015). Meanwhile, in New York, Giuliani followed up his comments with a recommendation that the city return to doing what his administration did for the homeless: 'You chase 'em and you chase 'em and you chase 'em' Giuliani told a reporter 'and they either get the treatment they need or you chase 'em out of the city' (Flagenheimer, Stewart and Navarro 2015).

Such a climate of official and popular opinion, which has been a fairly consistent climate for thirty years (even as it is crosscut at times with policies promoting and providing funding for shelters, and even rare moments of citizens rebelling against plans to criminalize homelessness, as with a 2012 vote in Berkeley, California refusing to enact a 'sit-lie' law, that would have made it illegal for homeless people to sit or lie on sidewalks) (Mitchell 2016). This makes the Department of Justice's intervention in Boise all the more important. It is the first federal recognition of a constitutional right to *be* on the streets. The Department's language in its court filing reads like it could have been taken right from the pages of a National Law Center on Homelessness and Poverty report, or from the pages of any number of scholars who have for years been seeking to describe and analyse the spatiality of homeless people's ability to *be* within the context of laws that seek to outlaw their very presence.

Yet the Department of Justice's action also raises three key questions: First, why is it that homelessness has persisted – and indeed continued to grow – in contemporary cities? Second, why has a major and consistent response been the criminalization of those made homeless? Third, what does it mean that the Department of Justice has intervened *now* and in this form? In this chapter, I will develop some answers to these questions and, as I trust will become clear, the answers are pertinent not just in the United States but around the world in an era of globally uneven capitalist development, since the answers hang on a more general theory of capital circulation, the shifting fortunes of class struggle and the changing role of the city *within* global capitalism.

In the analysis that follows, I will suggest that the first issue – why homelessness persists and why it grows – is *not* primarily or in the first instance a housing question as important as the question of housing is; rather, it is a fundamental question of how capital accumulation proceeds. The second issue – why anti-homelessness persists, why cities remain *revanchist* – is primarily *not* a question of social psychology (as critiques of urban revanchism often suggest), but is rather a very real, material response to very real, material processes of accumulation in the built environment, processes that require the production of city-space as abstract space. That the third issue, why the Department of Justice is intervening now, is not primarily a question of the state seeking to support human rights, but rather a result of concerted class struggle as well as a dawning realization that anti-homelessness just does not work as a method of safeguarding capital circulating through the built environment. Anti-homelessness is not always in the capitalist class's interest.

Capital circulation and the revanchist city

The best way to make such an argument is by working historically, by revisiting the early decades of the homelessness crisis and working forward. As Neil Smith (1996) made clear in *The New Urban Frontier*, the rise of homelessness was at the heart of contemporary revanchism in its birthplace, New York City. If the 1980s – the second decade of the urban crisis and the first decade of the homelessness crisis – in New York was a decade of struggle and ferment, then the 1990s was – similar to other cities – a particularly vicious decade. The 1990s are not usually thought of in this way. After all, the Thatcher-Reagan revolutionaries had ceded the pinnacles of power to the Clinton-Blairites. The 'authoritarian populism' that Stuart Hall (1979) identified as being at the heart of Thatcherite ideology (if not always in practice) had given way to what might be called the 'therapeutic authoritarianism' of Clintonism. In the 1980s, as 'the great U-turn' (Harrison and Bluestone 1988) of advanced capitalist economies was getting fully underway – with factory after factory being shut down, all pretence to social welfare being eviscerated, and whole cities being abandoned to their deindustrialized fate – the obvious ill and uneven effects called up concerted opposition. In the United States, as city streets became ever-more populated with the discards of postindustrialism – the new homeless (Hoch and Slayton 1989) – activist organizations around the country fought back. The class dimensions to both the wrenching political-economic changes and the remaking of cities were clear for all to see.

To be sure, activists, religious people and others scrambled to provide shelter and food, forcing cities to create homeless shelters in disused buildings, dedicating church halls to housing the poor and opening food pantry after food pantry and free meal program after free meal program. Yet they also fought politically, forcing Congress to pass and President Reagan to sign the McKinney-Vento bill that provided significant funding for homeless services.[3] Yet all of this activity, coupled with the professionalization of homeless services that McKinney-Vento promoted, seemed to have little effect on the number of people living on the streets. Homeless encampments ('Reaganvilles,' as they were called) became features of many American cities, sometimes becoming sites of significant organizing – places where homeless people came to recognize themselves in their commonality and to seek out ways to act collectively and in their own best interests.

At the same time, however, what pundits liked to call 'compassion fatigue' began to set in among many of the housed, as little in the official response to that point had alleviated the crisis of homelessness. Homeless people began to be widely seen less as a crisis and more as a scourge and a threat to others, to the city and even to themselves. Homelessness was redefined in the 1990s. It was now not an obvious crisis of *society* but an attribute of disordered and disorderly individuals. There's no such thing as society, only individuals and families, Margaret Thatcher famously declared. By the end of the 1980s and into the 1990s, more and more of these individuals and families were living on the streets, begging, seeking shelter under bridges, in river bottoms, and in parks, crowding doorways, and spending their days lugging their belongings from place to place in shopping carts. The scourge of these unproductive, wasteful and by-now undeserving individuals had to be addressed.

There was a dual response, and both worked to depoliticize homelessness and the homeless. First, there was a doubling down on the professionalization of homelessness services, including the codification of what is known as the 'continuum of care' model for addressing homelessness. This model declared homelessness not to be a condition of society but rather an attribute of individuals themselves. In this way of reasoning, people are homeless because they are mentally ill, or because they are alcoholics or drug addicts, or because they are uninterested in being a productive member of society. Such a position denies the class status of homeless people (as part of the permanent and latent reserve army of labour), as well as the political-economic conditions that produce homelessness, and instead focuses on how homeless people can be reformed or cured by inducement, interventions of care or compulsion: Here arises the therapeutic authoritarianism of the era. Homelessness is largely a moral failing on the part of individuals. What they need is either the proper incentives to get their lives together or what is invariably called 'tough love', a presumably compassionate, but ineluctably strict supervision (by charity, religious overseers or the police) so they can be 'set straight'.

Second, and working completely in tandem with the first strategy, was the invention of laws to penalize those who did not avail themselves of the tough love on offer and those who continued to sit on the streets, sleep in parks, beg at bus stations or build somewhat autonomous encampments. Anti-homeless laws worked fist in glove with the development of an archipelago of care in almost every American city. Since the continuum of care's therapeutic authoritarianism was available, laws criminalizing homelessness were understood by many not only to be needed – for society and for the good of homeless people themselves – but also to be *just*. They were a means of restoring order and civility on the streets, in the words of prominent anti-homeless law advocate Rob Tier (1998). They were the mechanism by which police were authorized to 'chase and chase and chase', and they still are.

The word 'restoring' is crucial here of course. For Neil Smith (1996), the paradigmatic case was New York City's efforts to 'reclaim' Tompkins Square Park in 1988. Tompkins Square, on New York's Lower East Side, had become home to a large encampment of homeless people – the 'evicted', as Rosalyn Deutsche (1996) accurately described them – in the midst of a neighbourhood gentrification was in the late 1980s just beginning to reach. Long an immigrant enclave, the Lower East Side had suffered decades of divestment but (despite something of a bombed-out appearance) still retained an interesting (and affordable) housing stock in an enviable location, much of it occupied by squatters. A radical guerrilla gardening movement had softened some of the Lower East Side's sharp edges, while creating space for organizing. Political organizing flourished throughout the neighbourhood, not least among homeless people. Symbolically, Tompkins Square Park represented what had to be reclaimed if the middle and upper class 'reconquest' of New York City was to be complete. It represented all that had gone wrong with the city in the 1960s and 1970s, from the counter- and drug-cultures, to the rise of new forms of radical politics, to the occupation of the streets and housing by 'unproductive' classes. It also made it startlingly visible that there were 'losers' in the new economy, losers who themselves seemed to be threat to the social and economic order.

The effort to reclaim Tompkins Square Park, which entailed violently evicting the homeless, closing the park entirely for several months, reopening it with a guarded fence around it, controlling access (limiting it primarily to neighbourhood children and their parents), totally re-landscaping it to make camping and occupation – or merely lying down – all but impossible, the writing of stringent new rules of use including bans on lying down on benches, still enforced to this day, was all part of a large cultural shift at the end of the 1980s that solidified the 'all-out attack on the social policy structure that emanated from the New Deal and immediate postwar era', as Smith (1996: 44) argued. The attack expressed a desire for 'revenge against minorities, the working class, women, environmental legislation, gays and lesbians, [and] immigrants…. Attacks on affirmative action and immigrant policy, street violence against gays and homeless people, feminist bashing, and public campaigns against political correctness and multiculturalism were the most visible vehicles of reaction' (Smith 1996: 44–45). The post-Keynesian city was a *revanchist city*.

Smith's argument, in its immediate historical context, was quite correct as Tom Slater argues in this volume, but naming the city 'revanchist' was also, in some ways a misnomer. 'Revanchism' means 'revenge', but what marked the city of the late 1980s and into 1990s was not so much revenge as a broad-based vengefulness. While there was a concerted effort to restore elite class power, and while there was a definite sense among many that the city and society more broadly had been 'lost' to liberals, minorities, gays, punks, homeless people and so forth, the political logic of *re*venge (by a particular class faction or an alliance of them) quickly gave way before a broader sense of historical and geographical entitlement by the rising classes of the new economy. All that stood in the way of this rising class had to be swept aside. The city was to be reclaimed with a vengeance.

But by whom and for what? Naming a city 'revanchist' does not directly answer this question, nor does it directly answer the first two questions I laid out earlier: Why is it that homelessness persists and continues to grow? Why are anti-homeless laws seen to be a reasonable response to this fact? To put it another way, it is a reasonable hypothesis that the reassertion of class power Smith identified had a reason, and we need to determine what it is. Such a determination begins by noting the obvious. The city has become – and is increasingly becoming – a significant, even primary site for capital accumulation. Gentrification, a particular form of urban development and especially of capital circulation, was a global class strategy, Smith (2002) argued, because accumulation through the built environment had become a global class necessity.

The outlines of the story are well known. The Keynesian reforms of the 1930s and World War II – forced in part by the manifest contradictions of liberal capitalism, in part by internal radical class struggle and the rise of powerful working class organizations and in part by the external pressure exerted by the socialist world – had, on the one hand, forced the socialization of significant aspects of reproduction, from housing to health care to education (with significant effects on the morphology and social geography of the urban landscape) and, on the other hand, wrested a larger share of capital surplus into working people's hands. By the 1960s, falling rates of profit and rising real wages (to say nothing of inflated military budgets) were creating structural pressure on capital accumulation, while those excluded from the general prosperity of the era (in the United States, primarily African-Americans), those at the blunt end of the bulldozers that reworked the urban landscape to make room for the mega-redevelopment projects of the modern era (again, primarily African-Americans), and those left to live amid the rubble of disinvestment that the fleeing newly prosperous white working class left in its wake (yet again, primarily African-Americans), fought back. The crisis of accumulation was coupled with an urban crisis. The resulting devastation left a social field in need of concerted policing and a formerly urban field ripe for reinvestment.

First, a new regime of accumulation had to be invented. The accumulation and fiscal crises linked to falling rates of profit had to be resolved. As test-run in post-coup Chile, the eventual solution was, of course, a massive deregulation of industry and finance, helping to create truly global processes of production and a worldwide market for finance capital, together with the evisceration of social welfare and the ratcheting up of the repressive forces of the state, especially against working class organizations (Harvey 2005, Mann 2013). Together these moves reset the value of labour power driving it down in ways far exceeding what could be accomplished through the socialization of social reproduction, while freeing up a significant portion of accumulated surplus to stalk the globe looking for productive sites for investment. The competitive pressure exerted by capital on capitals was sharply increased. A new, vicious round of accumulation accomplished through both increased absolute surplus value production (working days have been getting longer and longer since the 1970s and are now coupled with a growing use of precarious, 'zero-hour' labour) and increased relative surplus value production (we are all about to be made redundant by machines) has, near-constant economic crisis notwithstanding, reinvigorated capitalism, now in a neoliberal form (Harvey 2010).

Near economic crisis is, however, *not* ever notwithstanding. Crisis is a central, indispensable, part of the process. Capitalism is in a permanent crisis of overaccumulation. It has a massive realization and absorption process. Capitalist growth, as David Harvey (2014) has insisted, is and must be, compound growth. At historical rates of growth – about 2.2% globally – the size of the capitalist economy doubles every thirty years. Something like double the amount of capitalist value must be realized and something like double the amount of capitalist value must find productive outlets for investment today compared to 1985, the heyday of Reagan-Thatcherism. This same, roughly double amount will have to find productive outlets again in 2040.

Compound capitalist destruction: Abstract space and the persistence of anti-homelessness

The first thing to remember in relation to this fact is that the general law of capitalist accumulation never stops operating. The growth of capital at one pole, the law states, is the immiseration of ever growing numbers of people at the other pole. This is a logical – and historical – outcome of the capitalist process. Capital accumulation *must* produce misery (Marx 1987). Some portion of that misery has always been lived by the reserve army, both permanent and latent. With the rising organic composition of capital – which has been the purpose of, and accelerated by, the tech revolution we are all living through, more and more people are simply made redundant. Capital accumulation *necessarily* produces unemployment, poverty and homelessness. If these were ameliorated (not solved) through the housing, make-work, education, welfare and healthcare policies of the Keynesian era, then this is no longer the case.

The result is a massive problem of management as the *necessary*, and *necessarily growing* byproduct of capitalist accumulation litters city streets, seeks out housing by building up favelas, tent cities or squatting abandoned buildings, participates in black-market survivalism, wastes out its days hanging out by the corner stores or abandoned school yards, or takes over city centres in their indignation. Vigorous policing is one management solution. Anti-homeless laws, anti-social behaviour orders, stop-and-frisk policing, area bans, anti-squatting laws and policies, and kettling all find their rationale. Another management solution since social housing, universal education (formerly an important place to park a portion of the reserve army) or wage support seem unthinkable in an age of austerity (to the capitalist political class at least) is to rely on the charitable sector to succour those spit out of the great capitalist maw (David Cameron's Big Society; George Bush-the-Elder's Thousand Points of Light).

The second thing to remember about capitalism's necessary compound growth is that all that new-made surplus capital must go *somewhere*. Rising organizing compositions of capital do not just make people redundant; they also diminish the *rate* (if not the mass) of return on invested capital. Competitive pressures to increase the rate of return thus means that capital must seek out new sectors for investment. Commodity markets and production systems are therefore notoriously volatile, especially as the realization problem – the problem of selling all that is produced – never diminishes. Investment in production, and for that matter in retail, can quickly lose its lustre, but the built environment might be a different story. In the first place, capital and capitalism need a 'broadly suitable' landscape (as Harvey 1982 puts it), and suitability shifts as the logics and locations of capitalist production, circulation and accumulation shift. In order for some capital to circulate, other capital must be fixed in place, so new built environments must always be produced. Chances to make money abound. Not only are their delicious possibilities of rent, but also and perhaps especially (if counter intuitively), the relatively slow turnover times of capital fixed in the built environment make urban landscapes an enticing site for investment.

All this is, however, contradictory. Capital frozen – capital fixed in place – while an absolute necessity for circulation and accumulation, is also capital at risk. Production needs can and do a shift, innovations in transportation and communication make landscapes redundant. Hurricanes and earthquakes hit. Wars are launched. Landscapes are destroyed along with the investments in them. Destruction is a primary means of devaluation. Yet, in a finite world, a world where, as Neil Smith (1990) detailed, absolute geographical expansion of capital is no longer possible, a world where uneven development is now fundamental to the circulation and accumulation of capital, such destruction is also a necessary part of the dynamic. Destruction of capital is vital to the accumulation of capital. Compound capitalist growth entails compound capitalist destruction, destruction that prepares the ground for another round of accumulation. Such a dynamic is, of course, right at the root of Smith's theory of gentrification, which itself was inextricably bound to his theory of uneven development: The rent gap is a phenomenal form of compound capitalist destruction in particular places, given here the deceptively passive name 'disinvestment'.

Compound capitalist destruction – disinvestment, the rent gap – is not some sort of *deus ex machina*, but is rather fully and inescapably *social*. It is *lived*. This is true in four senses. Obviously, first, it is true for those living at the blunt end of the bulldozers or those living in housing being asset-stripped by its landlords, those toiling in abandoned factories, those in the paths of the levelling bombs, those flooded out by unmaintained levees. No matter how disinvested and how seemingly abandoned, these are landscapes of home, of life and of memory, and people will defend them while seeking to make and remake their lives in them anyway. This is especially true for those urban landscapes targeted for what Ruth Wilson Gilmore (2008) identified as 'organised abandonment' by capital and the state – planned devaluation – but within which this abandonment has not yet been fully realised.

Second, capital is fractionalised. Different capitals have different social and economic needs, desires and logic. While it is tempting to think of all capital as fully mobile, or fully footloose, only a fraction of capital is mobile in that way. Some capital is rooted in place: The preservation of *its* value – and its ability to spin off surplus value (or at least revenue) – requires the building up and preservation, rather than the destruction, of values all around it. It is in the interests of some fractions of capital to assure that the asset-stripping equity fund does *not* land in their neighbourhood. Intra-capitalist struggle is a vital part of the dynamic of capital circulation through the built environment. Place-bound capital seeks inward investment of capital perceived and hoped to be relatively stable, and hence all manner of inducements might be offered to attract it, from free land to tax breaks to the provision of elaborate, state-funded infrastructure. Mobile capital, to this fraction, is something to be deeply desired – and an existential threat.

What this looks like socially can be seen in any local development authority or chamber of commerce meeting, or at the mayor's office as she figures out how to attract the latest starchitect *and* repair the aging sewer infrastructure, or to manage the homeless who seem to be hanging around in ever-greater numbers and scaring residents, tourists and footloose capital away from the newly-gentrified warehouse district. For, third, the threat to local fractions of capital, and thus to local economies and lives, is not only existential and not only applied to certain fractions of society. Rather, it hunkers down and makes its presence felt in the tax coffers. As the neoliberal revolution rolled back taxes on high-income and high-wealth individuals and especially on corporations, localities have become even more dependent on revenues from sales and property taxes. Both are dependent on the 'success' of the urban landscape – on its development or redevelopment rather than its abandonment. The latter – property tax – is especially so dependent, not only because property values are relational, a function of the overall rent surface, but because increasingly the development tool of choice, in American cities at least, is tax-increment financing. With tax-increment financing, eventual payment for public investment in preparatory infrastructure – creating the conditions for inward investment – is derived solely from that portion of property value that is an *increase* over the values that had prevailed before redevelopment, not the total value of the properties. City revenues are dependent on continually increasing exchange values in the built environment. Anything that threatens such increases in exchange value must be confronted.

Fourth, and perhaps by now quite obvious, the dynamics of investment and disinvestment – the dynamics of capital circulation in and through – the built environment is a multifaceted struggle. It is a struggle over how abstract space – that is, space made commensurable or exchangeable – is to be produced. Abstract space is like abstract labour (Lefebvre 1991). In order to be useful, labour must be highly differentiated, but in order to be purchasable, labour must be commensurate. Marx's (1987) category of abstract labour helps us see the dialectic between difference and commensurability, and turns our attention to the question of how – through what sorts of social struggles and processes – differentiated labour is made exchangeable, how we know that X amount of janitorial labour is equal to Y amount of construction labour and Z amount of engineering labour. Lefebvre's (1991) category of abstract space performs the same function. In both cases these are concrete abstractions, which is to say, they are not just concepts but describe a real material basis. The built environment possesses value and what this value is, is a function of the degree to which that value is realized in relation to the values of other built environments. There is nothing abstract about struggles over abstract space. Precisely how these struggles get worked out – how space is abstracted and values are (or are not) realized – is never assured and never clear in advance. The struggles must be fought, and someone must win, or some sort of compromise and cease fire must be reached that properly sets and realizes value and does not entirely threaten the continued accumulation of capital, must be put into place (Mitchell 2016).

We now know why the class reconquest of the city has come with a vengeance. It is not only because certain people of certain classes – which may in fact include working and middle class homeowners – have come to think of the city as rightfully theirs (though this should not be discounted) but because it is theirs, both in concrete terms and in the abstract (the abstraction of abstract space). Their investment in property has given them a property right in investment, and they will do what they must to protect their investment and the value their property stores and which they expect to realise over time. They are opposed by other fractions of capital, or if not opposed, then certainly disciplined, since those other fractions might leave for greener pastures if not properly accommodated. Capital-strikes are an everyday occurrence and a non-stop threat to fractions of capital as well as working people, and the reserve army.

We now also have answers to our first two questions. Why has homelessness persisted? Because poverty is necessarily produced as part of the normal workings of capital accumulation and the sorts of ameliorations that marked Keynesianism have been eviscerated. At the same time, since cities are central sites of capital accumulation, built environments not put to their 'highest and best use' must be destroyed, remade or expropriated. Rents must increase, and people must be evicted. Why are anti-homeless laws a perennial solution? Because the visible homelessness that results is a threat to continued accumulation and to the processes of abstract space production. Charity and care have their role in managing the homeless population (Cloke, May and Johnsen 2010), but relying on philanthropy is simply inadequate. Anti-homeless laws are a class strategy (organized through the state); charity (at the whim of individuals) is not, at least not in this sense. 'Chasing, chasing, chasing' the homeless – working to assure they just 'leave' – begins to make social and seemingly political-economic sense.

The persistence of homelessness and the limits to anti-homelessness

Yet we still have not answered the third question: Why has the Department of Justice now intervened against the 'chase, chase, chase' strategy?

The answer is very straightforward. Anti-homeless laws don't work. Short of genocide, homeless people cannot be made to 'just leave'. Chasing them around does little beyond local good. Homelessness continues to be produced and homeless people continue to be spit out by the general law of capitalist accumulation. In America, at least, the limits to the criminalisation of homelessness as a class strategy are readily apparent. Anti-homeless laws have been around for thirty years, and as they have come to cover more and more of the American urban landscape they have done nothing to lessen the burden of homelessness, or the threat that homeless people are perceived to cause. Yet, like neoliberalism, which Smith (2008) famously declared to be dominant but dead, anti-homeless laws remain dominant but dead. Indeed, anti-homelessness is spreading. The European homelessness federation, FEANTSA, recently detailed the spread of punitive homelessness policies across Europe, from here in Barcelona to Budapest and from Poland to Belgium (Fernàndez-Evangelista 2013). Vicious legal attacks on homeless people are becoming as rife in Europe, as the austerity kick seeks to complete the dead and deadly neoliberal project, as they are in America. The reasons are the same. The landscape has become, and remains – crisis-ridden as it is – a primary site for capital accumulation in Europe. FEANTSA details the futility of the criminalization kick. The Department of Justice under the Obama administration, an administration fully beholden to the finance capital class fraction, similarly seems to be convinced of their futility and their expense. Much of their court filing in Boise is given over to the unnecessary expenses related to criminalization as opposed to other modes of intervention. The Department is less interested in the rights of homeless people and more in the fiscal strain they place on local and national states.

The Department of Justice has no interest in creating a City for All, to use the name of an impressive Right to the City-based political organization for homeless people and their allies in Hungary. City for All, according to one of its organizers, Balint Misetics (2013: 216), 'is a fierce critique of false and superficial solutions to homelessness, such as dominant government public policy responses which essentially consist of large, dormitory-style shelters, street social work, and further emergency measures in the winter' – all responses the Department of Justice in fact promote in their legal filing – but instead City for All 'emphasizes the root causes of homelessness: poverty and social inequality, the lack of affordable housing, the absence of an enforceable right to housing, the lack of an extensive system of social housing, or the underutilization of the small amount of public housing there is'. To push for this, City for All engages in class organizing in an effort to confront the class power that now dominates cities and the political economy.

By contrast, the Department of Justice is seeking to protect or safeguard class power and in finding new ways to safeguard accumulation even as the processes of capital accumulation necessarily create homelessness. In supporting the rights of homeless people to *be* in the city, by, perhaps inadvertently supporting homeless people's right *to* the city, the Department may have uprooted the dead plant of neoliberal capitalism to allow something else to grow in its place. That is the task the United States Department of Justice has inadvertently set us.

Notes

1 'Statement of Interest of the United States', *Bell et al v. City of Boise et al*, Civil Action No 1:09-cv-540-REB, August 6, 2015.
2 Press Release, 'Justice Department Files Brief to Address Criminalization of Homelessness', Office of Public Affairs, Department of Justice, August 6, 2015, emphasis added.
3 The discussion here reprises an argument made in Mitchell (2011).

References

Bellefante, G. (2015). 'Rudolph Giuliani's outrage on homelessness, and Richard Gere's' *New York Times*, August 30, p. MB1.

Cloke, P., May, J. and Johnsen, S. (2010). *Swept Up Lives? Re-Envisioning the Homeless City.* Oxford: Wiley-Blackwell.

Deutsche, R. (1996). *Evictions: Art and Spatial Politics.* Cambridge, MA: MIT Press.

Fernàndez-Evangelista, G. (ed.) (2013). *Mean Streets: A Report on the Criminalization of Homelessness in Europe.* Brussels: Brussels: Housing Rights Watch/FEANTSA/ Fondation Abbé Pierre.

Flagenheimer, M., Stewart, N. and Navarro, M. (2015). 'After playing down a homeless crisis, Mayor De Blasio changes his tune', *New York Times*, September 2.

Gilmore, R. W. (2008). 'Forgotten places and the seeds of grassroots planning', in C. Hale (ed.). *Engaging Contradictions: Theory, Politics, and Method of Activist Scholarship.* Berkeley: Global Area and International Archive, pp. 31–61.

Hall, S. (1979). 'The great moving right show', *Marxism Today*, January, pp. 14–20.

Harrison, B. and Bluestone, B. (1988). *The Great U-Turn: Corporate Restructuring and the Polarizing of America.* New York: Basic Books.

Harvey, D. (1982). *The Limits to Capital.* Chicago: University of Chicago Press.

Harvey, D. (2005). *A Brief History of Neoliberalism.* New York: Oxford University Press.

Harvey, D. (2010). *The Enigma of Capital and the Crises of Capitalism.* New York: Oxford University Press.

Harvey, D. (2014). *Seventeen Contradictions and the End of Capitalism.* New York: Oxford University Press.

Hoch, C. and Slayton R. (1989). *New Homeless and Old: Community and the Skid Row Hotel.* Philadelphia: Temple University Press.

Lefebvre, H. (1991). *The Production of Space.* Oxford: Blackwell.

Mann, G. (2013). *Disassembly Required: A Field Guide to Actually Existing Capitalism.* Edinburgh: AK Press.

Marx, K. (1987). *Capital, Volume 1.* New York: International Publishers.

Misetics, B. (2013). 'The city is for all from the "Human rights-based" perspective', in G. Fernàndez-Evangelista (ed.). *Mean Streets: A Report of the Criminalization of Homelessness in Europe.* Brussels: Housing Rights Watch/FEANTSA/Fondation Abbé Pierre, pp. 213–219.

Mitchell, D. (2011). 'Homelessness, American style', *Urban Geography*, 32, pp. 932–956.

Mitchell, D. (2016). 'People's Park again: On the ongoing history of the end of public space', *Environment and Planning A*,. Available at: doi:10.1068/a140435p.

Mitchell, D., Attoh, K. and Staeheli, L. (2015). 'Whose city? What politics? Contentious and non-contentious spaces on Colorado's front range', *Urban Studies*, 52, pp. 2633–2648.

Nir, S. (2015). 'San Francisco firefighters become unintended safety net for the homeless', *New York Times*, August 26, p. A10.

Smith, N. (1990). *Uneven Development: Nature, Capital and the Production of Space.* Oxford: Blackwell, 2nd ed.

Smith, N. (1996). *The New Urban Frontier: Gentrification and the Revanchist City.* New York: Routledge.

Smith, N. (2002). 'New globalism, new urbanism: Gentrification and global urban strategy', *Antipode*, 34, pp. 443–457.

Smith, N. (2008). 'Comment: Neoliberalism – Dominant but dead', *Focaal: European Journal of Anthropology*, 51, pp. 155–157.

Tier, R. (1998). 'Restoring order in urban public spaces', *Texas Review of Law and Politics*, 2, pp. 256–291.

10 Revanchism, ignorance and class struggle in austerity Britain

Tom Slater

Introduction

On 27th September 2012, a young man called Alex Haigh was sentenced to twelve weeks in prison for 'occupying an empty home without permission' in Pimlico, central London. He was the first person to be given a custodial sentence under Section 144 of the Legal Aid, Sentencing and Punishment of Offenders Act, which came into force on 1st September 2012. This is the law that makes squatting in residential premises in England and Wales a crime punishable by up to six months imprisonment and fines of up to £5,000. Haigh had travelled to London to seek work, and like so many he had found the costs of housing in the city prohibitively expensive. He initially sought shelter in a hostel, but left it because he did not feel safe living among people with serious drug addiction problems. The squat where Haigh was subsequently arrested was a housing association flat that had been empty for over a year, and one that had caused no problems to neighbours – police raided it only because they were searching for a criminal on the run. Upon doing so, they arrested all the squatters and many more people became criminals. On the day Haigh was jailed, a Ministry of Justice spokesperson remarked:

> For too long squatters have been playing the justice system and have caused homeowners untold misery in eviction, repair and clean-up costs. It is extremely encouraging that the new criminal offence of squatting in a residential building is enabling the police and other agencies to take quick and decisive action to protect homeowners against squatting.[1]

The architect of the law was the Conservative politician Grant Shapps, housing minister from 2010–12.[2] In the face of organised opposition from various housing activists in 2011, and against surveys suggesting that over 90% of people disagreed with the criminalisation of squatting, Shapps (2011) felt the need to write a column in *The Guardian* defending his position:

> Our reasons for criminalising squatting are crystal clear – we want to protect the rights of regular hard-working homeowners against the damage squatters can inflict on their homes, and the distress this causes in their lives.

This unapologetic defence of the interests of 'homeowners' – those fortunate enough to own property during a serious housing crisis – is particularly galling and bitterly ironic when considering the hard work that non-homeowner Alex Haigh was seeking in London. He was an apprentice bricklayer, living on a pittance whilst learning the craft of creating homes for people to own. Equally unsettling is the fact that *the housing crisis is not a housing shortage* per se (Dorling 2014): There are over 750,000 empty homes across the UK, and Haigh was jailed simply for bringing one back into use. Many more people experiencing housing precarity have since come up against the steady expansion of the UK penal state. As of April 2016, there have been 736 arrests (75% of those arrests occurring in London), 326 prosecutions and 260 convictions under Section 144 (SQUASH 2015). All of those prosecuted and convicted were aged between 21 and 41: the generation that is bearing the brunt of the affordable housing crisis. The majority of squat evictions and arrests occurred during the winter months (October – March), affecting the health and endangering the lives of those thrown out into the cold. Since 2012 there have been calls by politicians across the spectrum, as well as by mainstream media outlets, to extend criminalisation to commercial buildings, which threatens many more young people priced out of housing markets with the prospect of sleeping rough.

The criminalisation of squatting was just one of a series of punitive measures directed at people at the bottom of the class structure in David Cameron's Britain, justified by those at the top of the class structure as being in 'fairness' to 'hardworking' families/taxpayers/homeowners. Punitive policies directed towards the disabled, the unemployed and the working poor – the people most vulnerable in an economic crisis – have also proved to be very popular with voters (in England). The Conservative Party (Tories), campaigning by spreading both fear and ignorance, was re-elected with a small majority in 2015, gaining many more seats than in 2010. To make sense of it all, this piece takes inspiration and analytic guidance from the intellectual and political legacy of Neil Smith to expose, analyse and challenge the revanchism exhibited by ruling elites in austerity Britain.[3] For the purposes of clarity, I first recapitulate the concept of revanchism in its original form, summarise some recent critiques and then discuss its later extensions. I then scrutinise the emergence of revanchist political economy in Britain, with particular reference to what Hodkinson (2012a) has helpfully called 'the return of the housing question'. In order to explain how revanchism has become so ingrained in British society I focus on the *production of ignorance* via the activation of class and place stigma. I argue that this production and activation (especially by free market think tanks) carves an economic and political path for the opening and closure of the rent gaps underpinning gentrification on a scale never before seen in the UK. Given the prevailing social and political conditions, it seems highly appropriate to have an analytic lens that is focused on the 'class retaking of the city … [which] simultaneously involves a certain excommunication of working class people from their communities', one that 'represents a nexus of class memory with contemporary social violence' (Smith 2010: 25).

Revanchism: a theory of class struggle

Ludwig Wittgenstein (1937) once remarked that, 'concepts may alleviate mischief or they may make it worse; foster it or check it'. In order to understand the analytic utility of the concept of *revanchism*, it is instructive to look back to its birthplace: late 19th century Paris. Revanchists (from the French word *revanche*, meaning revenge) were a group of bourgeois nationalist reactionaries opposed to the liberalism of the Second Republic, the decadence of the monarchy and especially the socialist uprising of the Paris Commune, where the working-classes took over from the defeated government of Napoleon III and controlled the city for months. The revanchists (led by poet-turned-soldier Paul Déroulède and the *Ligue des Patriotes*) were determined to reinstate the bourgeois order with a strategy that fused militarism and moralism with claims about restoring public order on the streets. They hunted down enemies (the Communards) with a noxious blend of hatred and viciousness, intent on exacting revenge upon all those who had 'stolen' their vision of French society from them. In a series of provocative essays, Neil Smith identified a striking similarity between the revanchism of late 19th century Paris and the political climate of late 20th century New York City that emerged to fill the vacuum left by the disintegration of liberal urban policy. Whereas the liberal era of the post-1960s period was characterised by redistributive policy, affirmative action and anti-poverty legislation, the era of *neoliberal* revanchism was characterised by policies *against* the working class, the homeless, minorities, feminists, youth, environmental activists, gays and lesbians, and recent immigrants. In 1993, Rudolph Giuliani was elected mayor of New York City on the promise of offering a better 'quality of life' for 'conventional members of society'. In vivid detail, Smith illustrated that revanchism under Giuliani was sharpened by blaming the failures of earlier liberal policy on the marginalised populations that such policy was supposed to assist (Smith 1998). Giuliani identified homeless people (rebranded in policy documents as 'dangerous mentally ill street people'), panhandlers, prostitutes, squeegee cleaners, squatters, graffiti artists, reckless bicyclists and unruly youth as the major threats to urban order and the culprits of urban decay, and with police chief William Bratton by his side, instigated infamous 'quality of life' policing (later dubbed 'zero-tolerance policing') to harass marginalised categories in the metropolis. Just as the bourgeois order was perceived as under threat by the revanchists of 1890s Paris, in 1990s New York an exclusionary vision of 'civil society' was being reinstated with a vengeance – an attempt to banish those not part of that vision from the city altogether.

Smith's argument that revanchism was not something unique to New York or to American cities, but something common to the restructured urban geography of the neoliberal capitalist city – 'the ugly cultural politics of neoliberal globalization' (1998: 10) – invited many urbanists to test the applicability of revanchism in other urban contexts well beyond the US. This spawned an array of fascinating studies on revanchist attempts to cleanse cities of 'undesirables' in order to attract investment and/or tourists (e.g. MacLeod 2002, Atkinson 2003, Swanson 2007, Whitehead and More 2007, Uitermark and Duyvendak 2008, Mountz and Curran 2009, Jou, Clark and Chen 2016). An influential body of scholarship on homelessness in England and Wales, however, takes a rather different view of revanchism (e.g. DeVerteuil 2006, 2014, DeVerteuil, May and Mahs 2009, Cloke, May and Johnsen 2010, Johnsen and Fitzpatrick 2010). Whilst not denying that there are punitive policies that have serious consequences on the lives of homeless people, this literature attacks research into revanchism on numerous grounds. As well as calling it 'simplistic', 'dystopic', 'incomplete', 'narrow', 'inaccurate' and 'reductionist determinism' reflecting a 'spectacular triumph of structure over agency' where homeless people are 'stripped of their humanity', Cloke, May and Johnsen (2010) argue that the revanchist framework doesn't capture the 'negotiation' of punitive regulation by homeless people and ignores the existence of contexts where homeless people experience empathy and friendship in addition to (sometimes rather than) control and containment, referred to as 'spaces of hope'. They emphasise the work of faith-based (Christian) organisations in providing an ethic of care that, through 'extraordinary acts of kindness', empowers homeless people and facilitates their engagement in purposeful agency. Whilst there is much to admire in the fine-grained ethnographic portraits of the lives of homeless people provided by these authors, and in their original emphasis on the work of the voluntary sector (neglected for too long, and sometimes unfairly dismissed as enthusiastic foot soldiers of neoliberalism), it is unfortunate that they position their work so strongly *against* revanchist perspectives, rather than in analytic dialogue with them. This positioning reaches its extreme form when Cloke, May and Johnsen (2010) state that homeless people are 'convenient ciphers around which to build a wider critique of gentrification' (p. 18). A very substantial body of scholarship on the links between the erosion of affordable housing (something curiously off the radar of these scholars) and homelessness makes this claim at best unsubstantiated and at worst a rhetorical move (cf. Doherty 2011).

Whilst there is arguably validity in the observation that the language of some scholars working in a revanchist register is often thick in drama and thin in empirical details, the most significant weakness of the attacks on this register is that they fail to move beyond the microgeographies of homelessness to view revanchism as *a theory of class struggle*. Understood as Neil Smith (1996) intended, the revanchist city is an arena where those at the top of the class structure are determined to maintain and augment the privilege of their position via aggressive political tactics, institutional innovations, legal frameworks and policy experiments. Ever alert to history and geography, Smith later pointed out that this was a specific 1990s brand of revanchism that, although deeply troubling, was not as concerning as the brand of revanchism beginning to circulate in a new century – 'a far more dangerous revanchism stalks the world' (2009: 4) – an intensified revanchism associated with state-sponsored and private market violence at the global scale (including cities):

> Class reconquest, in the broadest sense – taking back the city, the nation or the planet from any whose presence or politics would challenge the narrow class and race ambition writhing in the verbiage of globalization and the discourse of neoliberalism – is the true glue that attaches local to global revanchism. [...] The brilliance of revanchism in its early twenty-first century incarnation is its universal ambition to terrify the populace into self-discipline and compliance. But it does not do so equally or evenly. [...] Whether one is revanchist or recipient has everything to do with existing power structures including especially, but by no means exclusively, class.
>
> (Smith 2009: 13–14)

As I will now argue, it is this new brand of revanchism that has appeared in Britain, and the intriguing research question is how and why such a class reconquest is unfolding.

A short sketch of the housing crisis under welfare reform

In 2010, a coalition government came to power in Britain following an election which did not yield a clear majority for any party. That coalition was a skewed alliance between the dominant Conservative Party (which campaigned using the language of compassion and social progress to shield the electorate from its rabidly right wing, ruling class and corporate ethos); and the subordinate Liberal Democrats (a small set of centre-right political lightweights without a coherent message or set of policies, nearly obliterated at the polls in 2015). David Cameron, and his Chancellor of the Exchequer, George Osborne (both members of the British aristocracy with substantial family fortunes, who surrounded themselves with many more such people) arrived in office during a global financial crisis, and were confronted by a substantial budget deficit which they argued was a consequence of irresponsible public spending by the previous Labour government, rather than what it actually was: the result of massive government bail-outs of the financial institutions responsible for the crisis. For Cameron and Osborne, two archdeacons of low taxation and low public spending, there was only one way to deal with this budget deficit: a vicious austerity package, which, conveniently, was also an opportunity to finish off the welfare state that 'Thatcher's children' of the Conservative Party so despise, and replace it with their dream of a thoroughly privatised and individualised society that would protect the sanctity of private property rights and a free market.[4] To continue the relentless pace of expanding global accumulation, British ruling elites set out to monitor and monetise more and more of those human needs that were not commodified in previous rounds of financialisation. Pensions, healthcare, education and especially housing have been more aggressively appropriated, colonised and financialised (Meek 2014). For Conservatives, the redistributive path – increasing taxation on corporations, land and property (London in particular is known as a tax haven for foreign investors in land and property) – is not a matter for public discussion, and an entire cadre of cultural-technical experts (chief among them economists, lawyers, think tank researchers and communications professionals) is in place to make sure the conversation does not head in that direction (more on this later). This ensures that it is generally not known that an estimated £120 billion a year is lost in the UK due to corporate tax avoidance, evasion and collection errors.[5]

Whilst welfare 'reform' (demolition would be a better characterisation) in the UK has encompassed far more than just housing, it is the cuts to housing benefit in particular that have become the flashpoint for a compelling and disturbing class struggle. Housing benefit in Britain became fully operational in 1988, and its objective was to subsidise the cost of rental accommodation for tenants in the social-rented and private-rented housing sectors (see Hamnett 2010). It is a means-tested benefit, which is administered by local authorities and paid to eligible tenants (to be eligible, a household's income has to be less than £16,000 a year). Entitlement is calculated by comparing current household needs and resources with household rent payments. Nearly 5 million people (about 10% of all adults in Britain) claim an average weekly rental subsidy of £84 – in London, the most expensive city, the average climbs to £130. Recognising the uneven geography of housing costs, the last Labour government in 2008 introduced the Local Housing Allowance, for new claimants living in the private rented sector: a flat-rate allowance for properties of different size in local rental markets. This scheme led to very high levels of payment in a very small number of cases, up to £2,000 per week for a four bedroom house in central London – entirely because of the scandalously high cost of housing there (which has nothing to do with the housing benefit claimant). Nonetheless, since 2010 the Conservatives have argued that it is 'unfair' that a 'hardworking household' is, via their tax payments, enabling a household where nobody works ('a feckless, work-shy household', according to Iain Duncan-Smith, infamously modified by George Osborne to 'shirkers') to live on the same street or same neighbourhood, and sometimes to live in a larger house in a more desirable district. The government therefore capped housing benefit at £400 per week for a four bedroom home, £340 per week for a 3-bed, £290 per week for a 2-bed, and £250 per week for a 1-bed.

Why is housing benefit being targeted like this? Notwithstanding right wing critiques of the implications of government 'handouts', the main reason is that the annual housing benefit bill is enormous, up from £6 billion annually in the mid-1990s to nearly £40 billion today: an easy target for a government wedded to austerity. There are crucial historical reasons why this bill is so high, however, which are conveniently ignored by those attacking housing benefit, and have everything to do with the explosion of the private rented sector in the UK. By 2014, for the first time since the late 1960s, there were more UK households renting homes privately than through social housing. Since 2000 this largely unregulated sector has doubled in size to well over 4 million households; buy-to-let mortgages have soared for nearly two decades, facilitated by insecure 'shorthold tenancies' and a get-rich-quick scheme for landlords, In mid-2015, subsidised by the state via tax breaks, 1.8 million households (and counting) are on the waiting list for social housing; homelessness is on the rise (there was a 31% increase in rough sleeping from 2010–2014), despite the existence of approximately 750,000 empty homes in the UK; 'affordable rents' in social housing (in England) are now set at 80% of market levels (up from 50%) – almost never affordable; 1.6 million households are spending over half their income on housing costs, and the very poorest households have approximately £60 per week left after housing costs are met (Lansley and Mack 2015).

These developments are a direct legacy of Margaret Thatcher's infamous Right to Buy scheme,[6] introduced in 1980 to encourage the direct sale of council housing at large discounts to tenants (to promote a 'property-owning democracy' at the expense of the public housing stock). Over the past 35 years, nearly three million publicly owned homes have been sold off under the scheme. Many working class people acquired an asset beyond their dreams,[7] at a steal, yet Right to Buy actually failed on its own privatising terms, as just under 40% of all those who exercised their Right to Buy then sold on to private landlords (Gallagher 2015), who rented the properties to tenants at double or triple the levels of council rent, *which required tenants to apply for housing benefit off the state.* So Thatcher's flagship policy actually ended up costing the state far, far more in housing benefit than it ever did in the maintenance and management of council homes. None of this appears to matter to the Conservative government: Right to Buy is now being extended to 1.3 million tenants in housing association properties, at discounts of more than £100,000 per property in London and £70,000 elsewhere.[8] Without question, the main long-term effect of Right to Buy was to rob future generations of affordable housing, diverting them into the landlord bonanza that is the private-rented sector. Since 1979, there has also been a substantial reduction in the amount of new social housing construction, which has led to the steady shrinkage of the sector. Senior Tories have spoken of their desire to rebrand social housing as 'taxpayer subsidised housing'.

Since 1st April 2013, the allocation of Housing Benefit has taken into consideration the number of rooms occupied by tenants in council and housing association accommodation, and restricts payments to allow for one bedroom per person or per couple. If it is deemed by the state that there are too many rooms in such a dwelling for the number of occupants, the 'under-occupancy penalty' is applied. Dubbed the Bedroom Tax[9] by its opponents, this reduces housing benefit payments by 14% for one extra bedroom and by 25% for two or more extra bedrooms. Children under 10 years of age are now expected to share a room regardless of their sex, and children of the same sex under 16 are expected to share. Tenants who use their 'spare' bedroom when recovering from an illness or operation are affected by this policy, as are disabled people living in adapted or specially designed properties. The only exemption made for disabled tenants is when they have a registered carer who needs to stay overnight. Politicians attempt to defend the indefensible by referring to the many thousands of tenants across all sectors 'languishing in overcrowded accommodation'; it is claimed that the 'reallocation effects' of the policy will reduce council housing waiting lists. This argument is immediately weakened once it is recognised that people being evicted from their homes due to rent arrears are not going to reduce council housing waiting lists. The evidence from people so far affected by the Bedroom Tax – either those evicted or those living in fear of eviction – is a catalogue of miserable experiences ranging from acute psychosocial stress to broken social networks and grieving for lost homes (Slater 2013).

It doesn't take a complex modelling exercise to conclude that the cuts to housing benefit outlined above mean that many thousands of people on low-incomes are no longer able to pay rent in cities across the UK, particularly those living in cities where the cost of housing is so exorbitant. This is leading to displacement on an epic scale, especially from London. In April 2015, it was revealed by reporters at *The Independent* newspaper that local authorities in London moved almost 50,000 families who had fallen into rent arrears out of the city between July 2011 and July 2014 – sometimes at a rate of nearly 500 families a week. In many cases, families were moved to cities far away from social networks in London: to Manchester, Bradford, Hastings, Dover and Plymouth, among others. When this is considered alongside the further cuts that were introduced in April 2016, when the *total* benefits that a family can receive (not just in housing) were reduced further, from £26,000 to £20,000 (£385 a week) and to £23,000 (£442) in London: Large parts of the south-east and south-west, and some northern cities, became unaffordable. Figure 10.1 offers a cartographic depiction of the implications for England, showing the areas where a family of two adults and two children living in a two bedroom home will be hit by the benefit cap.

The production of ignorance

In November 2010, when the implications of the cuts to housing benefit were first making national news, an article appeared on the BBC News website under the headline: 'Do the poor have a right to live in expensive areas?' (Kelly 2010). Guiding the article was the conservative view that housing support for the poor should not give access to housing that other households not on benefits cannot afford. The very existence of the article and its cold, brazen headline points to a phenomenon that is driving the diffusion of

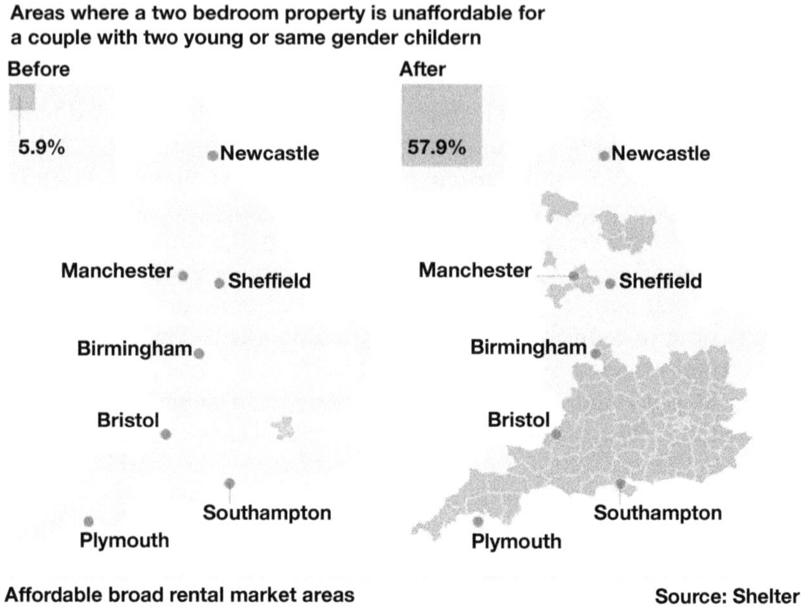

Figure 10.1 Housing affordability in England before and after the assault on housing benefit.

Source: Pennington (2015) in collaboration with *The Guardian* newspaper (Butler and Arnett 2015). Reproduced with permission of Shelter.

revanchism: the production of ignorance. As Mirowski (2013) argued in a swashbuckling critique of the economics profession in the build up to and aftermath of the 2008 financial crisis, one of the major ambitions of politicians, economists, journalists and pundits enamoured with (or seduced by) neoliberalism is to plant doubt and ignorance among the populace: 'This is not done out of sheer cussedness; it is a political tactic, a means to a larger end. ....Think of the documented existence of climate-change denial, and then simply shift it over into economics.' (Mirowski 2013: 83). Mirowski makes a compelling argument to shift questions away from 'what people know' about the society in which they live towards questions about what people do *not* know, and why not. These questions are just as important, usually far more scandalous, and remarkably under-theorised. They require a rejection of appeals to 'epistemology' and 'knowledge production', and instead, an analytic focus on ignorance production or *agnotology* (see Davies and McGoey 2012, for a timely intervention on 'rationalities of ignorance' vis-à-vis the 2008 financial crisis). This term was coined by Robert Proctor, a science historian, and refers to 'the study of ignorance making, the lost and forgotten' where the 'focus is on knowledge that could have been but wasn't, or should be but isn't' (Proctor and Schiebinger 2008: vii).[10]

It is telling that the BBC article included a side-column written by Alex Morton, then Director of Housing at the free market think tank *Policy Exchange* (labelled by the BBC as 'centre-right' when there is nothing centre about it at all), and now David Cameron's special advisor on housing. In an exemplar of think tank distortion, he wrote:

> Any 'right' for particular individuals would be based on government arbitrarily selecting certain people and taxing everyone else to pay the high costs necessary [...]. Government should [...] not take up time and energy working out how to select some lucky individuals who then receive overwhelming individual subsidy.
>
> (quoted in Kelly 2010)

What is striking here is what is not mentioned: the grotesquely high cost of housing. Instead the focus is shifted to appeals to fairness, the vague notion of desirability and the not-so-vague insistence that 'everyone else' should not be taxed to house 'certain people'. Also, contrary to what Morton would have us believe, the 'lucky individuals' are *not* those living in the desirable areas but the real beneficiaries of all 'overwhelming subsidies': landlords.

This example captures the extraordinary political influence and tactics of free market think tanks in the debate on welfare reform in the UK. What distinguishes *Policy Exchange* and the equally influential *Centre for Social Justice* (see Slater 2014) from the free market think tanks before them is the central involvement of (then opposition) frontline politicians in their formation. The former was the brainchild of the hard-right 'Brexiteer' Michael Gove in collaboration with other Tory politicians, and the latter that of Iain Duncan-Smith. These are two staunch Thatcherites and close allies of David Cameron, and the think tanks they created have not only become the go-to sources of comment and 'evidence' arming those of a neoliberal persuasion, but continue to provide the intellectual impetus behind the ongoing Conservative assaults on the public sector and on the welfare state in Britain. Supported by the press releases and noisy interjections of other free market think tanks such as *Reform* and *The Taxpayers' Alliance*, what Jones (2014) calls the 'outriders' have been instrumental in connecting the worlds of politics, business and the media to defend the neoliberal society they and their predecessors have helped to create, and to keep the national political conversation on terms/categories favourable to ruling elites. These think tanks always claim political independence (and seek respect by frequent appeals to their charitable or 'educational' status) but so evangelical is their devotion to market rule and loathing of big government that it serves as a honey pot for substantial financial donations from business interests. They are now so well financed (and thus well-staffed) that it makes it very challenging indeed for think tanks opposing free market logics and neoliberal nostrums to produce equivalent volumes of output.

When robust and overwhelming evidence against the neoliberal project mounts up, the think tanks scramble to produce 'evidence' to the contrary (usually based on small and rigged opinion polls that they have commissioned) defending the free market and attacking state intervention in economic and social affairs. Mainstream journalists are regularly knocking on their doors for comment, with the following effect:

> By a simple jujitsu move, the mechanical journalist's appeal to objective presentation of 'both sides' is turned into an imprimatur for the public to believe whatever their guts tell them is correct, at least until the marketplace of ideas delivers its final verdict. The aim of agnotology is not so much to convince the undecided, but to fog the minds of anyone lacking the patience [or time] to delve into the arguments in detail (which is pretty much anyone).
>
> (Mirowski 2013: 229)

At first sight it is absurd that the government should pay private landlords large sums of taxpayers' money simply to allow poor people to stay in a gentrified area. Yet if people did delve into the arguments in detail, without their minds being fogged by moralising around what is fair and unfair, they might begin to call into question the (always unstated) assumption that the deregulated market, buttressed by private property rights, is the force that determines the cost of housing – and the framing of the debate would change dramatically. 'Do the poor have a right to live in expensive areas?' would become 'Do the rich have a right to colonise historically poor areas?', and perhaps even 'What makes these areas so expensive and wages in them so unequal?': a rather different set of analytical and political concerns. The production of ignorance, something that free market think tanks have mastered, is thus critically important in the spread and survival of revanchist politics in the UK.

There is one final and central ingredient in the constitution of revanchist Britain that requires close scrutiny: stigma. One of the principal factors fuelling 1990s revanchism in New York City was (re)productions of paranoia and fear by the media that amplified and aggravated extant stigmatising sentiments among swathes of middle-class voters seeking to affix blame for their perceived lack of safety in urban public spaces. In austerity Britain, a similar revanchist dynamic is visible in the form of *the heightened stigmatisation of working class lives*, and of the places where working class people live. Recent years have seen the political activation of latent class hatred, where rogue images of welfare 'cheats' or 'scroungers' living in huge houses in expensive areas capture the attention of people on the look out for scapegoats for pervasive economic malaise (Mooney 2009). This has been the focus of important inquiries, notably those of Imogen Tyler (2013), who introduced the concept of *social abjection* to dissect the ways in which the mediation and figuration of stigma is employed to generate consent for neoliberal modes of governmentality in contemporary Britain. Drawing on a series of case-studies of marginalised and stigmatised categories, Tyler demonstrates that the production and mediation of stigma is not simply an effect of neoliberal ideologies and policies but rather a *core organ* of neoliberal governmentality: 'stigmatization operates as a form of governance which legitimizes the reproduction and entrenchment of inequalities and injustices.' (Tyler 2013: 212). Tyler's argument that stigmatisation is increasingly employed as a device to *procure consent* for punitive policies directed at those living at the bottom of the class structure (see also Jensen and Tyler 2015) resonates with Wacquant's account of stigma as a form of 'violence from above' (2008: 24) with clear emphasis on the 'virulence of its negative effects in the context of mass unemployment and political marginalization' (Wacquant 2008: 174). None of the punitive policies discussed in this essay would be possible without a political campaign to denigrate people in receipt of social assistance. Revanchist stigmatisation not only reinforces the beliefs held among those who have benefited the most from neoliberalisation of society: It also contributes to what Wacquant (2008) terms 'the symbolic splintering of the new urban poor' (p. 7), undermining solidarity as people internalise stigmatising views or project them onto their neighbours in strategies of lateral denigration.

A striking illustration was provided by Shildrick and MacDonald (2013) in their analysis of interviews with working class people in Middlesbrough, England. Despite material conditions of hardship, their respondents often denied being poor, and then castigated poverty among others around them as self-inflicted. Not only was this a coping strategy in response to the shame associated with poverty, unemployment and welfare, it also reflected the collapse of working class institutions and practices: 'Political orthodoxies about 'the undeserving poor' more easily take hold, even amongst 'the poor', in contexts of a diminishing politicized working class consciousness.' (Shildrick and MacDonald 2013: 302). In 1952, the architect of the National Health Service, Aneurin Bevan, famously stated that 'the whole art of Conservative politics' lies in finding answers to the question, 'How can wealth persuade poverty to use its political freedom to keep wealth in power?' Neoliberalism and the free market fanaticism that sustains it, in spite of its now obvious failures, relies on a considerable 'statecrafting' of the relationship between information and power.[11] The outcome is that people who are most affected by the brutal coupling of economic deregulation and welfare state retraction can end up supporting and even defending neoliberal beliefs and myths (chief among them that 'economic growth' is all that matters to a society, as wealth will supposedly 'trickle-down' and benefit everyone, and that success and failure are solely due to individual behaviour and personal responsibility). As Sayer (2014) demonstrates, one of the most dangerous and pervasive myths of all is that the largely unearned income of the super rich is actually being earned, and is only fair given their 'hard work' and talent as 'wealth creators'. Unearned income derives from control of an already existing asset, such as land, buildings, technology, or money, that others lack but need or want, and who can therefore be charged for its use. It is the production of ignorance that stops so many from seeing that the wealthiest in society extract wealth from the economy without creating anything in return.

Conclusion

The displacement of working class people from central city districts in Britain, particularly but not exclusively from districts in London and south-east England, should not be viewed as a consequence of the way the housing market works via the 'laws' of supply and demand, nor as a 'natural evolution' (Ball 2014), but as *calculated expulsion* being engineered by a revanchist national state in alignment with the interests of private capital. A principal motive is the consolidation and advancement of class power via the capture of rent from highly lucrative urban land markets. Just as Smith's concept of revanchism provides an understanding and critique of the class interests behind this motive, his theory of gentrification, the *rent gap* (Smith 1979a), offers an equally powerful explanation and critique of the class transformation of space. To conclude, then, I want to draw together the themes of austerity, revanchism, ignorance and stigma, and consider them in the context of that class transformation and within that theoretical register.

Due to the decades-long shrinkage of the manufacturing sector in Britain, capital has switched from its primary circuit of industrial production to its secondary circuit of accumulation, which is that of urban land and real estate markets, and this runs parallel to the primary circuit. In many capitalist economies, the secondary has supplanted the primary in terms of its overall importance, often accounting for over 40% of all economic activity. One illustration: 76% of all bank loans in Britain go into property (and 64% of that into residential mortgages), and 80% of all household debt is tied up in mortgages. David Harvey's concept of *fictitious capital* is crucial here: This is the driving force of the secondary circuit. Money doled out for a land title 'is equivalent to an interest-bearing investment, a claim upon anticipated future revenues, a future gain' (Merrifield 2014: 21). Thus, by fictitious capital, Harvey (1982) was referring to the anticipation of future surplus value production from land titles:

> The freer interest-bearing capital is to roam the land looking for future ground-rents to appropriate [...] and the more open the land market is, the more recklessly can surplus money capital build pyramid debt claims and seek to realize its excessive hopes through the pillaging and destruction of production on the land itself.
>
> (Harvey 1982: 368–369)

In urban Britain, we are seeing this phenomenon writ large. To address the crisis of continuous compound growth under long cycles of accumulation, capital has to devalue the existing capital fixed to the land, among other things, to reinvent investment opportunities for the absorption of a surplus (Harvey 2014). At times of crisis, speculation in land that is being devalued, or what Harvey (2010) calls *speculative landed developer interests,* becomes rife. In Britain, the institutional arrangements behind the distribution of housing incentivise rampant land speculation: the urban housing market in the UK (London especially) has now become a place for very rich people – especially investors from overseas – to park their money at an annual rate of return of around 10%.

Speculation means that more and more capital is being invested in search of rents and interest and future gains, rather than being invested in productive activity – a trend towards a *rentier* form of capitalism: a parasitic economy of unproductive extraction characterised by the marked escalation of unearned income. At its most extreme form what we are seeing is the *robbery* of value, rather than the production of value: asset pursuit and asset stripping, via land grabbing and evictions. The scholar-activists of Observatorio Metropolitano in Madrid explain as follows:

> The crisis is simply a lie: a sick joke were it not for the social consequences. The problem [...] is not the crisis...but rather the distribution of a bounty that years of abusive financial practices, government corruption, rampant deregulation and market ideology has left in the wrong hands.
>
> (Observatorio Metropolitano 2011: 66–67)

Rentiers have everything to gain from the global circulation of interest-bearing capital in urban land markets, and from the municipal absorption of surplus capital via all kinds of debt-financed urbanisation projects. Since 2008, it has become clear that vast sums of money in financial services are not made from big bets but from the *circulation* of capital. The relevance of the rent gap theory to campaigns and struggles against speculative landed developer interests is that it helps to

> redirect our theoretical focus toward the sphere of circulation...[where] we can trace the power of finance capital over the urbanization process, and the patterning of urban space according to patterns of profitable investment.
>
> (Smith 1979b: 24)

The function of rent in a capitalist mode of production is to underpin investment and reinvestment opportunity. The rent *gap* is the divergence between capitalised ground rent (the actual quantity of ground rent that is appropriated by the landowner, given the present land use) and potential ground rent (the maximum that could be appropriated under the land's 'highest and best use').

A recent example of the speculative rentier class attempting to exploit the rent gap in London is the struggle over the New Era housing estate, built by a charitable trust in the 1930s to offer working class Londoners affordable rental housing but, for many years, subject to disinvestment. Westbrook Partners is an investment firm based in New York City, which makes its billions by investing American pension savings in London land deals. Westbrook bought the New Era estate in March 2014 (initially a partnership deal with Benyon Estates, owned by the Tory MP Richard Benyon, until he had to pull out due to public shaming), and immediately notified tenants that rents would rise to market values: from £600 a month for a two-bed flat to £2,400 a month (Chakrabortty 2014). Rent gaps are actively produced by clearing out existing residents via all manner of tactics and legal instruments, such as massive rent increases, neglect (Westbrook has a history of building violations in property it owns in the US), harassment, eminent domain (or 'compulsory purchase orders' in the UK) and, as we have seen, welfare 'reform'. Closing rent gaps requires, crucially, separating people currently obtaining use values from the present land use providing those use values – in order to capitalise and gentrify the land to the perceived 'highest and best' use. Land value is not created from owning land: It is created from collective social investments in land, which landowners capture as unearned income via private property rights. Exploiting the rent gap requires *the expropriation of socially created use values*:[12] a form of structural violence visited upon working class people in contexts that are usually described these days as 'regenerating' or 'revitalising'. Instead of building shelter for people in need, the system encourages rentier capitalists to see who can best use their land-banking skills to anticipate the next housing bubble and survive the last one.

In December 2014, there was a significant victory for residents of the New Era estate, when they successfully fought plans by US land investors to evict dozens of families and more than double rents. Westbrook Partners, under huge public pressure because of a campaign against its profiteering motives (again led by mothers on the estate), sold the land estate to the Dolphin Square Charitable Foundation, an affordable housing charity committed to delivering low cost rents to Londoners on low to middle incomes. This struggle was part of an ongoing wave of protests in a capital city with grotesquely inflated housing costs across all tenures, and it generated massive attention to trigger a national debate on housing affordability and similar protests in other UK cities.[13] Above all, the movements demonstrated that when too many people are unable to afford shelter, it becomes a political problem, and a major basis for social mobilisation. What happens now will depend enormously on the abilities of these movements to resist the unrelenting stigmatisation they face, and the extent to which they can build alliances with other housing movements nationally and internationally to insist on the right to stay put *against* the forces shaping a revanchist planet, for 'nothing is more inconvenient, more unwelcome, than a multitude that decides to question everything.' (Observatorio Metropolitano 2011: 66)

Notes

1 Quoted here: www.bbc.co.uk/news/uk-england-london-19759056.

2 Shapps is most (in)famous for denying that he had a (prohibited) second job as a 'multimillion-dollar web marketer' under the pseudonym Michael Green for at least a year after he first became an MP.

3 It should be recognised that, due to devolution and huge political differences and historical geographies, there are considerable differences in housing affairs between Scotland and England especially, but welfare reform is UK central government legislation that is not devolved, and this has major implications for housing and for cities. So, whilst I acknowledge that the arguments of this article are skewed by the experiences of England, particularly south-east England, I think it would be a political mistake to jettison 'Britain' as a frame of reference, because of the impact of austerity, because of the accumulation/profit motives behind different policy approaches and because the circulation of capital in urban land markets (crucial in understanding the housing crisis) tends not to recognise national borders.

4 In case there is any doubt about intentionality, it is worth remembering that in April 2011, the Conservative MP Greg Barker (then Minister of State for Energy and Climate Change) told an audience in the US: 'We are making cuts that Margaret Thatcher, back in the 1980s, could only have dreamed of.' http://politicalscrapbook.net/2011/04/greg-barker-thatcher-video/.

5 The money through tax avoidance alone could pay for 25,000 nurses on a £24,000 a year salary for 20 years; could put 129,000 children through school from ages 5–18; and would allow the government to give every single pensioner in the UK an extra £65 a year: http://leftfootforward.org/2013/09/another-crackdown-on-benefit-fraud-yet-it-accounts-for-just-0-7-per-cent-of-welfare-budget/.

6 RTB has been the largest but by no means the only economic deregulation and privatisation scheme vis-à-vis housing in Britain. Many remaining council housing stocks have been transferred to housing associations ('stock transfer') as part of 'the state's desire to end municipal landlordism altogether, accounting for a further 1.5 million public dwellings' (Hodkinson 2012b: 510). Year upon year there is a dwindling public subsidy for housing associations, which have been forced to rely on commercial borrowing to balance the books, putting upward pressure on rents (ibid.).

7 As Hodkinson (2012b) has argued, Right to Buy 'played a decisive ideological role during the height of class conflict during the 1980s' (p. 513). The sudden emergence of homeownership had a 'diluting effect on council housing estates collectively trying to defend themselves' and lobby in solidarity for better maintenance and service delivery. Also, 'it represented the vanguard policy of Thatcher's 'popular capitalism' mantra' (ibid.) which aimed to pacify working class resistance to market logic and market rule. Millions of working class people found the opportunity to buy their homes cheaply from the state to be irresistible, and this helped legitimise the systematic assault on the welfare state in the 1980s. The view of conservative politicians was – and remains – that workers don't create as much fuss and trouble if they are suddenly immersed in the more 'secure' world of home ownership and mortgage repayments. Right to Buy was thus an electoral masterstroke, specifically targeting working class swing voters (Forrest and Murie 1988).

8 In a cynical appropriation of past language, the Tories have recently launched the 'Help to Buy' scheme to assist first-time buyers purchase residential property. It has thus far had two phases: (1) buyers contribute a 5% deposit and the government provides an equity loan for up to 20% of the property value, and

buyers must provide the remaining funds themselves (from a mortgage). This is available only for new-build properties under a certain price (less than £600,000) and the loan is interest-free for the first five years; (2) 5% deposit mortgages are available from ten different lenders with the government acting as a guarantor for the mortgage. This phase is not restricted to those buying new-builds: anyone wanting to buy any home in the UK worth less than £600,000 is eligible for the scheme. Even the IMF has warned George Osborne about this scheme creating another catastrophic housing bubble.

9 Following a two-week, five-city visit to the UK in 2013 to assess its achievements and challenges in guaranteeing the right to adequate housing, the UN's Special Rapporteur on housing Raquel Rolnik released a media statement that praised the UK's long track record in ensuring a right to housing, but then noted signs of retrogression: 'It is not clear that every effort has been made to protect the most vulnerable from the impacts of retrogression, indeed much of the testimony I heard suggests they are bearing the brunt. Housing deprivation is worsening [...] [P]eople appear to be facing difficulties to accessing adequate, affordable, well located and secure housing.' She concluded her statement with the recommendation that 'the so-called bedroom tax be suspended immediately and be fully re-evaluated in light of the evidence of its impacts on the right to adequate housing and general well-being of many vulnerable individuals.' Unsurprisingly, Rolnik's intervention sent Tories into white-knuckle rage, not least Grant Shapps, who fired back on several morning broadcasts, calling Rolnik's report a 'disgrace' and later triumphantly announcing that he had written to the Secretary-General of the UN calling for him to investigate how someone with 'apparent political bias' could be allowed to evaluate UK housing rights. He is still awaiting a reply.

10 It was while investigating the tobacco industry's efforts to manufacture doubt about the health hazards of smoking that Proctor began to see the scientific and political urgency in researching how ignorance is made, maintained and manipulated by powerful institutions to suit their own ends, where the guiding research question becomes, 'Why don't we know what we don't know?' As he discovered, the industry went to great lengths to give the impression that the cancer risks of cigarette smoking were still an open question even when the scientific evidence was overwhelming. Numerous tactics were deployed by the tobacco industry to divert attention from cancer risks, such as the production of duplicitous press releases, the publication of 'nobody knows the answers' white papers, and the generous funding of decoy or red-herring research that 'would seem to be addressing tobacco and health, while really doing nothing of the sort' (Proctor and Schiebinger 2008: 14). The tobacco industry actually produced research about everything except tobacco hazards to *exploit public uncertainty* (researchers knew from the beginning what they were supposed to find and not find), and the very fact of research being funded allowed the industry to say it was studying the problem.

11 Paul Gilroy is especially astute on this: http://dreamofsafety.blogspot.com/2011/08/paul-gilroy-speaks-on-riots-august-2011.html.

12 I am grateful to Peter Marcuse for this astute observation.

13 Recently, a 'Living Rent' campaign has emerged in Scotland, taking cues from the remarkable advances of the Living Wage movement to argue for urgent government action to ensure truly affordable rents in the private rented sector. This has resulted in the Scottish government announcing the introduction of rent controls in certain 'rent pressure areas' in Scottish cities, starting next year, and also to end the right of landlords to reclaim their properties from tenants without any reason.

References

Atkinson, R. (2003). 'Domestication by cappuccino or a revenge on urban space? Control and empowerment in the management of public spaces', *Urban Studies*, 40(9), pp. 1829–1843.

Ball, P. (2014). 'Gentrification is a natural evolution', *The Guardian*, 19th November. Available at: www.theguardian.com/commentisfree/2014/nov/19/gentrification-evolution-cities-brixton-battersea. (Accessed 26th January 2017).

Butler, P. and Arnett, G. (2015). 'Lower benefit caps "will exclude poor families from large parts of England"', *The Guardian*, 20th July. Available at: www.theguardian.com/society/2015/jul/20/lower-benefit-caps-exclude-poor-families-make-cities-unaffordable (Accessed 9th September 2016).

Chakrabortty, A. (2014). 'New Era estate scandal: Families at the mercy of international speculators', *The Guardian*, 19th November. Available at: www.theguardian.com/society/2014/nov/19/new-era-estate-scandal-london-families-international-speculators. (Accessed 26th January 2017).

Cloke, P., May, J. and Johnsen, S. (2010). *Swept Up Lives? Re-Envisioning the Homeless City*. Oxford: Wiley-Blackwell.

Davies, W. and McGoey, L. (2012). 'Rationalities of ignorance: On financial crisis and the ambivalence of neo-liberal epistemology', *Economy and Society*, 41(1), pp. 64–83.

DeVerteuil, G. (2006). 'The local state and homeless shelters: Beyond revanchism?', *Cities*, 23(2), pp. 109–120.

DeVerteuil, G. (2014). 'Does the punitive need the supportive? A sympathetic critique of current grammars of urban injustice', *Antipode*, 46(4), pp. 874–893.

DeVerteuil, G., May, J. and von Mahs, J. (2009). 'Complexity not collapse: Recasting the geographies of homelessness in a "punitive" age', *Progress in Human Geography*, 33(5), pp. 646–666.

Doherty, J. (2011). 'Review of "swept-up lives?" Re-envisioning the homeless city', *European Journal of Homelessness*, 5(2), pp. 225–234.

Dorling, D. (2014). *All That is Solid: The Great Housing Disaster*. London: Allen Lane.

Forrest, R. and Murie, A. (1988). *Selling the Welfare State: The Privatisation of Public Housing*. London: Routledge.

Gallagher, P. (2015). 'Right to buy: 40% of homes sold under government scheme are being let out privately', *The Independent*, 14th August. Available at: www.independent.co.uk/news/uk/home-news/right-to-buy-40-of-homes-sold-under-government-scheme-are-being-let-out-privately-10454796.html. (Accessed 26th January 2017).

Hamnett, C. (2010). 'Moving the poor out of central London? The implications of the coalition government 2010 cuts to housing benefits', *Environment and Planning A*, 42(12), pp. 2809–2819.

Harvey, D. (1982). *Limits to Capital*. Oxford: Blackwell.

Harvey, D. (2010). *The Enigma of Capital and the Crises of Capitalism*. London: Profile Books.

Harvey, D. (2014). *Seventeen Contradictions and the End of Capitalism*. London: Verso.

Hodkinson, S. (2012a). 'The return of the housing question', *Ephemera: Theory and Politics in Organization*, 12(4), pp. 423–444.

Hodkinson, S. (2012b). 'The new urban enclosures', *City*, 16(5), pp. 500–518.
Jensen, T. and Tyler, I. (2015). 'Benefits broods: The cultural and political crafting of anti-welfare commonsense', *Critical Social Policy*, 35(4), pp. 470–491.
Johnsen, S. and Fitzpatrick, S. (2010). 'Revanchist sanitization or coercive care? The use of enforcement to combat begging, street drinking and rough sleeping in England', *Urban Studies,* 47(10), pp. 1703–1723.
Jones, O. (2014). *The Establishment (And How They Get Away With It).* London: Verso.
Jou, S.-C., Clark, E. and Chen, H.-W. (2016). 'Gentrification and revanchist urbanism in Taipei?', *Urban Studies,* 53(3), pp. 560–576.
Kelly, J. (2010). 'Do the poor have a right to live in expensive areas?', *BBC News Magazine*, 4th November. Available at: www.bbc.co.uk/news/magazine-11674864. (Accessed 26th January 2017).
Lansley, S. and Mack, J. (2015). *Breadline Britain: The Rise of Mass Poverty.* London: One World.
MacLeod, G. (2002). 'From urban entrepreneurialism to a "revanchist city"? On the spatial injustices of Glasgow's renaissance', *Antipode,* 34(3), pp. 602–624.
Meek, J. (2014). *Private Island: Why Britain Now Belongs to Someone Else.* London: Verso.
Merrifield, A. (2014). *The New Urban Question.* London: Pluto.
Mirowski, P. (2013). *Never Let a Serious Crisis Go to Waste: How Neoliberalism Survived the Economic Meltdown.* London: Verso.
Mooney, G. (2009). 'The "broken society" election: Class hatred and the politics of poverty and place in Glasgow East', *Social Policy and Society,* 8(4), pp. 437–450.
Mountz, A. and Curran, W. (2009). 'Policing in drag: Giuliani goes global with the illusion of control', *Geoforum,* 40, pp. 1033–1040.
Observatorio Metropolitano (2011). *Crisis and Revolution in Europe: People of Europe, Rise Up!* Madrid: Traficantes de Sueños.
Pennington, J. (2015). 'The benefit cap: This changes everything', *Shelter Policy Blog,* 20th July. Available at: http://blog.shelter.org.uk/2015/07/the-lower-benefit-cap-this-changes-everything/ (Accessed 9th September 2016).
Proctor, R. and Schiebinger, L. (eds.) (2008). *Agnotology: The Making and Unmaking of Ignorance.* Stanford: Stanford University Press.
Sayer, A. (2014). *Why We Can't Afford the Rich.* Bristol: Policy Press.
Shapps, G. (2011). 'The idea squatting can be a solution to homelessness is false and cruel', *The Guardian*, 25th November.
Shildrick, T. and MacDonald, R. (2013). 'Poverty talk: How people experiencing poverty deny their poverty and why they blame "the poor"', *The Sociological Review,* 61, pp. 285–303.
Slater, T. (2013). 'Grieving for a lost home, revisited: The "bedroom tax" and displacement'. Available at: www.geos.ed.ac.uk/homes/tslater/bedroomtax.html. (Accessed 26th January 2017).
Slater, T. (2014). 'The myth of "broken Britain": Welfare reform and the production of ignorance', *Antipode*, 46(4), pp. 948–969.
Smith, N. (1979a). 'Toward a theory of gentrification: A back to the city movement by capital, not people', *Journal of the American Planning Association*, 45(4), pp. 538–548.
Smith, N. (1979b). 'Gentrification and capital: Practice and ideology in Society Hill', *Antipode*, 11, pp. 24–35.

Smith, N. (1996). *The New Urban Frontier: Gentrification and the Revanchist City.* New York: Routledge.

Smith, N. (1998). 'Giuliani time', *Social Text,* 57, pp. 1–20.

Smith, N. (2009). 'Revanchist planet', *Urban Reinventors,* 3, pp. 1–17. Available at: http://urbanreinventors.net/3/smith1/smith1-urbanreinventors.pdf. (Accessed 26th January 2017).

Smith, N. (2010). 'The evolution of gentrification', in J. Berg, T. Kaminer, M. Schoonderbeek and J. Zonneveld (eds.). *Houses in Transformation: Interventions in European Gentrification.* Rotterdam: NAi Publishers, pp. 15–26.

SQUASH (Squatters Action for Secure Homes). (2015). *Homes, Not Jails: 28 Month Review of Section 144 of the Legal Aid, Sentencing, and Punishment of Offenders Act.* Available at: www.squashcampaign.org/2015/04/homes-not-jails-april-2015-squashs-latest-report/. (Accessed 26th January 2017).

Swanson, K. (2007). 'Revanchist urbanism heads south: The regulation of indigenous beggars and street vendors in Ecuador', *Antipode,* 39(4), pp. 708–728.

Tyler, I. (2013). *Revolting Subjects: Social Abjection and Resistance in Neoliberal Britain.* London: Zed Books.

Uitermark, J. and Duyvendak, J.W. (2008). 'Civilising the city: Populism and revanchist urbanism in Rotterdam', *Urban Studies,* 45(7), pp. 1485–1503.

Wacquant, L. (2008). *Urban Outcasts: A Comparative Sociology of Advanced Marginality.* Cambridge: Polity Press.

Whitehead, J. and More, N. (2007). 'Revanchism in Mumbai? Political economy of rent gaps and urban restructuring in a global city', *Economic and Political Weekly,* 23rd June. 42(25), pp. 2428–2434.

Wittgenstein, L. (1937). *Culture and Value.* [Vermischte Bemerkungen]. Chicago: University of Chicago Press.

11 The class gap in gentrification

A political reading of the rent gap hypothesis

Marc Morell

Bartender's blues

Between 2010 and 2013, there was fierce conflict between fairly recently established residents and the owners of newly opened bars in Es Barri. This is a heavily gentrified neighbourhood on the eastern margins of the centre of Ciutat, also known as Palma, the capital city of Majorca and the Balearic Islands (Spain). By 2010, most of these bars had organised a pub-crawl scheme that is still popular nowadays. Under the name of *Ruta Martiana* [Martian/Tuesday Route], each Tuesday the bars offered a glass of beer and a tapa for €1. The idea quickly caught on and soon myriads of people swarmed the streets of Es Barri.

Although the newcomer residents of Es Barri were mostly organised through associations of neighbours (e.g. Canamunt, Banc de l'Oli), they actually used another platform to articulate their disgust at the noise and dirt that the Ruta Martiana caused: *sa Gerreria t'estim sense renou* [Gerreria, I love you without noise] being 'sa Gerreria' [Pottery] one of the many names of Es Barri. This was done because not everyone was against the bars, and especially since some of the bar owners were actually members of Canamunt. It was only a matter of time before the City Council started to look into this matter. In 2013, the council formally took sides with those against the pub-crawl. New bar openings were banned, and closing hours were established for those already in operation.

At first sight, this ethnographic sketch illustrates how gentrification is not exempt from its own internal contradictions. In its search for surplus value, the expansion of capital in the built environment adopts diverse conflicting strategies. Here, the uneven pattern of establishment set by a long-planned residential gentrification process was disturbed by a sudden commercial gentrification that fed upon nightlife bars (Ocejo 2011), but bars were not new to Es Barri. Well before the renewal scheme that enabled gentrification, the bars were the backbone of the neighbourhood. However, these bars were of a different kind. Linked as they were to the vice district of Ciutat, its stigmatised specialisations were eventually invoked to debase the value of properties in the area.

Rather than trying to determine the contradiction of capital in gentrification by looking at the different residential and commercial uses that are at play, I explore the relationship that capital and labour maintain in working out this same contradiction. By building on the urban productive forces that constitute the rent gap, Neil Smith theorised about, and by ethnographically illustrating these forces with the case of the bars of Es Barri, I contend that we can not only gain insight into the exploitative class relationship of the rent gap, the creation and appropriation of value, but also overcome the actual role of the state in gentrification by carving out a political project that is not severed from its material urban conditions and that it is based upon the efforts of those who labour the city. In this way, class fractions that otherwise seem to be opposites come together.

Rent gap, urban labour and the state

Back in 1979, Neil Smith published his groundbreaking article on the rent gap. By focusing on the distance found between the economic returns on the uses that potential ground rent and capitalised ground rent offered, he developed a consistent materialist explanation for gentrification. Smith found that whereas at the basis of the capitalised ground rent there is a 'present' use, at the basis of potential ground rent there is a 'highest and best' use. These different uses, argued Smith, are necessarily mediated by forms of collective social action.

Although Smith devised his hypothesis to explain how capital realised surplus value from the differentials found in the uneven development of the built environment, the argument about the 'uses' of land signals the further insights into the actual productive forces, and the extraction of surplus product, that are at stake in building the rent gap. This emphasis offers an urban material basis that not only contributes to the study of class formation, but also allows us to think of class struggle in political terms, that is, to imagine class action as a means towards the 'collective' use of land and its built environment.

Class was Smith's main concern. He dealt with class extensively, either in a more theoretical manner when advocating for the return of class to academia in order to reinvigorate cultural and identity politics (Smith 2000) or in a more descriptive way with his urban frontier project in which the revanchist city argument stressed the revengeful class-reaction of the well-off against the needy in their struggle to remain in the city (Smith 1996: 206–227). Yet class is inscribed in the rent gap (Morell 2015). Each moment of the rent gap is laboured by distinctive groups: the often disdainfully portrayed lumpen-class and the so-called middle classes, which otherwise prove to be politically irrelevant for change (De Angelis 2010).

These groups cooperate in the same chain of unwaged value production that I call 'urban labour'. There have been parallel advances in developing similar notions, either in the form of the symbolic labour of middle-class newcomers (Bidou-Zacharisian and Poltorak 2008) or in the 'collective labour' invested in maintaining the built environment by given communities (Harvey 2012: 68–88, also 139). However, none of these approaches take into account the labour required in destroying the value that usually precedes its creation from anew.[1]

The distance between the different use values of each group makes the gap, and hence the surplus product that a third party extracts, surplus value (local bosses, planners, builders, speculators, real estate, financial sector, etc.). Rent gaps are made by exploiting the everyday life of urban labour. Their most valuable outcome, improved housing, and its immediate urban environment, is sold as a commodity. Such an understanding does not irremediably draw us back to the question of housing classes (Aalbers 2016: 22–26 and 134–150), but it does acknowledge the material production and the realisation of surplus value by which class relations are shaped in the urban realm.

Class and the productive matter that shapes it are not alone, however, in their dealings with the inexorable expansion of capital in the urban environment. As Lees et al. (2008: 163–193) thoroughly explain, it is the changing role of the state that mediates the rent gap, and there are times in which this same state even leads the whole of the gentrification process. Smith was aware of this in his BA final dissertation:

> The role of the State was therefore twofold: to produce a distinct urban space for possession by the upper classes, and thereby to provide substantial investment opportunities for private capital.
>
> (Smith 1977: 20)

The idea that the state controls the production of space at large was already contended by Lefebvre's statist mode of production in his four volume opus on the state (Lefebvre 1976–1978), regardless of the form the state took. Since the fall of actually existing socialism and the rise of neoliberalism, recent approaches have emphasised the integration of the state system into the capitalist mode of production. Callinicos, for example, argues that state managers push state policy through micro-mechanic self-seeking actions in a direction that also tends to the promotion of capital accumulation (Callinicos 2009: 84–86).[2]

In the context of the Balearic Islands, capital accumulation comes in the form of tourism, and is constantly re-imagined by the state as a collective means for growth (e.g. Manera 2009). In reality, though, tourism feeds a powerful socioeconomic metabolism that permeates the whole of the archipelago's urban life (Murray 2012). The centre of Ciutat therefore serves as an extremely appropriate, and nevertheless specific, platform from which to research gentrification as a process that not only proves class inequality but actually maintains it.

The 'Dickensian' aspect

Due to the productive shift that tourism represented across the island in the second half of the 20th century, Es Barri witnessed a generalised flight of the working class from the manufacturing industries in which it had been employed. Meanwhile, property under-maintenance spread at the same rate as tenants with lower purchasing power entered (Morell 2015). Although Es Barri had always been host to it, by the 1980s it was considered to epitomise the full essence of what was commonly hold to be the despicable lumpen-class:

> [...] the catch-all for those who fall out [...] so that they are no longer functionally an integral part of society. To survive at all [...] they must adopt a parasitic mode of existence. The tendency toward illegality simply arises from the scarcity of other choices.
>
> (Draper 1972: 2309)

By the 1980s Es Barri maintained an impoverished elderly population that paid old cheap frozen rents, together with sexual labourers, and drug traffickers and users who had managed to move in thanks to low rents in shared accommodation.[3] In a clear exercise of territorial stigmatisation (Wacquant et al. 2014), in the 1980s and 1990s, the press fuelled the neglect of Es Barri by promulgating a biased view of who had the right to the neighbourhood: 'A neighbourhood for its *historic* residents, where passers-by can walk without fear of cars, delinquents or bad habits [...]' (Riera 1993). Yet this was not a ghost neighbourhood.

As Beltrán autobiographically writes in his graphic novels (Beltrán and Seguí 2011, 2014), Es Barri burst with a heterogeneous life throughout his youth, one that was killed off by the authorities and the well-to-do of Ciutat. Here, bars, bodegas and night clubs flourished. Far from polishing their nature, Beltrán and Seguí depict these sites as vice dens.[4] Although the stories to be found are not always suitable for all audiences, the authors take extreme care to show that these places held the neighbourhood together through informal networks of mutual aid, just as other small local businesses did.

By the mid-1980s the City Council announced the renewal of Es Barri (de Palma 1985). It would take more than a decade before the approval of the final partial scheme that was to plan the 'highest and best uses' of Es Barri (de Palma 1995), and even longer for additional EU funding (Equipo redactor del proyecto Urban-El Temple 1999). In the meantime, local bosses, both in government and in the opposition, made their own announcements of what renewal was supposed to bring:

> The scheme is [...] positive [...] for intervening in the most degraded area [by] generating businesses such as restaurants, bars and cafes [and] a

> substitution of people. [T]here is a need to [...] define the kind of people we want to fill the hole left by those who leave.
>
> (Carles Bona, municipal spokesman for urbanism in the opposition, in Oliver 1997)

There was a strong alliance among the political elite of Ciutat regarding Es Barri. The Social-Democrats supported the conservative-driven agenda for not only blatantly substituting inhabitants, but also for replacing the 'present uses' of businesses. Where there had been bars there would carry on being bars, but of a very different kind.

Bar-ri entrepreneurs

As the different planning schemes mentioned in the previous section show, the mediation of the state was decisive in paving the way for future private investment, yet the state by itself was not enough. There was also the need to entice other forms of collective social action such as strong grassroots organisations. Here, the ideal candidate was the *movement veïnal*,[5] and in particular the association of residents of Es Barri. This was founded in 1991 after the *Federació d'Associacions de Veïnats de Palma* [Palma's Federation of Neighbourhood Associations], together with other organisations, campaigned against the displacement of Es Barri's population during the organisation of fiestas:

> They are concerned with the high number of outcasts that will be left [...] in equal or worse conditions than the present, since they will be obliged [...] to leave for elsewhere where they will be equally marginal and badly housed.
>
> (R 1990)

This association of Es Barri was Canamunt – Ciutat Antiga. Its leadership was made of pioneer newcomers from the mid-1980s who, in the 1990s, worked hard to bring together the different groups in the neighbourhood. Not only did they lose the fight with the local authorities when these began to compensate evictions, however, they actually contributed to the displacement with the most thorough study ever on the neighbourhood demographics (Associació de Veïnats de Canamunt 1993). After the evictions and with the onset of the demolition outlined in the planning schemes, the few remaining members of Canamunt kept a low profile.

The association came back on the scene at full strength in 2007. The veteran members organised a meeting and more than fifty people attended. Most were new residents who had arrived from 2004 onwards. That was the year that a new bar opened and began to enliven the neighbourhood with their renamed celebrations. It did not take long before they joined efforts to bring together the opening anniversary of the bar and the resurrection of the Canamunt *fiestas*. In 2008, more than 3,000 people attended the main venue, and more newcomers arrived to fill the new and old housing of the spruced up neighbourhood.

By 2009 many other bar owners wished to join in, but Canamunt did not want to turn a grassroots organisation into a business agency. In addition, bars were overcrowding Es Barri by 2010. They were the only businesses that could afford the prices set by the owners of the not always eye-catching ground floors. Most of these bars put together the Martian Route pub-crawl mentioned at the beginning of the chapter but as a member of Canamunt remarked to a journalist, the bars had reached a turning point in which they were no longer bringing life to the neighbourhood (Zurimendi 2010). Other organisations present in the same neighbourhood had stronger views:

> [...] those of us who came to the old city many years ago see how the fruit of our labour is now being reaped by bar owners.
>
> (Associació de Veïnats Banc de l'Oli 2011)

Despite their differences, both sides saw the bars as a means of revitalisation that, in general, added value to the neighbourhood. Whereas for Canamunt the bars had accomplished the mission with which they had been entrusted, for Banc de l'Oli they represented a profiteering threat that could appropriate the efforts and life investments of its members. The ultimate fate of the bars has already been described, where state regulations put an end to the pub-crawl sprawl. What is yet to be explained is the hidden project for class politics the rent gap holds.

They say distance is oblivion

The title of this section corresponds to the first verse of the famous song 'La barca' written by the Mexican composer and singer Roberto Cantoral (1957). Written for a different purpose, it nevertheless helps to convey the idea that the wider the separation, the easier it is for us to forget experiences and practices that help to remain together and avoid and even overcome distance. Oblivion may well lead to work against one's own collective interest.

Take, for instance, Paton's recent work (Paton 2014). She elucidates how working-class Glaswegians consent to their own gentrification. She looks at their consumption choices and their way of life, one that supports relations of capitalistic accumulation. Her analysis sheds light on the problems raised by the translation of working-class consciousness to collective action, yet by focusing on individual consumption, the potential for class politics against gentrification is absent by definition.

It was Neil Smith who praised Henri Lefebvre in his prologue to the English translation of the *Urban Revolution*, for mastering the combination of 'hard-nosed critique and political desire' (Smith 2003: xii). Although I acknowledge that my political desire exceeds by far my skills in hard-nosed critique, I contend that it is only by exploring the productive fabric of gentrification that we may gain further insights into not only how gentrification works, but also how we ought to confront it. My double hypothesis on the competing forms of gentrification and on the urban labour involved in rent gaps aims to consider the limits of gentrification.

The thesis does so by examining the contradictory position of capital in its diverse quest for surplus value from within, while looking at how the objective productive forces of the rent gap, here embodied by a chain of production of value, become a potential class subject for social change. Creative practice, argued Williams, should be understood 'as the articulation and formation of latent momentary and newly possible consciousness' (Williams 1977: 212). What is possible, though, depends on the practice of a political present, one that can bridge distances:

> Distance (as in setting a safe distance) takes place under the sign of what is possible, in relation to what is current, present, real, existing; and not in the name of the past, or of an escape.
>
> (Lefebvre 2011 [1957]: 62)

By effacing the class gap between the class fractions that labour the rent gap, I hope to have critically contributed to such an appeal for new modes of consciousness that are nevertheless material and that therefore may contribute to relationally shaping class. Despite their past errors and bearing in mind that 'classes are simply not the transcendent, coherent, and empirically well delineated things they are too often taken to be' (Kalb 1997: 20), Canamunt might have started to bridge the class gap that rent gaps rely on:

> [...] it is now time to cook [...] both from the leftovers and from new ingredients, a finger licking good dish. This is our project: to put together what is left and what is lacking so as to build a lively and open neighbourhood [...]
>
> (Associació de Veïnats de Canamunt 2010: 1)

Since for rent gaps to be successful and find a profitable closure there is the need to maintain a class-gap as wide as possible upon which urban labour is founded, I maintain that any contentious class politics involving gentrification should do everything possible to combat oblivion by narrowing down the distance between the 'present' and 'highest and best' uses.

Acknowledgements

This chapter condenses the main argument of my doctoral thesis, the research for which was mostly funded by the *Programa de Beques Doctorals* 2006–2010 (*Direcció General de Recerca del Govern de les Illes Balears*). Since

Neil Smith has always been a source of inspiration, I thank Abel Albet and Núria Benach for their invitation and patience. Eleanor Lobb proofread the original draft. All translations of quotes and excerpts are my own. The usual disclaimers apply. Finally, I dedicate the chapter to my mother, Lynn Tipper, who, far from her parents' home, wisely knew how to instil in me their enduring working-class values.

Notes

1 After all, it was Marx (1980: 360–361) who, in an explosion of irony, apologetically conceived the crime of a criminal not only as producing criminal law but also the police, criminal justice, law degrees, etc.

2 In Morell and Franquesa (2011), we highlighted the ways in which concrete trends met abstract tendencies in producing the neighbourhood scale, one that was planned and geared towards gentrification by local bosses acting from different levels of state authority.

3 The pattern was the same in many Spanish cities. The vice district was known as Barrio Chino. See Fernández (2014) for an excellent study of the case of Barcelona.

4 Bars specialised to a great degree, such as the bar for transvestite sexual labourers described in detail in the realistic narrative masterpiece of the late Jaume Pomar, dedicated to the outlawed activities found in Es Barri (Pomar 2005 [1988]). Another author, Carlos Garrido, writing for a local newspaper, would later recall with nostalgia how the clubs of the area had served the young males of the city in their sexual initiations (Garrido 2004).

5 *Veïnat* means neighbour but *movement veïnal* translates best as Resident's Movement. This was a large social movement of the 1970s and 1980s that worked as a federated union of neighbourhood associations in each city across urban Spain. Its political leanings normally fell on the left of social democracy. For a detailed account of how it developed in Palma see Alemany (1986).

References

Aalbers, M.B. (2016). *The Financialization of Housing*. London: Routledge.

Alemany, G. (1986). *Les Associacions de Veïnats a Palma (1973–1984)*. BA Thesis. Palma: Universitat de les Illes Balears.

Associació de Veïnats Banc de l'Oli (2011). 'Palma: Propietat dels restauradors?', *Diario de Mallorca*, Thursday 10th October.

Associació de Veïnats de Canamunt–Ciutat Antiga et al. (1993). *Investigació sobre la realitat social i demogràfica de Sa Gerreria*. Palma: Patronat Municipal de l'Habitatge (unpublished 8th Ciutat de Palma Award, 3 vols.).

Associació de Veïnats de Canamunt–Ciutat Antiga (2010). *Es Brut de Can Amunt*, 1(1). Associació de Veïnats de Canamunt.

Beltran, G. and Seguí, B. (2011 and 2014). *Historias del barrio*. Bilbao: Astiberri (2 vols.).

Bidou-Zachariasen, C. and Poltorak, J-F. (2008). 'Le "travail" de *gentrification*: les transformations sociologiques d'un quartier parisien populaire', *Espaces et Sociétés*, 132–133, pp. 107–124.

Callinicos, A. (2009). *Imperialism and Global Political Economy*. Cambridge: Polity Press.

Cantoral, R. (1957). 'La barca', in *El reloj/La barca*. [45 rpm single] Promotora Hispanoamericana de Música.
De Angelis, M. (2010). 'The production of commons and the "explosion" of the middle class', *Antipode*, 42(4), pp. 954–977.
de Palma, A. (1985). *Pla General d'Ordenació Urbana*. Palma: Ajuntament de Palma.
de Palma, A. (1995). *Pla Especial de Protecció i Rehabilitació Integral de Sa Gerreria*. Palma: Ajuntament de Palma.
Draper, H. (1972). 'The concept of Lumpenproletariat in Marx and Engels', *Economies et Sociétés*, 6(12), pp. 2285–2213.
Equipo redactor del proyecto Urban-El Temple (1999). *Programa de rehabilitación integral Urban-El Temple. Gerreria-Calatrava*. Palma: Ajuntament de Palma.
Fernández, M. (2014). *Matar al 'Chino'. Entre la revolución urbanística y el asedio urbano en el barrio del Raval de Barcelona*. Barcelona: Virus Editorial.
Garrido, C. (2004). 'La muerte de "es Barrio"', *Diario de Mallorca*, Saturday 17th January.
Harvey, D. (2012). *Rebel Cities: From the Right to the City to the Urban Revolution*. London: Verso Books.
Kalb, D. (1997). *Expanding Class: Power and Everyday Politics in Industrial Communities, The Netherlands 1850–1950*. Durham: Duke University Press.
Lees, L. et al. (2008). *Gentrification*. New York: Routledge.
Lefebvre, H. (1976–1978). *De l'État*. Paris: Union Générale d'Éditions (4 vols.).
Lefebvre, H. (2011 [1957]). *Vers un romantisme révolutionnaire*. Fécamp: Nouvelles Éditions Lignes.
Manera, C. (2009). *L'eixam i les abelles. Per un nou model de creixement a les Illes Balears*. Barcelona: Abadia de Montserrat.
Marx, K. (1980 [1956] {1862/1863}). *Teorías sobre la plusvalía I. Tomo IV de El Capital*. México: Fondo de Cultura Económica.
Morell, M. (2015). 'When space draws the line on class', in J. G. Carrier and D. Kalb (eds.). *Anthropologies of Class: Power, Practice and Inequality*, Cambridge: Cambridge University Press, pp. 102–117.
Morell, M. and Franquesa, J. (2011). 'Playing snakes and ladders in Ciutat de Mallorca: An ethnographic approach to the neighbourhood scale', in M. Kousis, T. Selwyn and D. Clark (eds.). *Contested Mediterranean Spaces. Ethnographic Essays in Honour of Charles Tilly*, Oxford: Berghahn Books, pp. 195–220.
Murray, I. (2012). *Geografies del capitalisme balear: Poder, metabolisme socio-econòmic i petjada ecològica d'una superpotència turística*. PhD Thesis. Palma: Universitat de les Illes Balears.
Ocejo, R.E. (2011). 'The early gentrifier: Weaving a nostalgia narrative on the lower east side', *City & Community*, 10(3), pp. 285–310.
Oliver, A. (1997). 'La rehabilitació de Ciutat, el gran repte', *Ona*, 51, pp. 24–29.
Paton, K. (2014). *Gentrification: A Working Class Perspective*. Farnham: Ashgate.
Pomar, J. (2005[1988]). *Un dia o altre acabaré de legionari*. Palma: Hora Nova.
Reixac, J.R. (1990). 'Fiesta en Sa Gerreria por sobrevivir', *El Día 16*, Sunday 16 September.
Riera, J. (1993). 'La metamorfosis del "Barrio Chino"', *El Día del Mundo*, 5 February.
Smith, N. (1977). *The Return from the Suburbs and the Structuring of Urban Space: State Involvement in Society Hill*. BA Thesis. Saint Andrews: Saint Andrews University.

Smith, N. (1979). 'Toward a theory of gentrification. A back to the city movement by capital, not people', *Journal of the American Planning Association*, 45(5), pp. 538–548.
Smith, N. (1996). *The New Urban Frontier. Gentrification and the Revanchist City.* London: Routledge.
Smith, N. (2000). 'What happened to class?', *Environment and Planning A*, 32(6), pp. 1011–1032.
Smith, N. (2003). 'Foreword', in H. Lefebvre (2003 [1970]). *The Urban Revolution.* Minneapolis: University of Minnesota Press, pp. vii–xxiii.
Wacquant, L. et al. (2014). 'Territorial stigmatization in action', *Environment and Planning A*, 46(6), pp. 1270–1280.
Williams, R. (1977). *Marxism and Literature.* Oxford: Oxford University Press.
Zurimendi, A. (2010). 'L'experiment Gerreria. O com conviuen bars i veïnat', *Diari de Balears*, Monday 17 May.

12 The new urban frontier of everyday evictions

Contemporary state practices of revanchism

Vickie Cooper and Kirsteen Paton

In this chapter, we look critically at the political economic landscape of evictions within the United Kingdom context in relation to austerity. Our principle argument is that the current and unprecedented rise in evictions in the United Kingdom is not an 'organic' market-led occurrence, but has instead been actively orchestrated. Today, displacement is actively endorsed through state-level involvement via housing and welfare legislation and policy-making ushered in as austerity measures. As such, evictions have become daily occurrences in the United Kingdom. Evictions in neoliberal economies are fast becoming the most common form of displacement, with over 170 evictions per day in the United Kingdom (Ministry of Justice 2015), as well as an international rise in Ireland, United States, Spain, Germany and Greece (European Action Coalition 2015, Desmond 2016). Understanding the contemporary effects of gentrification and eviction requires scrutiny of the processes at large.

We begin this chapter by detailing the key ways that contemporary state-led gentrification has evolved, contributing to 'state-led eviction'. We situate evictions as a common form of displacement that underpins gentrification processes that have been under-researched in state-led gentrification discussions. We define state-led eviction as the cumulative impact of the government cuts in social support for housing and the promotion of housing market activities, which not only increase eviction risk amongst low-income households but, through various legal repossession frameworks, actively endorse it. We argue that state-led eviction comprises two aspects. The first relates to a transference of debt, from state level to household level, achieved through welfare cuts that directly affect housing benefit payments and ergo rent affordability. The second relates to a redistribution of wealth from the public to private sector. This includes the ways in which the state works in partnership with the private sector, including bailiff and enforcement companies that directly profit from widescale housing inequality and evictions processes. We explore this through secondary data analysis of welfare changes and legislative and court practices in the United Kingdom.

The evolution of gentrification

The ubiquity of gentrification today firmly casts aside the old 'first wave' culture/economic definitional tussles (see Ley 1996, Smith 1996). Gentrification is not simply a quixotic neighbourhood process (Smith 1996), but mediates urban restructuring towards neoliberal market capitalism. Gentrification is described as capitalism playing out in the landscape; it is essentially our economy's urban form (Brahinsky 2014), so as capitalism evolves, so, too, does gentrification. The most recent incarnation 'state-led gentrification', described by Neil Smith as being the urban frontier, names the active role that the state plays in supporting the market to drive gentrification. As a spatial manifestation of neoliberal urban policies, gentrification is not, then, a process exclusively driven by economic forces or the cultural and consumer choices of the middle classes or, as scapegoated today, hipsters (Kohmami and Halliday 2015). Institutionalised as urban policy, gentrification is deemed the panacea to the decline wrought by deindustrialisation (Lees 2003) and wholeheartedly subsumed into New Labour's Urban Renaissance strategy through the 2000s. State-led gentrification often occurred on brownfield sites of former industry, which meant that direct displacement, said to be the linchpin of gentrification processes, was not always a direct outcome. This led some to challenge surrounding the use of the term "gentrification" in this context (Lambert and Boddy 2002), while others claimed instead that it signified a wider processes of urban restructuring, that is, the creation of space for the more affluent user (Hackworth 2002). Given the protection afforded by social housing, displacement was claimed to not be the primary aim of state-led gentrification and instead was used to create the more affluent user through civilising the working class to be neoliberal consumers of private housing (Paton 2014). Either way, discussions of displacement and evictions waned in this context.

If gentrification was the spatial manifestation of neoliberal urban policies, then it is the management of individuals in these spaces which becomes the new frontier under neoliberal financial capitalism. This extends the 'revanchist thesis' as coined by Smith (1996), as an interpretation of how gentrification is used as a strategy by powerful groups in society to retake, at times by force, their territorial control over less powerful groups. Housing was still a largely protected state provision when this concept was articulated by Smith and others (Smith 1996, Mitchell 2003), so revanchist practices were identified as those that reclaimed prime *public* spaces from homeless people for the benefit of international capital. Today, the project of revanchism advances whereby the state is no longer the protector in housing and evictions, but the profiteer.

The very project of state-led gentrification was momentarily challenged by the financial crisis of 2007 (Harvey 2014), which was itself precipitated by a collapse in housing finance and real estate. Not long after, the financial system rebounded, finding new ways to extract profit and value without producing (Lapavitsas 2013). Generating profit in this manner involved a shift in the mode of governance, in which the imperatives of finance and private capital take precedence. This is characterised by the large-scale state withdrawal from the housing sector along with the simultaneous creation of policies that support a market-based housing finance model, which sees housing repositioned as a primary commodity and a global financial investment asset (Rolnik 2013). Thus, in the post-crash context, state-led gentrification performs a greater, rather than lesser, role in the political economy. This is evident in the post-crash United Kingdom housing policy's prioritisation of finance and private capital, no less through the retreat of welfare.

Previous to this global paradigm shift in housing policy, social housing played an important role in offsetting and, indeed, protecting tenants against eviction in the United Kingdom. Today, however, we are seeing an erosion of the protection originally offered through the historical development of municipal housing and the welfare state. In the aftermath of the financial crisis in 2008, austerity-driven housing and welfare policy made it increasingly difficult for individuals and families, who are foremost economically disadvantaged, to access affordable social housing. In 2012, the former Coalition government set out their plans for restricting access to social housing:

> Social housing is a scarce resource, and the Government believes that it is appropriate, proportionate and in the public interest to restrict access in this way, to ensure that, as far as possible, sufficient affordable housing is available for those amongst the local population who are on low incomes or otherwise disadvantaged and who would find it particularly difficult to find a home on the open market.
>
> (CLG 2013: 5)

The burden of the public debt and dominant global 'logic' of austerity have culminated in a series of welfare reforms and housing policies that have exacerbated housing inequality and amplified housing poverty for economically disadvantaged groups. These policy reforms bring the poorest households closer to private rented housing and market rent rates at a time when private sector rents are significantly more expensive than social housing rents. The private rental sector has become the dominant provider of private housing and social housing, and the state is no longer the protector in housing and evictions, but rather the profiteer. Lack of rent controls, lax regulation and unbridled transnational capital pumped into the property market have led to the creation of eviction 'hotspots' (Shelter 2014), where those in poverty and receiving welfare benefits face the greatest financial challenges in relation to rent and, thus, face the greatest risk of eviction.

Eviction as displacement

Displacement is defined as 'what happens where forces outside the household make living there impossible, hazardous or unaffordable' (Hartman, Keating and LeGates 1982: 3). Eviction is a form of everyday displacement and frequently plays out in processes of gentrification (Sims 2016). It is a key displacement technique that is used by landlords, housing associations and state authorities to forcibly remove working-class households from the area to free-up land value. Despite the centrality of displacement in gentrification processes, it has long been under-researched. Discussions of displacement have historically been overlooked in favour of theoretical debates aside from some notable United States studies in the 1980s from Marcuse, (1986), Hartman (1979), Hartman, Keating and LeGates (1982), and LeGates and Hartman (1986). Methodologically, displacement studies have been downplayed following the view that displaced people are 'hard to follow' (Newman and Wyly 2006) and that assessing scales of displacement can be like 'measuring the invisible' (Atkinson 2000). This 'neglect' (Chum 2015) has culminated in a cursory understanding surrounding the relationship between evictions, gentrification and market capitalism. Contemporary eviction studies (Chum 2015, Sims 2016) show that displacement can be captured, experiences documented and relationships with gentrification and market capitalism unveiled. While governments fail to accurately record the scale and cause of evictions, these studies show that eviction records can be sourced through other means, such as 'landlord associations, property management companies, civil courts, law enforcement, and state agencies' (Sims 2016: 30) and, furthermore, highlighting the concentration of evictions in one particular area can help to expose the relationship between evictions and market-driven forms of urban restructuring (Sims 2016).

In this next section, we will go on to demonstrate how the state now actively facilitates the displacement and eviction of people. The confluence of a retraction in welfare and rise of housing as a global financial product results in the displacement and eviction of tenants who now find themselves in rent arrears and too poor to live on land deemed high-value. We outline how austerity-driven welfare reforms have led to a hike in rent rates and affordability which has culminated in an unprecedented rise in rent arrears and 'eviction hotspots' (Shelter 2014) confounding revanchism in the housing sector which arguably constitutes state-led evictions.

Welfare cuts and housing policy

Government policies implemented to support the paradigmatic changes in housing have involved a redistribution of wealth and debt. In 2012, the Coalition Government introduced a historical set of welfare reforms. One of the most controversial and pernicious changes was undoubtedly the so-called 'bedroom tax': the spare room subsidy of tax on housing benefit levied at social housing tenants deemed to have a 'spare' bedroom. One year after its implementation in April 2014, research found that two-thirds of households in England affected by the bedroom tax had fallen into rent arrears, while one in seven families received eviction risk letters and faced losing their homes (NHF 2014). Around 6% of benefit claimants affected by the bedroom tax have moved as a result (BBC 2014). The more pervasive and perhaps most wide-reaching changes, however, were introduced through the 'benefit cap' (or Universal Credit). The benefit cap limits the total weekly income an individual or family can receive in welfare payments, with an estimated 58,700 households experiencing a reduction in Housing Benefit (45% in London). When a household exceeds the capped level, their benefit income is capped. This cap is administered primarily through housing benefit payments, therefore, household rent. It is estimated that 50,000 households have lost around £93 per week and 15% are losing around £150 per week (Shelter 2013). These households have to make up the subsequent shortfall they face in rent. Given that the benefit cap is administered primarily through housing benefit payments, it would be more useful to reconsider the benefit cap as a 'rent cap' as it automatically reduces a person's rent income.

Given this brief overview of austerity cuts affecting welfare recipients, it is not difficult to see how those in poverty and receiving welfare benefits face the greatest financial challenges in relation to rent and, thus, face a greater risk of eviction. The inability of households receiving welfare benefits to pay their rent has seen hundreds of thousands of households fall into rent arrears. From 2007–2013, the number of households in rent arrears increased by 130% (Money Advice Trust 2013) and resulted in unprecedented levels of repossessions and evictions. It is no coincidence that in 2013 we also saw 115,000 possession orders granted to evict tenants in private rented accommodation as a result of rent arrears. In 2015, across England and Wales more than 42,000 households in rented housing were evicted by County Court bailiffs, a 50% increase over the past four years, and the highest level since records began in 2000 (Ministry of Justice 2016). Nonetheless, these figures do not paint the full picture because they fail to highlight the number of renters who leave 'voluntarily', without challenging the initial possession claim. The realities of this are captured by Matthew Desmond's (2016) portrayal of the harrowing experiences of undocumented flight of families facing housing poverty in the United States as 'informal eviction'.

These changes in welfare and housing tenure are therefore distinct but interrelated modes of transferring debt to individuals and redistributing wealth to the private sector, and result in exposing the most vulnerable to a volatile housing market and the risk of eviction. There is more of this to come. In May 2016, the Conservative government advanced these austerity measures through their *Welfare Reform and Work Bill* that reduces the 'benefit cap' even further and expands Universal Credit, with even greater implications for rent affordability. Furthermore, the government's new

Housing and Planning Bill actively curtails housing choice by further privatising the housing sector. This bill proposes several new policies, including replacing the obligation to build homes for social rent with a duty to build starter homes; extending the Right to Buy to include Social Housing; forcing local authorities to sell 'high-value' housing; offering laxer planning permission; and introducing 'fast-track' evictions (Crisis 2015).

In light of these interventions made to housing policy and welfare, it is fair to say that evictions today are not solely market-driven. Conversely, since 2010, governments have consistently intervened and implemented a suite of hyper-austerity legislation and policies that primarily target households that are foremost economically disadvantaged. Specifically, where housing policy is concerned, and variously related welfare policies such as those described earlier, governments have stimulated private rented sector growth and exposed low-income earners to the vagaries of housing. Framed in this way, the state has been hyperactive in the amplification of housing poverty and insecurity, thus confounding the revanchist project.

The role of the state in evictions and circumstances leading up to it is the principal concern of this chapter. We reject the notion that the rise of the private rented sector economy, and the precarity it creates, is solely market-driven. Rather than conceptualising the demise of social housing and recent welfare reforms as a type of 'state withdrawal', or 'hollowing out' of state involvement (Cloke, May and Johnsen 2010), we see these events as a different type of state activity. These events are principally state-led, and what is more, they encompass strategies around enforcement and dispossession, supported by the legal repossession framework and private business. In this next section, we will show how legislative frameworks that prioritise housing ownership also enable an enforcement industry to prosper under these statutory powers: to repossess, recover debt and evict.

Accumulation by repossession

Just as the state intervenes in ways that increase the risk of eviction, it too endorses the enforcement of eviction by working in partnership with private enforcement businesses that supports the expansion of private industry profiting from displacement. This is a kind of accumulation by repossession, which builds from Harvey's (2005) notion of 'accumulation by dispossession' that describes how predatory capitalism operates, expands and accumulates at key points during economic recession. Where dispossession 'entails the loss of rights', this erosion also allows for greater flexibility around capital accumulation (Harvey 2005: 178). Wide-scale precarity precipitated by economic recession leads to dispossession, but that dispossession also allows for certain industries to accumulate and proliferate. In housing terms, accumulation by repossession reflects the ways in which rent arrears and the ultimate consequence of eviction, not only lead to the dispossession and displacement of tenants, but allow businesses to profit and expand in an industry that is centred around household debt, debt collection and eviction. As such, this activity cannot be explained simply as market-led. State intervention, including legislative frameworks that prioritise property ownership, have helped the expansion of an enforcement industry whose primary aim is to recover debt and/or evict tenants. Therefore, the industry profiting from eviction, we argue, comprises those private companies that profit from enforcing possession orders, as well as the legislative framework that protects them. These can include, for example, creditor companies as well as civil and high court bailiff and enforcement companies.

As the number of possession orders have increased, so too has the recruitment of civil and high court officials called upon to carry out the 'writ of execution'. In the last five years (Gayle 2015), bailiffs used by landlords for repossession have increased by 51%. Between 2014 and 2015, bailiff companies used by local authorities in London for the recovery of council tax debt increased by 50% (Ashton, Francis and Jarvie 2015) and HMRC recently spent £14.8 million on bailiff companies to recover unpaid taxes (Jones 2014). Business is booming. In 2015, one of the largest High Court enforcement and bailiff companies, The Sheriffs Office Ltd., was awarded *Enforcement Team of the Year* by the Chartered Institute of Credit Management (CICM), and further demonstrating the expanse of the industry, the Chief Executive of the Sheriffs Office Ltd claimed that, "we launched a number of specialist services and entered new markets, all leading to growth in all sectors of our business" (Credit Today 2016, np).

The growing involvement of private industry is crucial as we think about the complexity of state-led eviction. Evictions are often carried out by private bailiff and enforcement companies and those enforcement officers carry a range of legal powers as 'court officials', allowing them to forcibly remove people from their homes. Challenges to that authority can be met with the threat of prosecution. Furthermore, it is not uncommon for police to be present during tenant evictions, even though evictions are civil incidents, not criminal. In principle, the role of the police in these matters is to maintain independence, but their presence carries a symbolic effect, which leads tenants and communities to believe that residents being evicted have committed a criminal offence, where none has occurred. Police presence and the hybrid use of criminal and civil legislation in the eviction process present some challenges to resisting housing inequality.

No one is more familiar with these challenges than those activists fronting the housing resistance. In September 2015, 15 housing activists faced criminal charges (and some remanded in custody) for obstructing High Court Enforcement Officers evicting a tenant and housing activists from Sweets Way estate, North London. As rent arrears continue to grow and housing resistance intensifies, including occupation of empty homes and land, this is likely to trigger further action from private industry recruited to forcibly remove those tenants *and* activists in order to ultimately secure the development of capital accumulation and commodification of housing.

Conclusion

Smith and LeFaive (1984: 60) forewarned that '[c]apitalism is based precisely on its ability to displace the working class in all sorts of situations'. Today, under austerity, we see the punitive side of state-led gentrification really come into effect and, as a result, an advancement of revanchist practices. While the revanchism project originally spoke to the taking back of territorial power from public spaces, we see that under financial capitalism where housing is cast as a primary global asset, the state undertakes activity that extends revanchist practices to public housing. We have identified that state-led eviction advances revanchism because in transferring debt from state level to household level through welfare cuts, rent affordability is directly affected, and, moreover, state partnership activities effectively take back public housing through the increased involvement of bailiff and enforcement companies – companies that generate profit from eviction and repossession. These processes galvanise property ownership and effectively displace those who are not deemed 'valuable' from housing to free it up for more profitable use, while at the same time, making profit out of these processes. Welfare reforms that actively make households more precarious, ultimately leading to eviction, and the enactment of evictions themselves – whether through high court enforcement officers or by the presence of police in the point of evictions – reveal the state's central role. While the state has historically played a part in evictions, it is the confluence of these contemporary factors, whereby housing and rent are premium global assets, that has resulted in the removal of protection from residents. In this sense, Smith and LeFaivre's (1984) statement could not ring truer today. In the quest for profit under financial capitalism, the basic security of the home is under its greatest threat, and contemporary state policy and practice support rather than prevent this in new ways, indicating a grave new urban frontier. Evictions, we argue, are today the most common and visible side of housing insecurity and a significant marker in contemporary state-led gentrification processes.

References

Ashton, S., Francis, M. and Jarvie, M. (2015). *Too Poor to Pay: The Impact of the Second Year of Localised Council Tax Support in London*. CPAG/Zacchaeus 2000 Trust.

Atkinson, R. (2000). 'Measuring gentrification and displacement in Greater London', *Urban Studies*, 37(1), pp. 149–165.

BBC (2014). *Housing Benefits: Changes 'See 6% of Tenants Move'* [TV programme]. Available at: www.bbc.com/news/uk-26770727 (Accessed: 3 March 2016).

Brahinsky, R. (2014). 'The death of the city', *Boom. A Journal of California*, 4(2). Available at: www.boomcalifornia.com/2014/07/the-death-of-the-city/ (Accessed: 3 March 2016).

Chum, A. (2015). 'The impact of gentrification on residential evictions', *Urban Geography*, 36(7), pp. 1083–1098.

CLG (2013). *Providing Social Housing for Local People: Statutory Guidance on Social Housing Allocations for Local Authorities in England.* London: Department for Communities and Local Government (CLG).

Cloke, P., May, J. and Johnsen, S. (2010). *Swept Up Lives? Re-envisioning the Homeless City.* Oxford: Wiley-Blackwell.

Credit Today (2016). *High Court Enforcement Group Acquires Sheriffs Office*, 1st February. Available at: www.credittoday.co.uk/article/18168/online-news/high-court-enforcement-group-acquires-sheriffs-office- (Accessed: 3 March 2016).

Crisis (2015). *Crisis Submission to Housing and Planning Bill Public Bill Committee*, 5th November. Available at: www.crisis.org.uk/data/files/publications/Crisis%20 submission%20to%20Housing%20and%20Planning%20Bill%20Public%20 Bill%20Committee.pdf (Accessed: 3 March 2016).

Desmond, M. (2016). *Evicted: Poverty and Profit in the American City.* New York: Crown.

European Action Coalition (2015) 'Evictions across Europe', *European Action Coalition,* 26th May. Available at: www.housingnotprofit.org/en/articles (Accessed: 3 March 2016).

Gayle, D. (2015). 'Tenant evictions reach six-year high amid rising rents and benefit cuts fears', *The Guardian,* 14th May. Available at: www.theguardian.com/money/2015/may/14/tenant-evictions-reach-six-year-high-rising-rents-benefit-cuts (Accessed: 3 March 2016).

Hackworth, J. (2002). 'Post-recession gentrification in New York City', *Urban Affairs Review,* 37, pp. 815–843.

Hartman, C. (1979). 'Comment on neighborhood revitalization and displacement: A review of the evidence', *American Planning Association*, 45(4), pp. 488–491.

Hartman, C., Keating, D. and LeGates, R. (1982). *Displacement: How to Fight It.* Berkeley, CA: National Housing Law Project.

Harvey, D. (2005). *A Brief History of Neoliberalism.* Oxford: Oxford University Press.

Harvey, D. (2014). *Seventeen Contradictions and the End of Capitalism.* London: Verso.

Jones, R. (2014). 'Revenue doubles use of bailiffs to recover unpaid tax', *The Guardian,* 17th May. Available at: www.theguardian.com/money/2014/may/17/revenue-bailiffs-unpaid-tax-hmrc (Accessed: 3 March 2016).

Kohmani, N. and Halliday, J. (2015). 'Shoreditch Cereal Killer Cafe targeted in anti-gentrification protests', *The Guardian,* 27th September. Available at: www.theguardian.com/uk-news/2015/sep/27/shoreditch-cereal-cafe-targeted-by-anti-gentrification-protesters (Accessed: 3 March 2016).

Lambert, C. and Boddy, M. (2002). 'Transforming the city: Post-recession gentrification and re-urbanisation', in *Conference on Upward Neighbourhood Trajectories: Gentrification in the New Century*, 26–27th September, University of Glasgow.

Lapavitsas, C. (2013). *Profiting Without Producing: How Finance Exploits Us All.* London: Verso Books.

Lees, L. (2003). 'Visions of "urban renaissance": The Urban Task Force report and the Urban White Paper', in R. Imrie and M. Raco (eds.). *Urban Renaissance? New Labour, Community and Urban Policy.* Bristol: Policy Press, pp. 61–82.

LeGates, R. and Hartman, C. (1986). 'The anatomy of displacement in the United States', in N. Smith and P. Williams (eds.). *Gentrification of the City.* London: Unwin Hyman, pp. 178–200.

Ley, D. (1996). *The New Middle Class and the Re-making of the Central City.* Oxford: Oxford University Press.

Marcuse, P. (1986). 'Abandonment, gentrification and displacement: The linkages in New York City', in N. Smith and P. Williams (eds.). *Gentrification of the City.* London: Unwin Hyman, pp. 153–177.

Ministry of Justice (2015). *Mortgage and Landlord Possession Statistics Quarterly. England and Wales.* April to June 2015. Available at: www.gov.uk/government/uploads/system/uploads/attachment_data/file/453265/mortgage-landlord-possession-statistics-april-june-2015.pdf (Accessed: 3 March 2016).

Ministry of Justice (2016). *Mortgage and Landlord Possession Statistics Quarterly. England and Wales,* October to December 2015. Available at: www.gov.uk/government/uploads/system/uploads/attachment_data/file/499083/mortgage-and-landlord-possession-statistics-october-to-december-2015.pdf (Accessed: 3 March 2016).

Mitchell, D. (2003). *The Right to the City.* London: The Guildford Press.

Money Advice Trust (2013). *Rent Arrears the Fastest Growing UK Debt Problem,* Available at: www.moneyadvicetrust.org/media/news/Pages/Rent-arrears-the-fastest-growing-UK-debt-problem.aspx (Accessed: 3 March 2016).

Newman, K. and Wyly, E. (2006). 'The right to stay put, revisited: Gentrification and resistance to displacement in New York City', *Urban Studies*, 43(1), pp. 23–57.

NHF (2014). *Bedroom tax: The results are in*, 27th July. Available at: www.housing.org.uk/blog/bedroom-tax-the-results-are-in/ (Accessed: 3 March 2016).

Paton, K. (2014). *Gentrification: A Working-Class Perspective.* Farnham: Ashgate.

Rolnik, R. (2013). 'Late-neoliberalism: The financialization of homeownership and housing rights', *International Journal of Urban and Regional Research,* 37(3), pp. 1058–1066.

Shelter (2013). *Welfare Reform Bill – House of Lords Report Stage Joint briefing on Clause 94: The Overall Benefit Cap,* Available at: https://england.shelter.org.uk/__data/assets/pdf_file/0019/410590/Welfare_Reform_Bill_Clause_94_briefing.pdf (Accessed: 3 March 2016).

Shelter (2014). *Eviction and Repossession Hotspots 2013/14 Data*, London: Shelter. Available at: https://england.shelter.org.uk/__data/assets/pdf_file/0011/825176/Eviction_and_Repossession_Hotspots_201314_Data.pdf (Accessed: 3 March 2016).

Sims, R. J. (2016). 'More than gentrification: Geographies of capitalist displacement in Los Angeles 1994–1999', *Urban Geography,* 27(1), pp. 25–56.

Smith, N. (1996). *The New Urban Frontier: Gentrification and the Revanchist City.* London: Routledge.

Smith, N. and LeFaivre, M. (1984). 'A class analysis of gentrification', in J. Palen and B. London (eds.). *Gentrification, Displacement and Neighborhood Revitalization.* Albany: State University of New York Press, pp. 43–63.

13 Capturing urban rent through evictions

Home dispossessions in the historic centre of Palma (Majorca)[1]

Sònia Vives-Miró and Onofre Rullan

As highlighted by Charnock, Purcell and Ribera-Fumaz (2014), the bursting of the Spanish property bubble in 2008 effectively heralded the end of the 'Spanish Model' (López and Rodríguez 2010). This marked the beginning of a new urban crisis that, on the one hand, provided opportunities for accumulation structured around new forms of property speculation and financial engineering and, on the other hand, exacerbated urban precariousness and resulted in social conflict. Within this scenario, one of the most relevant processes from the perspective of Critical Urban Geography was the intensity of home dispossession, which manifested itself in the Spanish context through the eviction of tenants and mortgage foreclosures (with 70,000 per year from 2008 onwards) (CGPJ 2016). In his rent gap theory, Smith (1979) described the process of capturing urban rents through devaluation and the subsequent reinvestment in the urban space. Following this argument, the present text seeks to open a discussion about the role that home dispossession, through the eviction of tenants and mortgage foreclosures, plays in the process of devaluing and revaluing urban space and the creation of the rent gap. To do this, we propose examining a number of different questions: How do the evictions of tenants and mortgage foreclosures affect the production of urban rent? Do they devalue or revalue urban space? What role have they played in the emergence of new processes of gentrification?

In order to examine these questions, we start with the argument that evictions form part of a strategy for producing urban rent and are not simply a consequence of the present crisis. To support this stance, we put into conversation two frames of theory: on the one hand, those that have analysed the concepts of the rent gap, gentrification and displacements and, on the other hand, those that have studied the financialisation of the built environment. In order to ground our analysis, we have centred our study on the process of devaluation/revaluation and its relationship with the eviction of rental and mortgage foreclosures carried out in the Historic Centre of Palma (Mallorca). To conclude, we have sought to conceptualise these logics within the framework of the new geographies of inequality that have emerged as a result of the production and capture of urban rents associated with the financialisation of housing in areas with high property values.

Evictions as a strategy for producing and capturing urban rent

The great wave of evictions observed in Spain from 2008 onwards is often interpreted as a simple consequence of the crisis. It is seen as the collateral damage associated with a more general economic crisis, with narratives referring to 'effects' and 'impacts' in the form of the 'loss of housing' and 'the most vulnerable sectors of society finding it difficult to access the housing market', etc. In this chapter, we seek to argue that evictions, along with the other mechanisms used for home dispossession, are not just consequences of the crisis, but also form part of the very strategy of producing and capturing urban rent. In other words, we are not speaking about an effect or impact, but rather about an integral part of the phenomenon itself; it forms part of the very logic which lies at the root of the urban crisis.

The phenomenon of evictions must be understood within the wider framework of processes of accumulation by dispossession (Harvey 2003) involving the urbanisation of capital and, more precisely, through the process of home dispossession. Harvey (2003) uses the concept of accumulation by dispossession to refer to the present-day equivalents to the original processes of accumulation that Marx highlighted when he referred to historical actions of enclosure as a form of primitive accumulation (Marx 1990 [1864]) based on the separation of the producers of land from their means of production. Although the process of enclosure has been present throughout history, the crisis of Fordism, and the subsequent restructuring of the capitalist system, brought about a new round of enclosures in which, reaching beyond the means of production, the means of social reproduction were also enclosed (Midnight Notes Collective 1990). The sectors subject to dispossession within the social reproduction of the countries of southern Europe are currently education, health and housing, and since the 1990s, the main modus operandi used for plundering them within the city has been via different forms of neoliberal urbanism (Smith 1996).

Evictions and mortgage foreclosures represent the dispossession of housing within processes of accumulation of urban rents. We must therefore understand them as an essential, inseparable and inherent part of the process of urban rent accumulation. Without one part, the other cannot exist; without the dispossession of housing, it is not possible to produce, appropriate, capture or accumulate this type of urban rent. Although the concept of accumulation through housing has already received considerable attention, we feel that the integration of home dispossession as an inherent part of this logic has, to date, remained relatively neglected.

Evictions as displacements due to the rent gap

The rent gap has been described as 'the disparity between the potential ground rent level and the actual ground rent capitalised under the present land use' (Smith 2012: 126). The potential land rent of a given site is determined only by its 'highest and best' use, whereas the actual land rent is also a function of that site's current intensity and type of land use (Clark 1995: 1490). The rent gap is therefore 'the disparity between the potential ground rent level and the actual ground rent capitalised under the present land use' and 'is produced primarily by capital devalorisation (which diminishes the proportion of the ground rent able to be capitalised) and also by continued urban development and expansion (which has historically raised the potential ground rent level in the inner city)' (Smith 2012: 65). In this sense, following Smith (2009), the rent gap constitutes the central mechanism of the process of gentrification. When neighbourhoods suffer disinvestment, the land rent that can be obtained in the area falls, as do property purchase and rental prices. As disinvestment continues, the abyss between the ground rent and the rent that could be obtained from a potential redevelopment grows, until the point at which reinvestment begins to become profitable (Smith 2009). When the rent gap is sufficiently wide for promoters to be able to capture a satisfactory surplus, gentrification takes place (Smith 2012).

Even so, the concept of gentrification requires displacements that reach beyond the process of rent appropriation. These displacements constitute home dispossession via the creation of a rent gap. In other words, gentrification is a process of accumulation by dispossession in which the accumulation manifests itself in the form of appropriated urban rents via the rent gap, and dispossession takes the form of home dispossession via displacements.

According to Slater, however, studies of displacements produced by gentrification had practically disappeared until very recently (Slater, 2015). This void has been explained by the difficulty involved in its quantification (Newman and Wyly 2006) and because it effectively implies 'measuring the invisible' (Atkinson 2000). Recovering the approach originally used by Marcuse, Slater observes that 'if the pain of displacement is not a central component of what we are dealing with in studying gentrification [...] we are missing the central point that needs to be addressed.' (Marcuse 2010: 187) and thinks that 'it is vital to do something about this in order to restore critical perspectives' (Slater 2015: 131).

Related to this need, we have conceptualised the eviction of tenants as part of the displacements that occur through gentrification. By doing this, we can, on the one hand, look more deeply into the specific characteristics of this type of displacement and, on the other hand, justify an understanding of evictions as a fundamental part of the process of accumulation by dispossession. Even so, within this conceptualisation, we can only consider evictions of tenants, because although those relating to mortgage foreclosures essentially follow the same logic, they form part of other mechanisms to which we shall refer later on.

In this sense, evictions of tenants are covered by what Marcuse (1985) called 'direct displacements', whether of the 'last resident' or 'chain' type.[2] Using these two categories, we included tenants in the group of those displaced for economic reasons: that is to say, when the tenant cannot pay their rent because it has been increased, leading to their eviction. Although indirect displacements (displacement by exclusion or due to pressure) also form part of the process of home dispossessions and housing enclosure, they are not the object of this study.

In summary, evictions of tenants form part of the process of direct displacement due to economic motives. In this way, evictions form part of the process of the production of the rent gap and can therefore be considered part of the strategy for the production or capture of urban rent. This is a conceptual integration that reinforces our understanding of the phenomenon of evictions within the logic of capturing urban rents.

Evictions within the logic of the financialisation of the built environment

The financialisation of housing constitutes another of the mechanisms of the logic of accumulation of urban rent through the home dispossession. In the literature, the financialisation of housing is included in studies of the financialisation of the built environment (Clark, Larsen and Lund 2015). The point of contact between financialisation (Krippner 2005) and the urban space can be found in the conversion of land into a financial asset and its inclusion, along with its associated urban rent, in the circulation of different types of international financial capital. In this sense, we recover the frequent use of the concept of 'land as a financial asset' (Harvey 1982). The fact that land has become a financial asset implies that what is bought and sold is no longer the land itself, but rather the expectant, fictitious right to appropriate its supposed potential urban rent (Harvey 1982).

In this way, the growth in financing has led to increasing capital investment in built environments. In urban areas, however, the aim of this investment has largely been to 'improve' the exchange value of this land rather than the value of its use. Following this logic, financialisation has implied an increase in its potential rental value. According to Sayer (2012: 171) and Clark, Larsen and Lund (2015: 22–23), under the sway of investments, what is expected of an allocational efficiency is that the expected rates of financial return should be as high as possible. Along these lines, the increase in potential rent generates an increasing disparity between this and the capitalised rent; in other words, it increases the rent gap. Thus, as Clark and Lund have observed, 'the making and taking of rent gaps at increasing scales is a principal form of accumulation by dispossession [...]. These processes have intensified and sprawled with burgeoning financialisation, increasing competition over land [...] and increasing inequalities associated with the global wave of neoliberalisation' (Clark and Lund 2012).

Looking beyond the rent gap, the financialisation of housing, which also implies its growing revaluation, also entails the use of other mechanisms for capturing rents through home dispossession. In countries like Spain, this has often been translated into property speculation bubbles in which financial entities have played a fundamental role by extending the availability of cheap credit and where the mortgage loan has become an essential condition for obtaining access to home ownership (López and Rodríguez 2010). In Spain, widespread indebtedness became a new mechanism for social governance: In order to comply with their mortgage obligations, those in debt have had no choice but to accept the precarious conditions imposed on the labour market (Colau and Alemany 2012: 35). Along these lines, revaluation constitutes a new form of housing enclosure, which is intensified when salaries fall and which makes it impossible to sustain the increasing cost of housing (Vives-Miró and Rullan 2014).

It is precisely in this scenario that mortgage foreclosures form an inherent part of the process: Without the appropriation of the rents of debtors that mortgage foreclosures provide, it would not be possible to conceive a phenomenon of such magnitude. Once again, this entails a logic of capturing urban rents through home dispossession.

The eviction of tenants and mortgage foreclosures both form a highly important part of dispossession within processes of accumulation generated by the financialisation of the built environment. Evictions are due to both the increase in the rent gap and the more general process of urban revaluation and mortgage foreclosures as part of the expansion of credit required to capture rents. The different waves of evictions that have taken place in Spain since 2008 are no more than a reflection of the dynamics of the 'Spanish Model' (López and Rodríguez 2011), which has been based on financial and property-based accumulation by dispossession.

Home dispossession and property revaluation in the historic centre of Palma

In order to answer the questions posed in the introduction, we have analysed the relationship between evictions and processes of devaluation and revaluation in the historic centre of Palma (Mallorca): the area located within the old, pre-industrial, walled city. In this space, which has a fully constructed surface area of around 130 hectares with a high heritage value, we have mapped both the evolution of tenant evictions and mortgage foreclosures and the price of housing.

Between 2003 and 2014, there were 890 cases of home dispossession, of which 810 (91%) corresponded to the eviction of tenants but only 80 (9%) to mortgage foreclosures; 62% of this total were recorded since the onset of the crisis (Table 13.1 and Figure 13.1). For the whole study period (2003–2015), this implied a density of 6.25 direct home dispossessions per hectare due to economic motives. The predominance of tenant evictions (91%) over mortgage foreclosures shows that it has been the population renting accommodation that has suffered most from home dispossession. Rents are more susceptible to increases as a result of pressure in favour of the urban revaluation in neighbourhoods than the conditions of mortgage contracts.

Table 13.1 Evictions and mortgage foreclosures in the Historic Centre of Palma (2003–2014)

	2003–2007		*2008–2014*		*Total 2003–2014*	
	Number	*%*	*Number*	*%*	*Number*	*%*
Tenant evictions	318	35.73	492	55.28	810	91.01
Mortgage foreclosures	17	1.91	63	7.08	80	8.99
Total	335	37.64	555	62.36	890	100.00

Source: Own data.

On the other hand, at the beginning of the 1990s, the geography of property prices[3] in the Historic Centre of Palma showed that there was a clear asymmetry between the eastern (10 €/m^2) and western (14–15 €/m^2) parts of the city. With the entrepreneurial turn of Palma (Vives-Miró 2013), its local government began to promote both public and private European investments in the eastern sector, through two urban reform plans,[4] and also several other interactions associated with the Urban Plan (1997–99). This marked the start of a process of gentrification in part of the city (Franquesa 2010, Vives-Miró 2011). Within this context, and as part of Palma's own particular take on entrepreneurial initiative, there was also the promotion of 'touristisation', or of uses linked with leisure and tourism, which was particularly evident in the central and western parts of the seafront.

This logic implied an important change in the price map for residential uses within the built environment of the historic city. This began to emerge in the first decade of the 21st century and by the second showed a clear north/south gradient that contrasted with the east/west gradient evident prior to the last property bubble. In this way, the central-eastern front, the La Calatrava (Franquesa 2010) neighbourhood, became the area in which property revaluation was most intense between 2006 and 2015.[5]

With respect to this transformation of the geography of revaluation, however, perhaps we must ask ourselves what the geography of evictions has been like. As we can observe from the map (Figure 13.1), the home dispossessions analysed here do not have their epicentre in the most revalued areas (opposite the eastern port), but rather in areas corresponding to the second level of revaluation (located in the central-eastern part of the city). This fringe corresponds to the La Gerreria neighbourhood, an area of bars and restaurant establishments, which in recent years has been promoted by the business community and by Palma City Council.

Comparing the most revalued areas and the geography of dispossessions, it is also possible to observe a clear link (though one that is not always direct) between revaluation and the density of evictions (Figure 13.1 and Table 13.2).

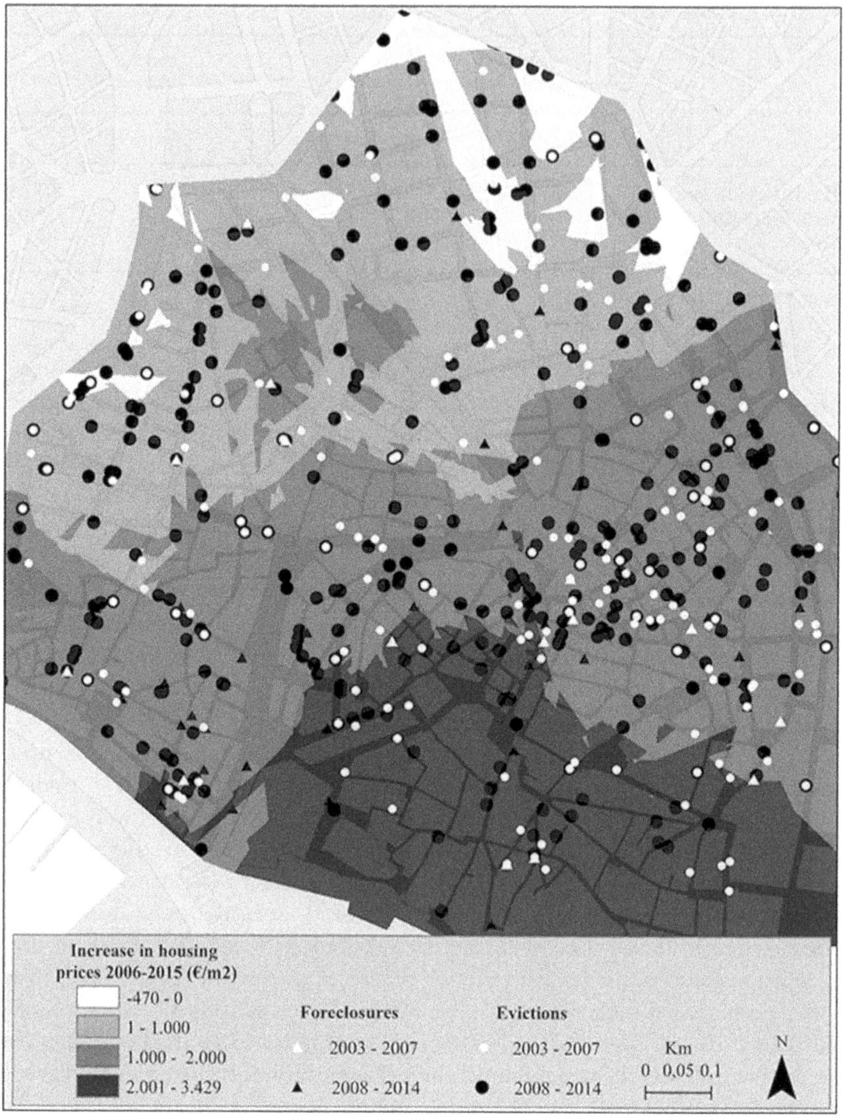

Figure 13.1 Tenant evictions, mortgage foreclosures and revaluation in the Historic Centre of Palma.

Source: Own data.

Figure 13.2 shows how the greater the revaluation is, the greater is the density of dispossessions, but only to a certain point, after which the number of dispossessions suddenly drops off. In fact, the area where density of dispossessions is lowest (2.84 D/ha) corresponds to that in which revaluation has been greatest: over 2,000 €/m^2.

Table 13.2 Revaluation and densities of dispossessions in the Historic Centre of Palma

Revaluation (€/m²)	*Dispossessions*	*Hectares*	*D/ha*
Between −470 and 0	21	4.97	4.22
Between 0 and 1,000	465	68.00	6.84
Between 1,000 and 2,000	318	39.16	8.12
Over 2,000	86	30.26	2.84
Total	890	142.39	6.25

Source: Own data.

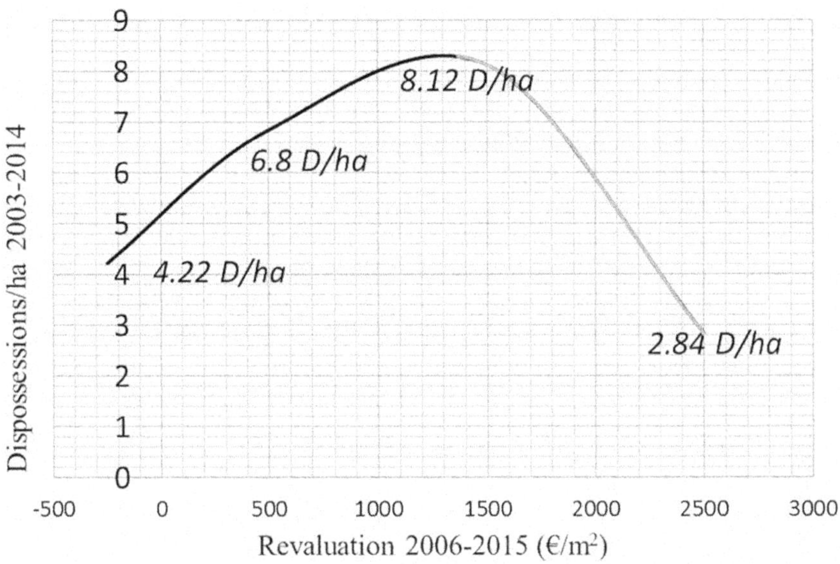

Figure 13.2 Density of dispossessions and urban revaluations in Palma (2003–2014).
Source: Own data.

The low density of direct dispossessions in the most revalued area must be interpreted in light of the chronology that we have previously analysed: the period since 1990s. This was due to the fact that urban transformations had taken place in this area prior to our study period, resulting in the direct expulsion of tenants from this area, thereby excluding their displacement (Marcuse 1985: 206), as this had already taken place (Franquesa 2010). In fact, the displacements that took place in the La Calatrava neighbourhood in the 1990s were carried out thanks to collaboration between the Neighbours' Association and the City Council, who worked hand-in-hand with investors to displace the tenants. The neighbours' association 'facilitated the most difficult part of the rehabilitation process: that of coming to an understanding with the tenants [and] looking after the social question by working on a door-to-door basis' (Franquesa 2010: 130). This explains the relative lack of data about dispossessions for the area corresponding to the Historic City of Palma compared to other areas in which this process is less mature.

Conclusions and discussion

In the analysis of the historic centre of Palma, we have observed a clear relationship between evictions and revaluation. The greatest density of direct dispossessions was found in the area nearest to that with the highest level of revaluation, which subsequently became the area of greatest pressure in terms of potential urban rent. On the other hand, the most revalued areas, which are those in which the gentrification process is most consolidated, presented lower densities of evictions. This was because displacements from these areas would have taken place in periods previous to our study. Our argument therefore focuses on the fact that as the revaluation wave advances, there is also an advance, in the same direction, in the wave of evictions and this helps to force up the rent potential of the area concerned and, in other words, to increase its rent gap.

Based on what we have observed, we can present a model for discussion which derives from this relationship between evictions and the rent gap within a context of high heritage values (see Figure 13.3). As shown in the graph, the line showing evictions starts to increase just before the predominant tendency in the neighbourhood changes from devaluation to revaluation. This line suddenly increases in parallel with capitalised rent

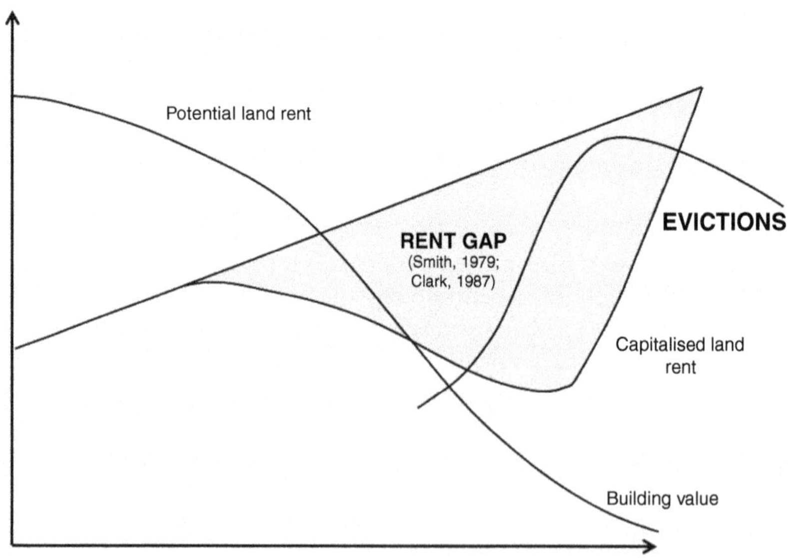

Figure 13.3 Evictions and the rent gap in the Historic Centre of Palma.
Source: Own work based on Smith 1979 and Clark 1987.

until a point at which the process of revaluation has become fully consolidated. It is then that it begins to descend, although with a gentler slope than during the initial rise.

In conclusion, in response to the questions posed, we think that evictions in urban areas with high heritage values can increase their potential rent and thereby exacerbate the rent gap. They therefore become a mechanism that can be used to produce urban rent and foster the revaluation of the urban space in which they are concentrated. In this way, waves of evictions could be one of the gateways to gentrification.

Notes

1 The research that led to this work forms part of the National Plan for R&D of the Ministry of Economy and Competitiveness: 'Crisis and vulnerability in Spanish island cities: transformations in social reproduction spaces' (CSO2015-68738-P).
2 The direct displacement of the last resident describes low-income residents displaced by physical or economic mechanisms, while direct displacement in a chain looks beyond the 'last resident' displaced and considers households that may have been forced to move during a particular period (Blanco et al. 2015).
3 All of the prices that we refer to in this text correspond to Euro equivalents in 2015.
4 This refers to the redevelopment plans for the La Calatrava neighbourhood in 1989 (modified in 1993), and La Gerreria neighbourhood (previously the city's red-light district), in 1995.
5 According to the newspaper *El País* (08 June 2007), in 2007, this area included what was the most expensive street in Spain, with prices reaching as high as 17,000 €/m^2.

References

Atkinson, R. (2000). 'Measuring gentrification and displacement in Greater London', *Urban Studies*, 37(1), pp. 149–65.

Blanco, J., Apaolaza, R., Bosoer, L. and González, A. (2015). 'Movilidades, desplazamientos y territorios: algunos aportes para el debate de la gentrificación'. *Contested Cities* (Working Paper).

Charnock, G., Purcell, T. and Ribera-Fumaz, R. (2014). *The Limits to Capital in Spain: Crisis and Revolt in the European South.* London: Palgrave Macmillan.

Clark, E. (1987). *The Rent Gap and Urban Change: Case Studies in Malmö 18601985.* Lund: Lund University Press.

Clark, E. (1995). 'The rent gap re-examined', *Urban Studies,* 32(9), pp. 1489–1503.

Clark, E. and Lund, A. (2012). 'Financialization, rescaling rent gaps and land grabbing'. Available at: http://backdoorbroadcasting.net/2012/05/eric-clarke-and-anders-lund-hansen-financialization-rescaling-rent-gaps-and-land-grabbing/.

Clark, E., Larsen, H. G. and Lund, A. (2015). 'Financialisation of built environments: A literature review', *FESSUD Working papers series*, 114.

CGPJ (2016). *Consejo General del Poder Judicial.* Available at: www.poderjudicial.es/cgpj/es/Temas/Estadistica-Judicial/.

Colau, A. and Alemany, A. (2012). *Vidas hipotecadas. De la burbuja inmobiliaria al derecho a la vivienda.* Barcelona: Cuadrilátero de libros.

Franquesa, J. (2010). *Sa Calatrava mon amour. Etnografia d'un barri atrapat en la geografia del capital*. Palma: Documenta Balear.
Harvey, D. (1982). *Limits to Capital*. London: Verso.
Harvey, D. (2003). *The New Imperialism*. Oxford: Oxford University Press.
Krippner, G. (2005). 'The financialization of the American economy', *Socio-Economic Review*, 3, pp. 173–208.
López, I. and Rodríguez, E. (2010). *Fin de ciclo. Financiarización, territorio y sociedad de propietarios en la onda larga del capitalismo hispano (1959–2010)*. Madrid: Traficantes de sueños.
López, I. and Rodríguez, E. (2011). 'The Spanish model', *New Left Review*, 69, pp. 5–29.
Marcuse, P. (1985). 'Gentrification, abandonment and displacement: Connections, causes and policy responses in New York City', *Journal of Urban and Contemporary Law*, 28, pp. 195–240.
Marcuse, P. (2010). 'On gentrification. A note from Peter Marcuse', *City. Analysis of Urban Trends, Culture, Theory, Policy, Action*, 14(1), pp. 187–188.
Marx, K. (1990 [1864]). 'La llamada acumulación originaria' in *El capital*, volumen 1.
Midnight Notes Collective (1990). 'Introduction to the new enclosures', *The Commoner*. Available at: www.commoner.org.uk/02midnight.pdf.
Newman, K. and Wyly, E. (2006). 'The right to stay put, revisited: Gentrification and resistance to displacement in New York City', *Urban Studies*, 43(1), pp. 23–57.
Sayer, A. (2012). 'Facing the challenge of the return of the rich', in W. Atkinson, S. Roberts and M. Savage (eds.). *Class Inequality in Austerity Britain*. Basingstoke: Palgrave Macmillan, pp. 163–179.
Slater, T. (2015). 'La expulsión de las perspectivas críticas en la investigación sobre gentrificación', in Observatorio Metropolitano (ed.). *El mercado contra la ciudad. Sobre globalización, gentrificación y políticas urbanas*. Madrid: Traficantes de sueños, pp. 107–144.
Smith, N. (1979). 'Toward a theory of gentrification. A back to the city movement by capital, not people', *Journal of the American Planning Association*, 45(4), pp. 538–548.
Smith, N. (1996). *The New Urban Frontier: Gentrification and the Revanchist City*. London: Routledge.
Smith, N. (2009). 'Revanchist planet', *Urban Reinventors*, 3, pp. 1–17.
Smith, N. (2012). *La nueva frontera urbana. Ciudad revanchista y gentrificación*. Madrid: Traficantes de sueños.
Vives-Miró, S. (2011). 'Producing a successful city: Neoliberal urbanism and gentrification in the tourist city – The case of Palma (Majorca)', *Urban Studies Research*, 2011, pp. 1–13.
Vives-Miró, S. (2013). *L'espai urbà del capitalisme. La construcció del projecte neoliberal de Palma*. PhD. Available at: www.tesisenred.net/handle/10803/283230.
Vives-Miró, S. and Rullan, O. (2014). 'La vivienda en el tránsito de la reproducción ampliada del capital a la acumulación por desposesión en España', *Territorios inconclusos y sociedades rotas*. Madrid: Asociación de Geógrafos Españoles, pp. 1–16.

PART IV

Policies and strategies

14 Financialised rent gaps and the public interest in Berlin's housing crisis

Reflections on N. Smith's 'generalised gentrification'

Laura Calbet i Elias

Introduction

In Berlin, housing prices began to increase in 2009 as the real estate boom in many parts of the world came to an (at least temporary) halt and the financial crisis spread. In this context, as in other geographies in which the new housing question has emerged, Neil Smith's gentrification theory (Smith 1979, 1982) has received particular attention. This is not surprising, since the concept has the virtue of connecting questions of capital accumulation with an accurate analysis of uneven urban development. In his work from 2002 onwards, Smith developed the concept of 'generalised gentrification' which can be seen as an inspiring starting-point for understanding a significant transformation of ways in which urban production is currently taking place. This paper takes the concept of 'generalised gentrification' as a start, and then relates the question of the rent gap to worldwide processes of housing financialisation and the role of real estate investment in current definitions of public interest. Both of these concepts are illustrated using the city of Berlin as an example. Therein, Smith's different conceptions of gentrification and the rent gap are presented briefly, illustrating them simultaneously with different periods from Berlin's recent housing history. Further, building upon the concept of generalised gentrification, the term financialisation is related to the concept of planetary rent gaps. Finally, the focus is put on the question of interurban competition in finance-oriented accumulation and the discussion of its consequences.

Gentrification and rent gaps – revising concepts

In their article 'The changing state of gentrification', Jason Hackworth and Neil Smith (2001) described three waves of gentrification in New York. According to them, until the 1970s, gentrification occurred sporadically and usually focused on the rehabilitation of run-down buildings. During the second wave, in the 1980s, gentrification became a strategy to attract investment into inner-city areas. The third wave of gentrification began in the first half of the 1990s. Referring to this third wave, one year later Neil Smith coined the term 'generalised gentrification' (2002).

> The generalization of gentrification has various dimensions. These can be understood in terms of five interrelated characteristics: the transformed role of the state, penetration by global finance, changing levels of political opposition, geographical dispersal, and the sectoral generalization of gentrification.
>
> (Smith 2002: 441)

What is particularly inspiring here is that the reference to global finance as a characteristic of generalised gentrification helps to raise questions of social and urban dynamics in the financialisation discussion. Furthermore, considering the financialisation of housing in terms of gentrification helps link it to the concept of the rent gap. 'Rent gap' defines the difference between the true achieved rent (capitalised ground rent) and the potential rent that could be realised from a property 'under the land's "highest and best use"' (Smith 1979: 543). Despite the extreme difficulties in operationalising and measuring rent gaps, at the theoretical level the difference between the capitalised rent and the prognosis as to what rent could be achieved from a property after its redevelopment or under new conditions of exploitation still seems to be crucial for understanding housing investment, and price dynamics, also in a financialised housing production system.

In successive phases of his work, Neil Smith conceptualised the emergence of the rent gap in different manners. In his first definition of gentrification, he considered reinvestment processes in run-down areas. Thus, gentrification would take place when 'depreciation produces the possibility of profitable reinvestment' (Smith 1979: 542). Later, Hamnett and Randolph (1986) adapted the concept to cases in which tenure transformations led to the possibility of closing the rent gap: particularly, when rental apartments were capitalised through their conversion to owner occupancy.

These processes took place in Berlin following German unification, especially after 1996. The years following the fall of the Berlin Wall were characterised by an enormous real estate bubble. When the bubble burst in 1996, it left oversupply of housing that resulted in the limitation of real estate investment to a few neighbourhoods. This was particularly felt in the run-down and deprived inner-city quarters of former East Berlin which became the target of public rehabilitation subsidies offering middle-class home buyers the possibility of closing the existing rent gap and achieving high-rent returns. The gentrification process was also boosted by the restitution policies of unification, which restored properties in the German Democratic Republic (GDR) from public ownership back to their pre-war owners and converted rental apartments into owner-occupied units (Holm 2006).

In the rewriting of his rent gap exposition for the book *The New Urban Frontier* (1996), Neil Smith specified a further situation in which the rent gap arises:

> We have focused here on the widespread situation in which a cycle of capital devalorization, brought about by disinvestment, accounts for the emergence of a rent gap. But it is also possible to conceive of a situation in which [...] the potential ground rent is suddenly pushed higher, opening up a rent gap in a different manner. This might be the case, for example, when there is rapid and sustained inflation, or where strict regulation of a land market keeps potential ground rent low, but is then repealed.
>
> (Smith 1996: 65)

This was the case for public and/or social housing in Berlin. The aforementioned bubble fostered the dismantling of public social-housing programmes and rent controls in Berlin. In 1993, the city government decided to privatise approximately 15% of available public housing. About 230,000 housing units were sold between 1990 and 2011. The most important sales took place in 2004 when thousands of units were sold *en bloc* to high-risk private equity funds. The rent gap was opened up by the commodification of a previously highly regulated housing provision system, as suggested by the previous quote. Furthermore, 'penetration by global finance' played an important role in the process. The privatisation procedure and the high returns aspired to by the municipality meant that opportunistic investors were the only players able to acquire the housing sites, as Sabina Uffer (2011) has remarked. The appearance of international institutional investors has been regarded as the starting-point for housing financialisation in Berlin.

The real estate market in Berlin remained, however, largely unaffected by the worldwide real estate price increases until 2007. Since the beginning of the crisis, in contrast, the city has become an objective for financialised investments, leading to consequent price increases. Today the issue of housing is not limited to privatisation processes, but also to an increase of housing-related indebtedness and the emergence of a new kind of demand, by both speculative foreign private investors and actors related to the financial market, usually investment funds. For these reasons, I consider that housing financialisation in Berlin has become generalised since 2009. The five characteristics of generalised gentrification can also be observed in Berlin at present: Aside from the penetration by global finance, gentrification has spread geographically and sectorally. Although political opposition is increasing, repression has prevented an effort comparable to the 1980s squatter movement. Other questions related to the transformed role of the state are discussed later in this article.

Penetration by global finance: financialised rent gaps

Since the advent of generalised gentrification seems to be related to a financialised accumulation system, it is relevant to analyse how rent gaps emerge in a financialised urban development. Similarly, Loretta Lees, Tom Slater and Elvin Wyly insist on the need for research

> to measure how the concepts of potential and capitalized ground rent themselves are altered when a significant fraction of housing market activity involves buyers and sellers working or moving across international boundaries. Is potential ground rent itself, for instance, becoming globalized as local property transactions are tied into world financial markets?
>
> (Lees, Slater and Wyly 2010: 62–63)

The relationship between global finance and housing provision has been the focus of the majority of urbanisation-related financialisation research (Aalbers 2008, 2012; Harvey 2011). Housing financialisation can be defined as the phenomenon according to which local housing provision is characterised by the exploitation of real estate in an accumulation system 'in which profits accrue primarily through financial channels rather than through trade and commodity production' (Krippner 2005: 174). Financialisation research examines how the immobile and long-term character of land and real estate has been transformed by financial instruments, regulations and rationalities, creating 'liquidity out of spatial fixity' (Gotham 2009). As a result, real estate investments have become one principal source of revenue for the global finance market. If we acknowledge the increasing role of real estate as a financial asset, it is plausible that the financialisation of housing shapes the (exchange) value and thus housing prices, resulting in a rise in capitalised rents as well as in those that could potentially be achieved.

In my view, there are at least four reasons for rent gap increases as a result of financialisation processes, which might be known, using Tom Slater's term, as 'planetary rent gaps' (2015):

1. The internationalisation of real estate markets may trigger price prospects that are not based on the local demand and price levels, but on housing markets in other (sometimes disparate) contexts and geographies.
2. The predominance of investors with high access to capital results in greater price competition and higher price prospects.
3. The predominance of institutional investors with high return expectations has the consequence of price prospects which are orientated towards the exchange values of not solely real estate but also financial investments.
4. The increasing prices, and also the fictive basis of a financialised accumulation system, tend inherently to produce speculative spirals.

Two brief examples illustrate the creation of planetary rent gaps in the case of Berlin. Housing prices in Berlin are, despite the recent increase, lower than in other German and European cities. In the aftermath of the financial crisis, this has resulted in greater international demand, usually seeking a safe harbour for investment or just an opportunity to speculate.

> Berlin is still cheap compared to Hamburg and Munich. [...] Berlin has become an investment destination not only for German investors. Foreign investors are increasingly discovering the capital as well.
>
> (cash.online 2015, author's translation)

In this context, price increases are framed by discourse presenting the process as a convergence towards housing prices in other cities. The fact that housing prices in Berlin remained at low levels for a long time is presented as the 'undervaluation' of Berlin's housing market compared to other geographies. The enormous real estate bubble worldwide that shaped the housing price increase, now perceived as 'natural', remains unmentioned in these sources. These 'normal prices' express a potential rent.

> Rising [housing] prices in Berlin still represent a regional adaptation. The Berlin real estate market has long been considered as undervalued and is now rapidly adapting to fundamental factors.
>
> (Investitionsbank Berlin 2014: 14, author's translation)

Therefore, prices are judged in comparison to other cities, although they are above the purchasing power of most of Berlin's population, characterised by low wages and high unemployment rates. Similar to other parts of the world, the potential value of urban rent increases to magnitudes that are mainly linked to financial returns, and can only be achieved through high degrees of private indebtedness. Thus, a rent gap might emerge between the local, 'undervaluated' ground rents and an abstract international rent level.

In addition, as another example, financialisation may open a rent gap due to the intrinsic characteristics of new investors. Institutional investors' high access level to capital, as well as their return prospects, influences both competition within the real estate market and prices. Thus, the potential rent rests on the activity of these actors in the real estate market. In my analysis of housing developers, I have observed how the funds used to finance the projects are linked to new financialised instruments and actors with higher profit expectations and shorter run times than in previous decades. In some cases, long-standing developers in Berlin who originally financed their projects with public funds and were engaged in social housing amortised through 30-year subsidies now create joint ventures with international financial companies that concentrate capital from both private and institutional investors. While small investors in the 1980s saw investment as an

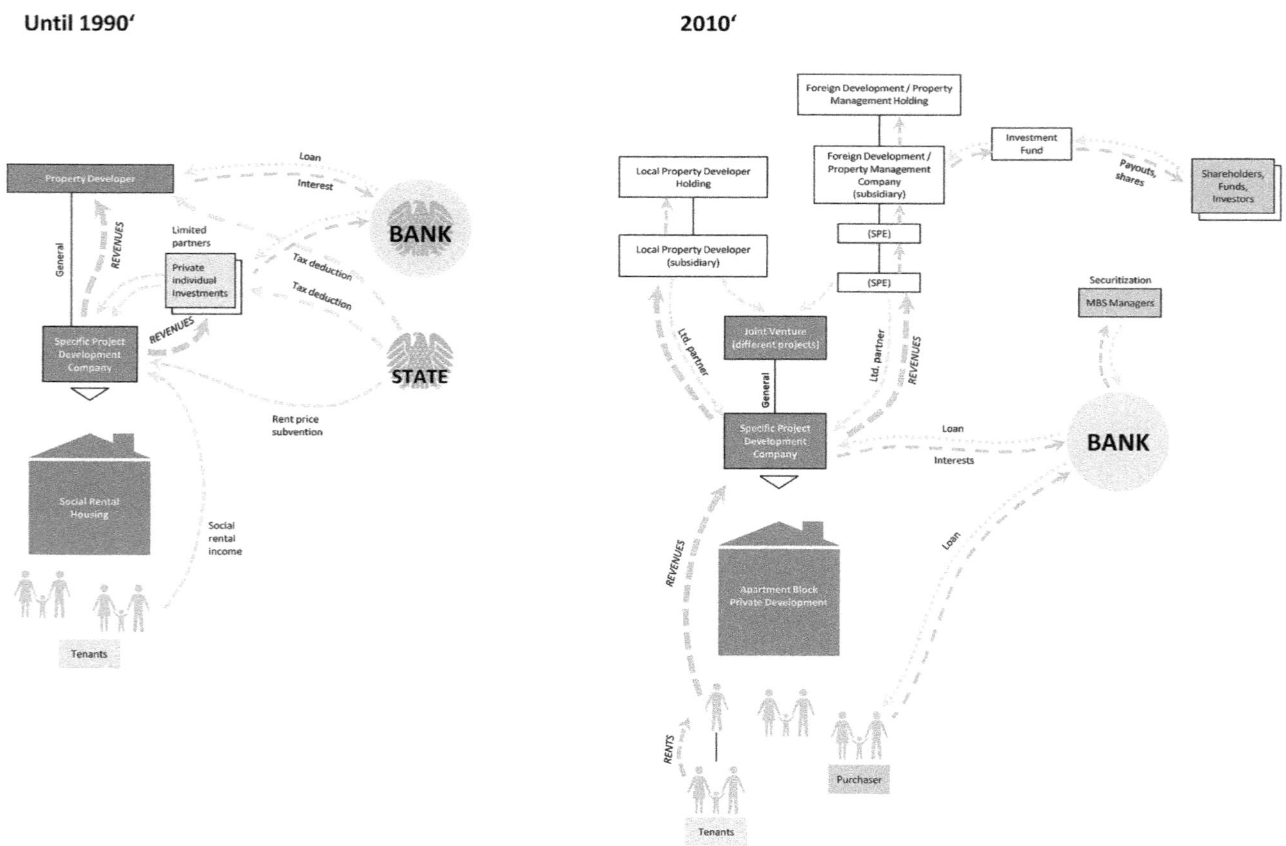

Figure 14.1 Developers' finance and business strategies until the 1990s and at the present. The graphic represents the specific case of one existing developer related to closed end funds. In the first case, amortisation times comprised 30 years, and the annual returns were approx. 4%–7%. In the current case, the amortisation times are 3–4 years, and the expected returns are approx. 12% for institutional investors participating in the fund.

Source: Laura Calbet.

opportunity for tax reduction and for savings with returns of c. 4%–7%, infomercials for real estate funds today suggest returns of approximately 12% per annum for institutional investors and 7% for private investors, with amortisation times of 3 to 4 years. Their projects are usually highly-priced apartment blocks, which are sold rapidly to owner-occupiers and minor landlords. Thus, the value of the potential rent no longer relies on a 'highest and best *land* use' but, rather, is based at least partly on 'the highest and best *financial asset*' (Figure 14.1).

Hence, the financialisation of housing means even more than just 'penetration by global finance'. It results in an estimation of investment in relation to other geographies and assets that might result in the appearance of rent gaps relying on the perception of the market from a global perspective.

Real estate investment as a strategy of urban competition

Having exposed the reasons for understanding current price increases as the result of planetary rent gaps, this next section considers the transformed role of the state, and particularly the relation between real estate investments and interurban competition. Neil Smith depicted 'generalised gentrification' as a stage in which

> urban real estate development – gentrification writ large – has [...] become a central motive force of urban economic expansion, a central sector of urban economies. Urban regeneration is a central strategy in the global competition among different urban agglomerations in the context of a newly globalized world.
>
> (Smith 2006: 207)

Research on gentrification has often analysed the role of municipal policies supporting urban regeneration processes that result in (and sometimes explicitly intend for) the displacement of social groups with lower income levels. These neoliberal policies aim to improve the economic base of a city, strengthening location advantages for, say, logistics companies or new technologies, fostering what Neil Brenner called 'Standortpolitik' (2003). Neil Smith's statement, however, seems not to be strictly limited to such principles. By saying that 'urban real estate development has become a central motive force of economic expansion', Smith's quote connects instead with Mark Purcell's characterisation of neoliberal urban policies. According to Purcell, neoliberal urban policies are framed within 'the imperative of competitiveness, the inviolability of property rights, and the primary importance of the exchange value of urban land' (2009: 148). Following Purcell, a central instrument for the legitimation of neoliberal policies has been to present 'competition' as a necessity, and to present policies facing competitiveness as beneficial to everyone. The growth imperative, essential in capitalism, has been coupled in neoliberalism with the necessity of competitiveness. From this point of view, it is assumed that competitive advantages will lead to economic success and will increase social wealth in general. Consequently, competitiveness becomes a fundamental aspect of the neoliberal definition of 'public interest'.

In 1995, Donald Krueckeberg related the term 'public interest' to property issues: 'The taking issue refers to the conflict between private and public interests in the use of a piece of land' (1995: 304). Usually, the planning discipline is presented as a neutral instance dealing with differing interests. Nevertheless, the

> problem with our planning tradition is that it assumes, to begin with, that what is public, what is private, and who owns what can all be accepted as clear. That is not a good beginning, because it begs the question of who has a right to what.
>
> (1995: 301)

Subsequently, Krueckeberg quotes John Christman's *The Myth of Property* (1994), in which two categories of property rights are defined: those rights linked to the right to use, related to use values; and those rights that allow deals with the aim of economic benefit, related to exchange values. In a context in which competition is supposed to be beneficial to everyone, if real estate investments represent an essential form of capital accumulation, then property development necessarily has to become a political priority. Interurban competition for real estate investments, no matter whether these investments focus on the realisation of exchange value to the detriment of the rights of use, becomes part of the public interest. By doing so, real estate investment becomes a matter of consensus, a question to which there are no alternatives. By employing this logic, the concept of the 'housing crisis' acquires a new, post-political dimension. It refers not just to the question of deprived living conditions, but also to the impossibility of counteracting them without challenging the imperative of competitiveness, which is equivalent to contesting neoliberalism. Discussions about post-politics that criticise the lack of alternatives in the political debate and highlight the consequences of consensus-based politics (e.g. Swyngedouw 2007) seem to be helpful for developing these ideas.

I will try to illustrate this by again using the case of Berlin. As explained above, the increase in housing prices in Berlin seems to be closely related to financialisation, especially in the wake of the financial, economic and sovereign debt crises. Nevertheless, official discourses put a remarkable emphasis on denying a speculative bubble and link the rise in prices to increased demand, referring merely to a use-related demand. In this way, the rise in housing prices is perceived as merely an inevitable side-effect of a situation presented as positive.

> Despite the many opportunities [comprised in the current city growth] there are also some side effects, especially for the housing and real estate market: growth makes prices rise.
>
> (Senatsverwaltung für Stadtentwicklung 2013: 2, author's translation)

Indeed, the population of Berlin is increasing, partially as a result of immigration from those European countries most affected by the crisis. However, it is clear that this demand does not possess high purchasing power. Actually, there is no contradiction between a real estate bubble and growing levels of (use-related) demand, but rather, presumably, most bubbles take place parallel to processes of demographic growth.

Corresponding to the dominant interpretation, the main solution proposed to the apparent housing shortage is to increase supply, as classical economic theory would suggest. Thus, the emphasis of housing policies in Berlin is on the construction of a large number of homes, supported by measures to facilitate the approval and building processes.[1] Between 15,000 and 20,000 housing units a year are to be constructed mostly by private developers, regardless of the fact that they mainly produce high-priced apartments. Simultaneously, numerous protests against new constructions have arisen as a result of the diminishing quality of life in the neighbourhoods, and particularly as a result of the rising rental prices related to the quarters' revalorisation. Yet the opposition to this housing policy is considered illicit, as the assertion of the representative for urban development illustrates:

> In this context, we cannot say: Please do not change, build anywhere but next to our property, not in our neighbourhood. [...] This so-called 'not in my backyard' (NIMBY) attitude is also problematical for the common good.
>
> (Geisel 2015: 6, author's translation)

Thus, while housing construction becomes a matter of public interest, the rejection of high-priced housing developments is seen as particularistic, defending selfish interests that differ from the common good. Even the rejection of the building plans for the perimeter of the new Tempelhof Park by the majority of Berlin's citizens in a referendum, representing not a NIMBY-position but a general rejection of recent housing policies, is perceived as contrary to Berlin's wealth. The regional head of the Social Democratic Party at the time of the referendum expressed his discontent as follows:

> Growth also leads to more growth, which means new jobs, more income, new companies, higher tax revenues. Without investment, growth can shrivel and die. If Berlin does not understand its potential for growth

> and cannot absorb migration, then growth will move to Hamburg, to Frankfurt am Main, to the Ruhr region.
>
> (Stöß 2014, author's translation)

Thus, the dissent regarding current housing policies is interpreted as a threat to urban competitiveness, while real estate investments are assumed to catalyse social wealth. This confirms Neil Smith's depiction of a generalised gentrification, according to which

> real-estate development becomes a centerpiece of the city's *productive* economy, an end in itself, justified by appeals to jobs, taxes, and tourism. [...] [T]he construction of new gentrification complexes in central cities across the world has become an increasingly unassailable capital accumulation strategy for competing urban economies.
>
> (Smith 2002: 443)

Therefore, the new role of the state in generalised gentrification goes beyond 'Standortpolitik' as characterised in the academic debate on policies of urban competitiveness, encouraging investments in real estate as a means of capital accumulation in a financialised economy. By doing so, the public interest is assumed to derive from property investments based on the exchange value of land, to the detriment of policies that defend the human need for housing.

Conclusions

Summarising, the concept of generalised gentrification is an inspiring starting-point for improving the connection between the debates on housing financialisation and ideas regarding a post-political urban regime, and for a better understanding of processes of uneven development. It helps to relate the question of financialisation to the concept of the rent gap, enabling us to identify at least four ways in which the potential rent can increase as a result of the current asset character of housing. It is therefore apparent that not just the arrival of cross-border financial investors but, especially, the concept of planetary rent gap will become crucial for attaining globally operating real estate markets. It also allows the changing terms of interurban competition in the context of a finance-oriented accumulation system to be taken into consideration. The assertion that real estate investment becomes an end in itself, defined as the common good against the background of the competitiveness imperative, allows us to illustrate how the 'housing crisis' has become a post-political question that requires us to challenge the (neoliberal) frame in order to contest it.

Note

1 Besides strengthening new construction, further measures have been implemented, for example a more restrictive regulation of rental price increases. Nevertheless, new construction is still presented as the most important strategy to counteract the current housing price increases.

References

Aalbers, M. (2008). 'The financialization of home and the mortgage market crisis', *Competition & Change*, 12(2), pp. 148–166.

Aalbers, M. (2012). *Subprime Cities: The Political Economy of Mortgage Markets*. Chichester/Malden: Wiley-Blackwell.

Brenner, N. (2003). 'Standortpolitik, state rescaling and the new metropolitan governance in Western Europe', *DISP*, 152, pp. 15–25.

cash.online (2015). 'Berlin: Kaufpreise steigen schneller als Mieten', *cash.online*, 6.5.2015 Available at: www.cash-online.de/immobilien/2015/wohnimmobilien-berlin-4/249460 [Accessed 18 September 2015].

Christman, J. (1994). *The Myth of Property*. Oxford: Oxford University Press.

Geisel, A. (2015). 'Wohnungen bauen, heißt zukunft gestalten', *Stadtforum Berlin. Welche Wohnungsneubau braucht Berlin? Dokumentation 15.4.2015*, Berlin, pp. 6–7.

Gotham, K. F. (2009). 'Creating liquidity out of spatial fixity. Crisis: the secondary circuit of capital and the subprime mortgage', *International Journal of Urban and Regional Research*, 33(2), pp. 355–371.

Hackworth, J. and Smith, N. (2001). 'The changing state of gentrification', *Tijdschrift voor Economische en Sociale Geografie*, 92(4), pp. 464–477.

Hamnett, C. and Randolph, B. (1986). 'Tenurial transformation and the flat break-up market in London: The British condo experience', in N. Smith and P. Williams (eds.). *Gentrification of the City*. Boston: Allen & Unwin, pp. 121–152.

Harvey, D. (2011). 'The urban roots of financial crises: reclaiming the city for anti-capitalist struggle', *Socialist Register, 2012: The Crisis and the Left*, 48, pp. 1–35.

Holm, A. (2006). *Die Restrukturierung des Raumes: Stadterneuerung der 90er Jahre in Ostberlin: Interessen und Machtverhältnisse*. Bielefeld: Transcript.

Investitionsbank Berlin (2014). *IBB Wohnungsmarktbericht 2013*. Berlin.

Krippner, G. (2005). 'The financialization of the American economy', *Socio-Economic Review*, 3(2), pp. 173–208.

Krueckeberg, D. A. (1995). 'The difficult character of property', *Journal of the American Planning Association*, 61(3), pp. 301–309.

Lees, L., Slater, T. and Wyly, E. K. (2010). *The Gentrification Reader*. London: Routledge.

Purcell, M. (2009). 'Resisting neoliberalization: communicative planning or counter-hegemonic movements?', *Planning Theory*, 8(2), pp. 140–165.

Senatsverwaltung für Stadtentwicklung (2013). *Gut und sicher wohnen: Soziale Wohnungspolitik und bezahlbare Mieten*. Berlin.

Slater, T. (2015). 'Planetary rent gaps', *Antipode*. doi:10.1111/anti.12185.

Smith, N. (1979). 'Toward a theory of gentrification: a back to the city movement by capital, not people', *Journal of the American Planning Association*, 45(4), pp. 538–548.

Smith, N. (1982). 'Gentrification and uneven development', *Economic Geography*, 58, pp. 139–155.

Smith, N. (1996). *The New Urban Frontier: Gentrification and the Revanchist City*. London: Routledge.

Smith, N. (2002). 'New globalism, new urbanism: gentrification as global urban strategy', *Antipode*, 34(3), pp. 427–450.

Smith, N. (2006). 'Gentrification generalized: from local anomaly to urban 'regeneration' as global urban strategy', in M. S. Fisher and G. Downey (eds.). *Frontiers of Capital: Ethnographic Reflections on the New Economy*. Durham: Duke University Press, pp. 191–208.

Stöß, J. (2014). 'Die besten Jahre für Berlin kommen noch. Kommentar des SPD-Landeschefs nach der Tempelhof-Entscheidung', *Der Tagesspiegel,* 30.5.2014 Available at: http://www.tagesspiegel.de/berlin/nach-der-tempelhof-entscheidung-spd-landeschef-die-besten-jahre-fuer-berlin-kommen-noch/9967576.html.

Swyngedouw, E. (2007). 'The post-political city', in Bureau for Architectural Theory (ed.) *Urban Politics Now: Re-imagining Democracy in the Neoliberal City.* Rotterdam: NAi Publishers, pp. 58–76.

Uffer, S. (2011). *The Uneven Development of Berlin's Housing Provision.* PhD Thesis. London School of Economics and Political Science.

15 Beyond the 'revanchist city'

When public policies softly support gentrification in the name of social mix. The case of Inner Paris

Anne Clerval

As early as 1979, Neil Smith was probably the first to highlight the role of State and municipalities in gentrification (Smith 1979). During the 1990s, he analysed the public policies of New York and saw them as *revanchist* against working-class, ethnic and LGBT minorities (Smith 1996). His idea of *revanchism* is inspired by the pre-fascist context in France after the Commune and the defeat against Germany at the end of the 19th century. *Revanchist* urban policies are fully part of class struggle dynamics on different scales. I consider, like Neil Smith, that gentrification is a form of class struggle in the urban space that leads to working-class eviction. Following this point of view, the role of public policies in this process is a real issue, both in the urban field and in social science research. In fact, Slater (2006) pointed out the eviction of critical perspectives from gentrification research along with the development of state-led gentrification as 'revitalisation' of inner cities. Wacquant (2008) interpreted this eviction as being the result of the general marginalisation of the working classes in political, media and scientific discourse, and also of the underestimation of the part played by public authorities in the dynamics of gentrification. During the 2000s, several studies have pinpointed the importance of this influence, showing how gentrification can even, in some instances, become a full-scale local political strategy (Smith 2002, Atkinson and Bridge 2005). In their introduction to a special issue of the journal *Urban Studies* devoted to the role of public policies in the phenomenon of gentrification, Lees and Ley (2008) reposition these local policies within the more general context of neoliberalism and international competition between cities. The 'third wave' of gentrification, symbolised by the conversion of the London docks into a new world-class business quarter, appears to federate researchers, who hitherto had disagreed over the causes of gentrification. Lees and Ley (2008) align themselves on the analysis by Smith (1982), who argued the key position of gentrification in world capitalism at the time of the neoliberal turning point. They also recall how the consensus-making objective of a social mix can be used today to favour the gentrification of working-class quarters.

This paper aims to contribute to analyse the role of public policies in gentrification with the case of Paris that has, paradoxically, not been widely studied in this respect. The city of Paris (Inner Paris) is interesting because it has been experiencing a late gentrification process under the French Left-wing Socialist Party since 2001 (the party returned to power in France in 2012). The urban public policies they led in Paris were very different from *revanchist* New York under Rudy Giuliani. In Paris, 'social mix' is the key-word, based on strong public policy to re-launch social housing production. After a presentation of the main features of gentrification in Paris, I will show how these urban policies are unable to curb the eviction of the working-class from Inner Paris and even foster gentrification in several quarters.

The context of gentrification in Paris

The specific features of Paris compared to other metropolises

In Paris, gentrification appears as a process occurring late in comparison to London or New York. There are several reasons to explain this. Some emphasise the fact that the French capital has been a place of residence for the ruling elite for a long time, and this is indeed still true (Préteceille 2007). The heart of the Paris agglomeration did not experience the white flight noted in American cities where middle- and upper-class white people left the city centres for the suburbs in the 1950s and 1960s. Only the Marais quarter of Paris has seen a desertion and later return of the wealthier classes over the course of its history, the first type of trend to be described as gentrification, in reference to numerous Victorian areas of London (Glass 1964). Nevertheless, the extent of areas of older buildings in working-class Paris represents a considerable potential for gentrification. From the 1960s, the central quarters on the Left Bank of the Seine (*Rive Gauche*) were gradually transformed, extending the process of removing working-class quarters that began with the work of Haussmann. Indeed, the quality of the old buildings in Marais led to policies for rehabilitation or upgrade that fostered the return of the wealthier classes to the area.

As is the case with cities in the south of Europe, the French capital stands out on account of the existence of a considerable mass of old buildings and the heritage that they represent. Unlike the case of London, these buildings were largely spared from the destruction brought about by the Second World War. The post-war period did not experience city planning alterations on the scale seen in the City in London, nor indeed on the scale of American cities such as Philadelphia in a different context (Smith 1979). In addition, rent in Paris was tightly controlled for a long period, which slowed speculation in real estate, or at least restricted it to certain market sectors. Thus, the takeover of working-class quarters by affluent classes occurred in a gradual manner in Paris. It was only in the 1990s that gentrification became truly visible and that the demolition of old, run-down housing was officially abandoned in favour of upgrading.

Actors and factors in gentrification

Gentrification in Paris thus occurred mainly through private initiative, via the rehabilitation and upgrading of working-class housing (Clerval 2013). The people subsequently moving in were often artists and architects looking for professional premises, and they took over former craft and industrial workshops in the eastern part of Paris towards the end of the 1970s, sometimes in the wake of the squats that characterised that decade. More widely at the same time, however, middle-class households (among whom cultural professions were over-represented) acquired and rehabilitated homes in the working-class quarters. This trend was supported by estate agents and by the banks, who lowered mortgages which partially compensate for the rise in house prices. The trend was also fostered by promoters who undertook rehabilitation and upgrades for split sales. These promoters move in when prices are sufficiently high in a given area, this being particularly characteristic of the large artisan or industrial *cours* (enclosed yards housing workshops and living premises) of Eastern Paris, in some cases involving former factories.

As elsewhere, the gentrification of the working-class areas of Paris can be explained by several structural factors in the spheres of housing and employment. The most obvious of these is the continuous decrease in the number of working-class jobs in Île-de-France (the Paris region) and Paris from the 1960s. This was compounded by a decrease in the number of unskilled jobs in the services sector in Paris from the 1980s, while in the same period the number of managerial, professional and intellectual jobs increased. This change in the structure of the job market in Île-de-France was the result of the redistribution and reorganisation of production worldwide, a phenomenon that was accelerated by neoliberal macro-economic policies from the 1980s. Deregulation and the international integration of the economy increased competition among unskilled workforces on a global scale, and facilitated the globalisation of industrial production, while the metropolisation that resulted led to a concentration of qualified jobs in Île-de-France. Nevertheless, the changes in the structure of the job market are not sufficient to explain the increasing social selection occurring in the residential space in the inner city of Paris. Here the patterns of housing and the structure of the housing market accentuated the contrasting dynamics between professional, intellectual and working-class categories. The earlier working-class housing was deteriorating under the effects of strategies aiming at short-term financial returns for landlords, or else as a result of their intention to sell their properties (this being particularly true for complete buildings). The policy of mass construction of 'social' housing in city outskirts in the 1960s and 1970s led to the departure of large numbers of working-class people from Paris to its suburbs, generating a phenomenon of vacancy in the working-class quarters. These were then partly taken over by more precarious populations, often immigrants, alongside gentrified households. The partial vacancy and the poor state of repair of housing created a rent gap (Smith 1982) in these central quarters; after the first gentrifiers moved in, so did trendy retail stores, meaning numerous quarters gained lucrative potential for investors.

Earlier public policies played a considerable part in this process. From the 1960s to the 1990s, the operations consisting in the 'renovation' of areas by way of demolition and reconstruction contributed to de-structuring working-class quarters, but at the same time led to the building of a large amount of social housing. In addition, the first gentrifiers were opposed to the demolition of old housing, which demonstrates the lasting hiatus between the public authorities and the gentrification process. The de-regulation of rent in the late 1980s, however, encouraged real estate speculation, while at the same time several prestige operations gave impetus to the upgrading process in working-class areas for the benefit of the middle and upper classes. This can be seen, for instance, in the demolition and development of Les Halles (originally a large covered market), the construction of the Georges Pompidou Modern Art Centre (Beaubourg) in the 1970s, the rehabilitation of the Marais area in the same period, and the creation of certain large facilities to the east and north of Paris (see Figure 15.1).[1]

Spatial dynamics: diffusion, outposts and bypass

Figure 15.2 makes it possible to apprehend the spatial dynamics of gentrification based on statistics analysis and fieldwork (Clerval 2011). What can mainly be seen is a diffusion from the wealthy quarters (*les Beaux quartiers*) in the West. On the Left Bank of the Seine, gentrification is longstanding and based on demolition and reconstruction operations such as the Beaugrenelle development (15th *arrondissement*[2]) on the *Front de Seine*, on new private constructions in peripheral quarters and on the rehabilitation of older housing in the quarters that had become fashionable, such as La Butte-aux-Cailles (13th) in the 1990s. Today, the Left Bank displays an almost exclusively upper-class profile, the only distinction being between old and new buildings. On the Right Bank, the large crescent of former working-class areas extending from the Batignolles (17th) to the Faubourg St Antoine (11th, 12th) was progressively taken over and renovated by middle and upper-class occupants, moving outwards from the centre to the periphery. This was a continuous process of occupancy turnover, with gentrifiers becoming increasing wealthy. This explains why today the central and immediate peri-central *arrondissements* have a wealthier social profile than those that are more peripheral, although these were also in the process of becoming gentrified. In space, this diffusion process is not completely continuous. Gentrification possesses a sort of a pioneer fringe, with outposts, particular spaces such as the Butte Montmartre, the canals, or the

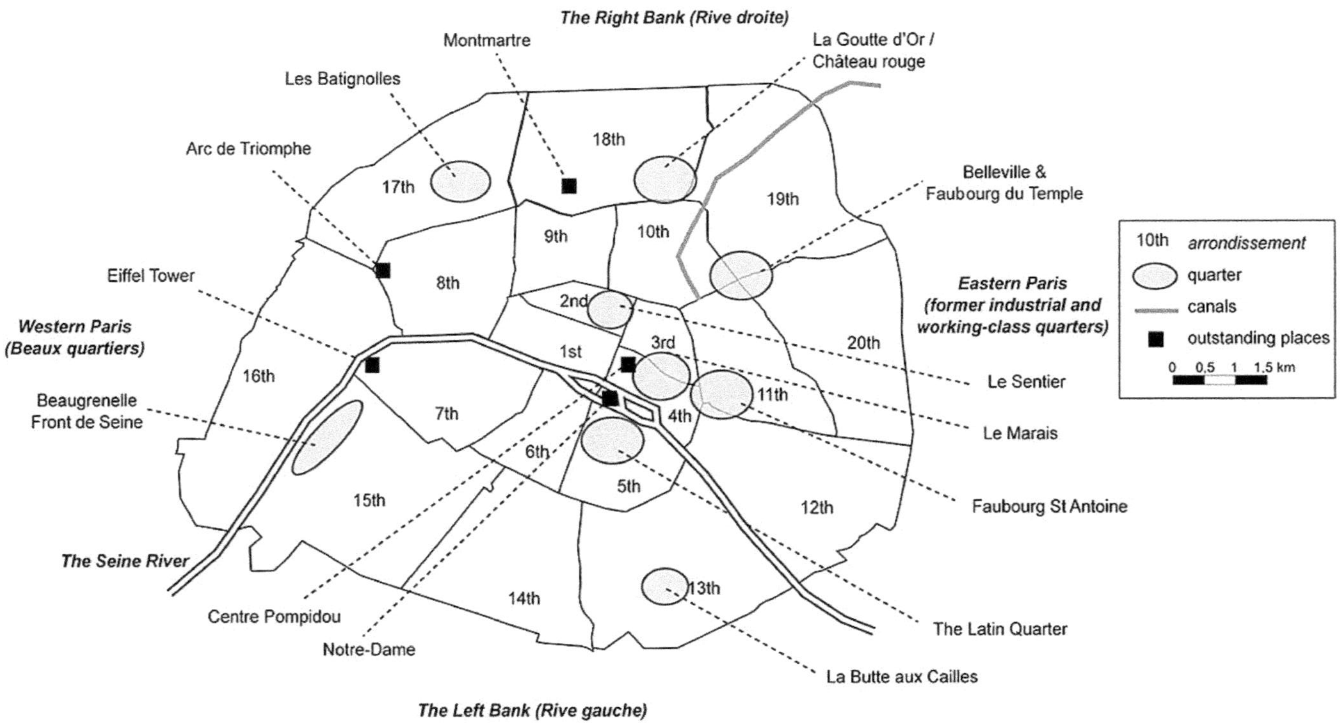

Figure 15.1 Paris: study area.
Source: Map produced by Anne Clerval with an APUR base map.

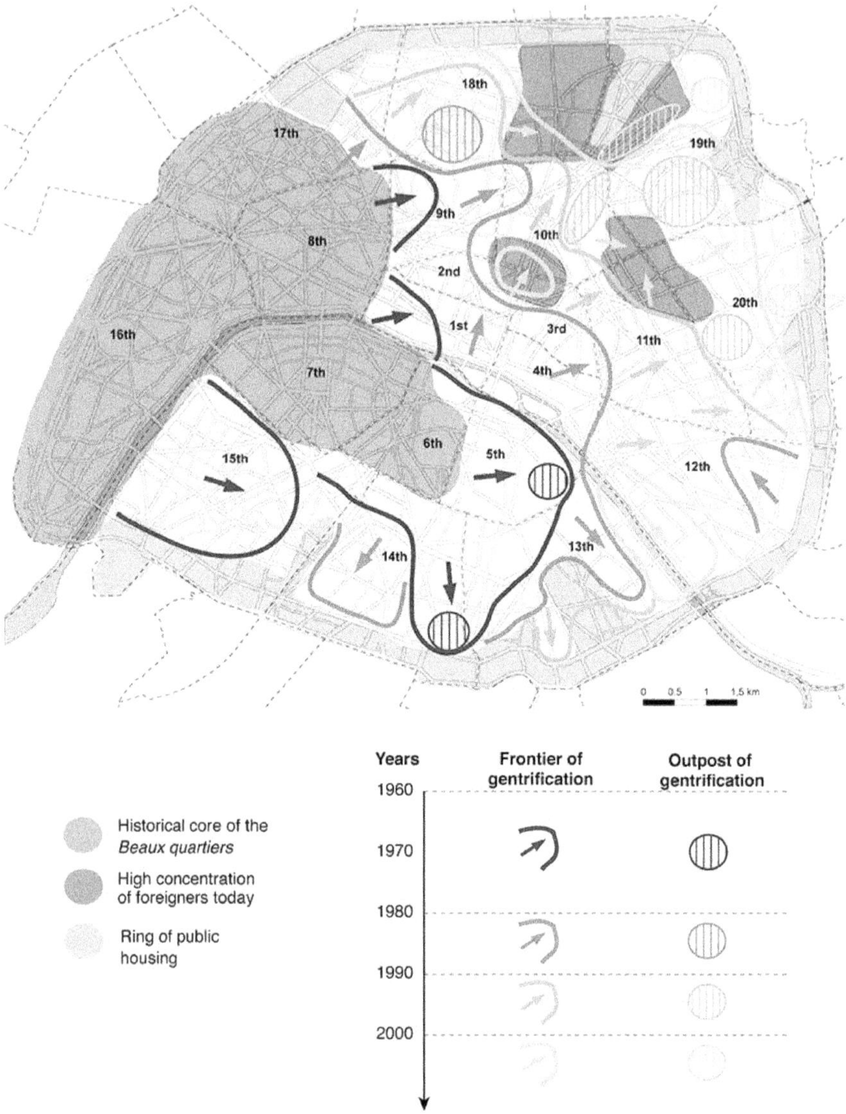

Figure 15.2 The gentrification of Inner Paris: a frontier since the sixties.
Source: Map produced by Anne Clerval with an APUR base map.

green areas, and bypass effects, in particular around spaces with marked proportions of immigrant populations, such as the centre of the Right Bank (around Le Sentier in the 2nd) or Belleville (20th) and the Faubourg du Temple (10th, 11th). These quarters appear as pockets of resistance bypassed by gentrification, although not completely exempt from the process.

How public policies led by the Left since 2001 subtly support gentrification

In 2001, the municipal elections brought a Socialist majority to power in the Paris council (*Mairie*) for the first time. The new Mayor, Bertrand Delanoë was re-elected in 2008 and replaced in 2014 by Anne Hidalgo with the same majority. The left power does not seem to be against gentrification. Indeed, almost all of the gentrifier households I met during my fieldwork (35 persons) voted for the Socialist party and were satisfied with their urban policies. Since 2001, the Socialist majority in Paris has never stopped promoting social mix and a shared city where all types of people live in harmony. It insists on the re-launching of social housing and urban public intervention in general, which seems to be very different from revanchist urban policies, especially neoliberal state-led gentrification (Davidson and Lees 2010). As we saw, however, the gentrification frontier continues to extend at the same pace as before, sparing no area of Inner Paris.

Urbanism and housing: preserving older buildings and re-launching the production of social housing

With the abandoning of demolition and reconstruction in 1996 by the Right in Paris, the production of social housing collapsed. The re-launching of the production of social housing in Paris by the Left has been particularly highlighted in Council communication. The increased production of social housing relies on diversification in the modes of production and in modes of financing. New constructions were implemented alongside the rehabilitation of old buildings (acquired by the municipality) and in some cases purchasing and commissioning of buildings without any work being performed. The two latter novel forms of production of social housing enabled the target of 3,000 approved homes per year to be reached as early as 2001. Production reached 7,000 homes a year in 2010 and 2013 via the resumption of new construction since 2007.[3] Since 2001, 76,000 new social homes have been approved. The objective of 20% of main residence homes being social housing (provided for in the SRU – *Solidarité et renouvellement urbain* – law in 2000, and raised to 25% in 2014) is about to be reached. In 2014, the proportion of SRU social housing had reached 18.5% in Paris (versus 13.4% in 2001).

These figures, however, mask the structural limitations of this endeavour. Indeed, available building plots are increasingly scarce in Paris, and new constructions are most often replacements of old buildings too run-down to be rehabilitated. Thus, 'official' social housing has in fact replaced *de facto* social housing, without enabling the housing of as many individuals, both because the homes are larger, and because the ratio of the number of occupants to the number of rooms is much lower in the first instance than in the second. Thus, the production of social housing needs to be positioned within the wider context of the housing pool in Paris overall. This housing, as we have seen, is markedly affected by gentrification. The process contributes to restricting the private housing rental offer, in particular, the offer affordable to working-class households, i.e. *de facto* working-class accommodations. Thus, the production of social housing is insufficient to replace the decrease in *de facto* social housing and the main residence pool overall remains more or less stable.

Thus, to reach the objective of 20% social housing, if the rest of the housing pool – which is private – is occupied by the middle and upper classes, this equates to maintaining only 20% working-class households in Paris, in other words half as many as the proportion they actually represent in the structure of the Île-de-France workforce.

In addition, not all SRU social housing is allocated to working-class households. In particular, a social housing funding system, the *Prêt locatif social* (PLS), was instated in 2001, producing homes intended for households that were 30% above the income level for obtaining PLUS accommodation (*Prêt locatif à usage social*), the classic social housing provision. The PLUS also leads to a degree of social mix, since it allows for 10% of the housing to be allocated to households with resources above the maximum level. The re-launching of the production of social housing is nevertheless grounded in this diversification of types of funding: Thus, PLS housing merely requires State approval without funding, which enables the State to withdraw from the production of social housing. Nearly 30% of the social housing produced in Paris since 2001 comes under the PLS system, while this does not correspond to the social structure of the demand: Among the 130,000 applicants for social housing in Paris, between 70% and 75% have incomes that are below the PLAI housing limit (*Prêt locatif aidé d'intégration*), intended for the lowest income category, while only between 5% and 10% could apply for PLS housing (depending on the year). In fact, this 30% PLS housing reflects the proportion of Parisian households with incomes allowing them to apply for this type of accommodation. This can be read as a policy aiming to reorient part of the social housing pool towards middle-class households, while at the same time working-class households are experiencing ever-increasing difficulties in finding housing in Paris, whether in the social sector or in the private offer.

The diversification of these types of financing, and hence of the social status of households qualifying to apply for social housing, also serves the strategy aiming to rebalance the housing pool geographically and to enhance the social mix promoted by the Socialist municipality. The idea is to create social housing in *arrondissements* where it is lacking, in particular 'very' social housing (for the lowest incomes), and to promote PLS housing in areas where there is already 20% social housing, or more generally in working-class quarters. Thus, the Municipality has acquired buildings on Avenue Mozart (16th *arrondissement* – a traditionally wealthy area) to turn them into PLAI (low-rent) housing, and has created PLS housing in the working-class areas of Château Rouge (18th) or lower Belleville (10th, 11th).

While the process of gentrification is already pronounced in Paris, after accelerating in the period of rising property prices from the end of the 1990s up to now with only a pause in 2008–2009, these immigrant quarters, as seen earlier, are those in which the process is the least advanced today. Thus, in areas where this process could still be controlled – for instance by increasing the offer of genuinely social housing targeting the working class – the municipality contributes, on the contrary, to accompanying gentrification, replacing only a minority of *de facto* social housing, via demolition or upgrading, and via its conversion into social housing with particular emphasis on the PLS type.

To summarize, the Left's housing policy in Paris only led to improvements in the housing conditions of a minority among the working class that can stay in Paris. Without strict rent control or any other action in the private sector, it can neither stop nor curb gentrification. Since 2015, a rent control law has applied only to the city of Paris: It is a step backwards because this law should have applied to many urban areas in France where the housing market is under pressure. Additionally, rent control applies only for the new leases and depends on market prices, which are already too high for the working class in modern-day Paris. Even if this law has been sharply attacked by the Right and by the Socialist Party itself, it would be insufficient to curb gentrification.

Small businesses, public space and cultural policies in working-class quarters

In Château Rouge (18th), this is accompanied by an attempt to restrict or remove the so-called 'exotic' retail trade, frequented by African populations, and by the overall upgrading of public spaces, vigorously called for by the new homeowners in the quarter. Some of them created an association in the early 2000s named *Droit au calme* [right to tranquillity] against working-class and immigrant use of the streets. In fact, Château Rouge is an old immigrant quarter housing people from Maghreb prior to the 1980's and nowadays mostly people from sub-Saharan African countries. The area became an immigrant commercial centrality and attracts customers from the whole metropolitan area and beyond. The gentrifiers from *Droit au calme* wanted a 'normal area', which is to say, without such a concentration of immigrant retail and trade (halal butchers, African hairdressers, call shops, fabric merchants), to instead contain more traditional French shops (wine cellars, cheese shops, independent supermarkets). In fact, these traditional shops are declining across the city because of the competition from chain supermarkets, aside from some special market areas like organic shops or high-quality and expensive businesses. Since then, *Droit au calme* has disappeared, but the Municipality uses a local semi-public company (the SEMAEST) to preempt small-scale business premises to install selected new small shops (i.e. wine cellars, small supermarkets or flower shops).

This comes along with the eradication of run-down housing, beautification of public spaces, rehabilitation and creation of new green areas and cultural facilities. One of the most prominent examples is situated in the North of Paris, where gentrification needs public help to progress. A 35,000 m^2 area near the railway tracks of the Gare de l'Est was set aside for contemporary art, a restaurant, a café and shops. It was inaugurated in 2008 on 104 rue d'Aubervilliers in the building that previously housed the city funeral enterprise. With an investment of €102,000,000, equivalent to almost a quarter of the 2008 social housing budget, the 'Cent Quatre' (104) project constituted the largest project in Delanoë's first term of office. On the same street a new green area was created, Le Jardin d'Éole, and two blocks of run-down housing were demolished to create private (yet regulated) housing for the middle classes.

Alongside the creation of new facilities, the public authorities are also investing more and more in the organisation of cultural manifestations taking the form of events in public spaces. In particular, certain major events are gaining wide media coverage, such as *Paris Plage* [Paris Beaches] or *Nuit Blanche* (literally a 'sleepless night'), and also festivals like *Paris quartier d'été* or *Cinéma au Clair de Lune.* These events take diverse forms, very often integrating music and dance, but also increasingly contemporary art. Most are designed as events based on some form of scenography, and some entail non-permanent fixtures. These events are, all in all, the tools of a cultural policy that has been seeking to reach a wider public audience since the 1960s. They are used to promote the new urban planning implemented by the Municipality.

The geography of these policies is revealing of the way in which they accompany gentrification: The working-class quarters taken over by the middle and upper classes are those that have benefitted from the largest number of new green spaces, new sporting facilities and new cultural facilities. The extension of the itinerary for the *Nuit Blanche* and the shift of *Paris Plages* to the northeast constitute explicit public support for the arrival of the gentrifiers in quarters that are still marginal in relation to this process.

Indeed, the absence of reference to the process of gentrification that is underway, the manner in which it is progressing, and its spatial dynamics, casts light on the motives for embellishing the city and promoting culture within its boundaries. The establishment of large cultural facilities – despite the fact that Paris was not lacking in these – in the northeast of Paris clearly contributes to gentrification by way of culture, as in Los Angeles (Davis 1990) or Bilbao (Vicario and Martínez-Monje 2003). Culture is not only instrumentalised in order to upgrade the image of the city and draw wealthier occupants, tourists and private investments; the creation of cultural facilities is also addressed to professionals in the cultural field, and it has been seen that these are the avant-garde of the gentrification process. The intellectual middle class, cultural professionals and highly-qualified managerial staff are targeted by the policies implemented by the municipality, which is in turn composed of elected members from these same social groups. In practical terms, they are in a better position to make use of the local councils (*conseils de quartier*), the new sporting, educational and cultural facilities, or the infrastructures set aside for what is known as *circulation douce* ('soft transport' such as bicycles and roller skates). Ideologically, it is their vision of the city that dominates aesthetics, heritage and recreation: Present public policies go a long way towards creating urban landscapes for 'consumption' by the middle and upper classes (Smith 2002). The orientation of urban policies in favour of the middle class is often operated under the cover of common interest, the middle class being considered as a clear majority, along the ideological lines of the averaging-out society in the post-war boom, despite the fact that this has been completely contradicted by the tendency for the income gap to widen (Bihr and Pfefferkorn 1999). In reality, the Parisian middle class is generally well above the median for income in France. Among middle-class voters, the gentrifiers largely vote for the Socialist or Green parties, so that the municipal policies implemented in Paris are also intended to meet the approval of an electorate. The re-election of the Socialist Party in 2008 and 2014 shows the success of this strategy: In particular, the takeover of environmental concerns generally promoted by the Greens eventually benefitted the Socialist party, since it strengthened its hegemony across the political left at the expense of the Greens.

Conclusion

It thus appears that the specific features of public policies implemented in Paris in the past, as well as the maintaining of rental controls up to 1986, and the production of social housing up to the 1980s, all partly explain the late occurrence of gentrification in Paris. Since Haussmann, nevertheless, public authorities have sought to upgrade or add value to the inner city, fostering gentrification in various modes. Even today, public policies in Paris show this trend, although they cultivate a certain degree of ambiguity as to their objectives and the populations targeted. The regeneration of public action promoted by the Socialist Left in Paris from 2001 has been unable to halt the process of gentrification in housing, and, on the contrary, has contributed to accompanying the trend by the development of public spaces and by its cultural policies.

With a high level of public action, and especially the promotion of social housing, the municipality do not lead a *revanchist* policy, and do not use the tools of neoliberal urbanism. Yet, public authority does, in a way, what the private sector does in other cities (for example CCTV development, or cultural promotion). Thus, despite a pervasive discourse about 'social mix', the different policies taken as a whole appear to foster gentrification in Paris. Indeed, this promotion of 'social mix' is more linked to social Catholicism and class collaboration than Marxism and class struggle and it is used upside down in working-class quarters to gentrify them, even if it's strongly denied by the Municipality.

Alongside the revanchist public policies, this paper is a contribution to show the 'soft' ways to promote gentrification, as they are not only soft but *concealed* and apparently generous (like promoting 'culture' and 'social mix'), and they often paralyse opposition and resistance.

Notes

1 For instance, the following can be quoted: The Cité des sciences et de l'industrie opened in 1986 on the site of the former La Villette abattoir, or the Opéra Bastille opened in 1989 on the edge of the Faubourg St Antoine which was in the process of marked gentrification.
2 Paris is divided into *arrondissements* that are numbered from the centre spiralling outwards.
3 A new mode of calculation included extensive rehabilitation in the category of new constructions.

References

Atkinson, R. and Bridge, G. (eds.) (2005). *Gentrification in a Global Context: The New Urban Colonialism*. London: Routledge.

Bihr, A. and Pfefferkorn, R. (1999). *Déchiffrer les inégalités*, 2nd edition. Syros: La Découverte.

Clerval, A. (2011). 'The spatial dynamics of gentrification in Paris: a synthesis map', *Cybergeo: European Journal of Geography,* 553 Available at: http://cybergeo.revues.org/23745.

Clerval, A. (2013). *Paris sans le peuple. La gentrification de la capitale.* Paris: La Découverte.

Davidson, M. and Lees, L. (2010). 'New-build gentrification: Its histories, trajectories and critical geographies', *Population, Space and Place*, 16, pp. 395–411.

Davis, M. (1990). *City of Quartz: Excavating the Future in Los Angeles.* London: Verso.

Glass, R. (1964). 'Introduction' in Centre for Urban Studies (ed.). *London, Aspects of Change.* London: Macgibbon & Kee, pp. XII–XLI.

Lees, L. and Ley, D. (2008). 'Introduction to special issue on gentrification and public policy', *Urban Studies*, 45(12), pp. 2379–2384.

Préteceille, E. (2007). 'Is gentrification a useful paradigm to analyse social changes in the Paris Metropolis?', *Environment and Planning A*, 39(1), pp. 10–31.

Slater, T. (2006). 'The eviction of critical perspectives from gentrification research', *International Journal of Urban and Regional Research*, 30(4), pp. 737–757.

Smith, N. (1979). 'Gentrification and capital: Practice and ideology in Society Hill', *Antipode*, 11(3), pp. 24–35.

Smith, N. (1982). 'Gentrification and uneven development', *Economic Geography*, 58(2), pp. 139–155.

Smith, N. (1996). *The New Urban Frontier: Gentrification and the Revanchist City.* New York: Routledge.

Smith, N. (2002). 'New globalism, new urbanism: gentrification as global urban strategy', *Antipode*, 34(3), pp. 427–450.

Vicario, L. and Martínez-Monje, P. M. (2003). 'Another "Guggenheim Effect"? The generation of a potentially gentrifiable neighbourhood in Bilbao', *Urban Studies*, 40(12), pp. 2383–2400.

Wacquant, L. (2008). 'Relocating gentrification: the working class, science and the state in recent urban research', *International Journal of Urban and Regional Research*, 32(1), pp. 198–205.

16 Urban regeneration, rent and labour

Insights from Barcelona's 'knowledge district'

Greig Charnock, Hug March, Thomas Purcell, and Ramon Ribera-Fumaz

Neil Smith's *The New Urban Frontier: Gentrification and the Revanchist City* (1996) was an enormously influential book. Myriad studies have since examined urban regeneration with a focus on gentrification processes and revanchist city policies, understood – thanks to Smith – as moments of the uneven development of capitalism on a global scale. As the present book attests, Smith-influenced research has without doubt shed much needed light on the dynamics of urban transformation across the world, and for the benefit of scholars and activists alike.

Central to Smith's argument was the coordinating role of rent in the urban process (cf. Harvey 1982). For Smith, the notion of the 'new urban frontier' captured the rapaciousness with which global and domestic capital sought to take advantage of the opportunity to regenerate decaying post-industrial sites and neighbourhoods with the emergence of a financialised and service-oriented 'new' economy. In this chapter, we look at Smith's original focus on how rent performs a crucial coordinating role at 'the frontier'. To do so, we explore the more recent regeneration of the city of Barcelona realised under the official guise of promoting knowledge-based competitiveness.

As Smith (2002: 439) once noted, any 'back-to-the-city movement' of the part of capital usually involves governmental, corporate and corporate-governmental partnerships as the agents of urban renewal in which land use for new productive activities is heavily linked to highly speculative forms of rentier profit-seeking. As we show, in this regard the transformation of Barcelona has been no different. The flagship redevelopment of the particularly post-industrial district, Poblenou, in Barcelona, has both reproduced and intensified an urban process that is highly dependent upon 'the power of finance capital over the urbanisation process' and 'the patterning of urban space according to patterns of profitable investment' (Smith 1979: 163). As such, Barcelona provides ample evidence of how contemporary strategies to engineer urban competitiveness and sustainability at the urban frontier continue to be premised upon speculative rentier practices by key agents. We also highlight how such practices fuelled the destructive process of crisis formation in Barcelona and, more generally, in Spain. Finally, we show how the unevenness of, and limits to, knowledge-based urban transformation in Barcelona has been manifest not only in economic crisis but also in the sudden and everyday visibility of very different working subjects to that of the 'creative class' of human capital or 'talent' that were supposed to populate the new 'city of knowledge': the *sans papiers*.

Barcelona: rents and the knowledge-based economy[1]

From 1998, Spain experienced a domestic property boom that lasted the best part of a decade. It was an expression of various interrelated processes (see Charnock, Purcell and Ribera-Fumaz 2014b): the reform of land and planning policy and associated shifts in banking behaviour at the national level, the introduction of the Euro, which allowed low to negative real interest rates for a sustained period, and a general trend towards capital switching (Smith 2008: 168), as well as the financialisation of urban assets at the global level that made Spanish land and real estate a particularly attractive investment opportunity. Additionally, in 1998, the Barcelona City Council (BCC) and town planning agencies were considering how best to initiate a second phase of post-Olympics urban regeneration, focusing upon the district of Poblenou in line with the recent reform of the city's General Metropolitan Plan (GMP) (Oliva 2003). The '22@' project, which re-branded Poblenou as a 'knowledge district', was eventually unveiled in 2000.[2] Under the new GMP, real estate developers could build higher than was previously permitted (densification went from 2.2 to 3 m^2) and were allowed to mix commercial and residential developments. The redevelopment aspect of the project was on a considerable scale, covering 198.26 ha, and leading to the estimated transformation of 1,159,626 m^2 of existing industrial land and the potential creation of around 3,200,000 m^2 of new construction and involving a total infrastructure investment of €180 million – the largest project of its kind in Europe at that time.

What really marked out the 22@ project, however, was its presentation by BCC as a project tailored towards making the district – and the city – a leading node in the global 'knowledge economy'. This entailed a focus on the public sector-led engineering of the 'hard' and 'soft' conditions for future competitiveness (Pareja-Eastway et al. 2007), and on exploiting the competitive advantages supposedly offered by Barcelona's compact urban morphology and its attractiveness to a globally mobile 'creative class' of knowledge-based entrepreneurs and workers (Charnock and Ribera-Fumaz 2011). Crucially, at that time, the overarching rationale for the project, as denoted by its branding as an innovation district, was from the start rooted in BCC's rhetorical embrace of contemporary economic theories that highlight particular factors which come together to allow specific locations to innovate in highly knowledge-intensive productive sectors, not least on the basis of concentrating a pool of creative 'human capital' or 'talent', and to compete globally on that basis (Leon 2008). The promotion and management of the 22@ project have been overseen by the municipal company 22@ bcn Inc. Created in 2000 by BCC, this is the agency tasked with financing the transformation of the urban environment, channelling funds into ICT-focused employment and training schemes and promoting new businesses in the area both domestically and internationally. Part of the rationale of the institutional design of the project is to link these activities with the start-up incubators housed in BCC's Poblenou premises. This new remit was representative of the 22@ project's broader rationale: to reengineer the competitiveness of the city on the basis of received wisdom regarding urban competitiveness in the knowledge-based economy.

At the heart of the reform of the GMP and the redevelopment of Poblenou is an urban redevelopment financing model that is dependent upon the generation and capture of revenues by the municipal company 22@ bcn Inc., major real estate developers and private investment fund managers. The model adopted by BCC as a means of leveraging capital to redevelop Poblenou is known as 'Value Capture Finance' (VCF), and the Urban Land Institute has cited the 22@ project as a best practice example of its operationalisation (Huxley 2009). VCF is precisely the kind of supply-side regeneration financing mechanism characteristic of entrepreneurial urban governance (Harvey 1989), in which the local state minimizes the debt it takes on from large-scale redevelopment whilst incentivizing the private sector – developers, investors and ultimately companies – to invest, with the potential to appropriate rents through facilitated access to public lands (see Ingram and Hong 2012).

As such, 22@ can be considered an exemplary form of urban regeneration marked by the intensification of partnerships between private capital and the local state, in which the search of highest returns – and/or tax receipts – overrides other planning and developmental objectives (Smith 2002). High returns have been generated by BCC through the exchange of urban plans and building permits. The plans were a mixture of public and private initiatives and, in exchange for building permits, BCC routinely claimed between 10% and 30% of the planned development, or the equivalent in monetary terms, and charged a levy of €80/m^2 of land developed (an annually updated charge). In addition, BCC has acted as a brokering agent between local institutions and financial capital, facilitating the circulation of fresh capital through the built environment.

From its inception until the advent of recession in Spain in 2009, this strategy appeared to be successful. In 2001 alone, and at the height of the property boom, Barcelona received €800 million of investment in real estate – 70% of which was of foreign origin at a time when global pension funds, in particular, targeted the office real estate market as a stable source of financial investment (*Europa Press* 2002). At the time, and given extremely low interest rates and the relative lack of other investment opportunities, large investment funds targeted single flagship developments of €20 million and above – the kind of properties that until the emergence of 22@ were of scarce supply in Barcelona. The subsequent appearance on the market of large single developments with rates of return of above 7% became particularly attractive to large investment fund managers (Moya 2002).

The process through which properties were developed in Barcelona, more for the rent they might yield rather than for their constitutive role in engineering knowledge-based competitiveness, was realised through so-called 'turnkey' transactions. These typically involved the delivery of a property by a developer to an investor according to a contractually-defined cost and time period. As Smith (1979: 546) notes, a distinguishing feature of gentrification is that the capital advanced in the form of loans is productively consumed for the creation of new property that can be capitalised into a stream of rents for the investor. Most developments in 22@ were based upon advanced sales, and in order to qualify for full development rights within the 22@ district, developers were obliged to reserve 20% of the final built space for BCC. This allowed the latter to capitalize on Poblenou's central location and a lack of existing large-scale development in the city more generally which formed the basis of demand and rising prices.

From this point onwards, it is telling that 22@ became the city's epicentre for office-based real estate demand (Sarrià 2005). Capital investment via VCF into the 22@ district attracted city-centre firms looking to relocate in search of cheaper office spaces – often incentivised by public funds and/or an eagerness to capitalize on anticipated rises in 22@ land prices. By 2006, they represented 75% of newly-located firms in the district (García 2006). This pattern highlighted the extent to which rents served a key role in co-ordinating the transformation of the 22@ district, and in spite of the stated intentions of city planners whose original plans for engineering urban competitiveness at least rhetorically centred around incubating or attracting small to medium-sized firms specialised in knowledge-intensive innovation and production. Instead, 22@ quickly became the preferred location for larger, well-established companies and mass tourism-related businesses looking to profit from new cost and location advantages; while real estate developers quickly came to favour such clients as buyers and tenants. For its part, and as *de facto* landlord, 22@bcn Inc. profited from VCF arrangements whilst continuing to propagate an ideological discourse of enhancing competiveness through the branding of 22@ as a space for innovation and the attraction of an international 'creative class' of knowledge workers.

If you build it, they will come…

Perhaps the clearest sign of the failure to create a new space for knowledge and people concerns the district's newest residents. In theory and rhetoric, 22@ was designed to promote quality of work and life in the district, defined in terms of aiding knowledge entrepreneurs and supporting vocational training and labour market entry in knowledge-intensive sectors. BCC also placed great emphasis upon attracting internationally mobile 'talent', making the most of the district's proximity to the city's famed beaches and entertainment centres. The 22@ blueprint therefore incorporated the production of housing and spaces of consumption for a young, cosmopolitan, university-educated, and predominantly male 'creative class' (Charnock and Ribera-Fumaz 2011; cf. Peck 2005).

Ironically, the district has succeeded in attracting a new class of creative and talented young people, albeit not quite the people the planners had in mind. At the fringes of 22@, abandoned former industrial sites have become the makeshift homes of (predominantly Sub-Saharan African) migrants, most of whom now submit to the harsh daily grind of collecting and recycling scrap metal, which they transport through the district in shopping trolleys. According to BCC, by 2013 in Barcelona, there were over 60 similar settlements with squatters (Rodríguez 2013). Probably the most well-known example was a factory at Puigcerdà, 127, located midway between the two major urban redevelopment centres: 22@ and Diagonal Mar. It was here that around 300 migrants sought to make a living under very difficult conditions and with the constant threat of eviction, which was eventually executed by court-ordered police on 24 July 2013. This is only one example of many such cases, although it deserves attention, first, because it was the largest informal settlement of migrants in Barcelona in recent years and, second, because it publicly exposed for thc first time the issue of how a particularly vulnerable and hitherto invisible group was coping in the fallout from the economic crisis and the collapse of the construction economy.

These migrants are certainly not the 'globally mobile creative class' of knowledge entrepreneurs and workers that 22@ aimed to attract yet they can be seen as an internationally mobile group of people responding creatively to meet basic subsistence needs. Furthermore, since a large part of the recycled material the migrant workers collect is being purchased and shipped internationally, they are effectively labouring within in a real global economy.

As Smith stressed in all his work on gentrification and urban regeneration, the move back to regenerated urban centres is too often on the basis of 'a self-interested appeal that the white middle and upper-middle classes retake control of the political and cultural economies as well as the geography of the largest cities. Probing the symptomatic silence of who is to be invited back into the city begins to reveal the class politics involved' (Smith 2002: 445). In the case of Barcelona, behind the empty rhetoric of the construction of a city for knowledge and talent, Smith's warnings about the capital's disinterest in anything other than turning a profit from real estate development – and the class politics that are symptomatically silenced in the process – still ring true.

Notes

1 This section is an updated summary of Charnock, Purcell and Ribera-Fumaz (2014a).
2 Land exclusively designated for industrial use under city planning was formerly given the classification '22a'. Regulations now cover '@ activities': i.e. new information and communications technology industries, R&D, publishing, multimedia and database and knowledge management.

References

Charnock, G., Purcell, T. and Ribera-Fumaz, R. (2014a). 'City of rents: the limits to the Barcelona mode of urban competitiveness', *International Journal of Urban and Regional Research*, 38, pp. 198–217.

Charnock, G., Purcell, T. and Ribera-Fumaz, R. (2014b). *The Limits to Capital in Spain: Crisis and Revolt in the European South*. Basingstoke: Palgrave-MacMillan.

Charnock, G. and Ribera-Fumaz, R. (2011). '"A new space for knowledge and people"? Henri Lefebvre, representations of space, and the production of "22@ Barcelona"', *Environment and Planning D*, 29, pp. 613–632.

Europa Press (2002). 'La inversión inmobiliaria en Barcelona alcanzará los 800 millones al cerrar al año', 8 December. Available at: www.europapress.es/economia/noticia-inversion-inmobiliaria-barcelona-alcanzara-800-millones-cerrar-ano-gracias-tiron-extranjero-20021208181628.html. [Accessed 4 May 2016].

García, J. (2006). 'Sólo el 30% de las 226 empresas instaladas en el 22@ se dedica a la comunicación', *El País*, 23 March, p. 40.

Harvey, D. (1982). *The Limits to Capital*. Oxford: Blackwell.

Harvey, D. (1989). 'From managerialism to entrepreneurialism: the transformation in urban governance in late capitalism', *Geografiska Annaler: Series B, Human Geography*, 71, pp. 3–17.

Huxley, J. (2009). *Value Capture Finance: Making Urban Development Pay Its Way*. Washington DC: Urban Land Institute.

Ingram, G. K. and Hong, Y-H. (eds.) (2012). *Value Capture and Land Policies*. Cambridge: Lincoln Institute of Land Policy.

Leon, N. (2008). 'Attract and connect: the 22@Barceolona innovation district and the internationalisation of Barcelona business', *Innovation: Management, Policy & Practice*, 10, pp. 235–246.

Moya, F. J. (2002). 'Barcelona se afianza como plaza rentable para invertir en oficinas', *el Periódico de Catalunya*, 3 February, p. 145.

Oliva, A. (2003). *El districte d'activitats 22@bcn*. Barcelona: Aula Barcelona.

Pareja-Eastway, M., Turmo Garuz, J., Pradel i Miquel, M., García Ferrando, L. and Simó Solsona, M. (2007). 'The city of marvels? Multiple endeavours towards competitiveness in Barcelona: pathways to creative and knowledge-based regions', *ACRE report 2.2*, Amsterdam: AMIDSt, University of Amsterdam.

Peck, J. (2005). 'Struggling with the creative class', *International Journal of Urban and Regional Research*, 29, pp. 740–770.

Rodríguez, P. (2013). 'Puigcerdà, 127: a l'espera d'aigua corrent i una vida digna', *elDiario.es*, 4 April. Available at: www.eldiario.es/catalunya/aigua-assentaments_a_barcelona-poblenou-carrer_puigcerda_0_107439294.html [Accessed 4 May 2016].

Sarrià, S. (2005). 'Las nuevas áreas de negocio de Barcelona se consolidan', *el Periódico de Catalunya*, 5 July, p. 27.
Smith, N. (1979). 'Toward a theory of gentrification a back to the city movement by capital, not people', *Journal of the American Planning Association,* 45, pp. 538–548.
Smith, N. (1996). *The New Urban Frontier: Gentrification and the Revanchist City.* London: Routledge.
Smith, N. (2002). 'New globalism, new urbanism: gentrification as global urban strategy', *Antipode*, 34, pp. 434–457.
Smith, N. (2008). *Uneven Development: Nature, Capital and the Production of Space*, 3rd edition. Oxford: Blackwell.

17 A disappearing world

The ever-expanding 'frontier of gentrification' through the eyes of Porto's historic centre long-time residents

João Queirós

Introduction

This paper renders a tribute to Neil Smith by highlighting the creative and heuristic nature of his work, which has been key in framing my research on urban and social change in inner city Porto, Portugal (Queirós and Rodrigues 2005, Queirós 2007a, 2014, 2015). Smith's work on the nature of the capitalist city and his pioneering theory and critical stance on gentrification (1979a,b, 1996, 2002) have influenced my own view on the evolution and consequences of urban and housing policies and state action within my favoured research site (see Box 1; see also Pereira and Queirós 2012 for a synthesis on urban and housing policies as seen in Porto throughout the past century), and supported some of my most relevant research options regarding these subjects.

This chapter shows how my initial work on urban renewal policies and urban and social change in inner city Porto tried to empirically illustrate and analytically specify Smith's (2002) hypothesis on 'gentrification as global urban strategy', and how his reflection on the emergence of the revanchist city and the difficulties social forces opposing gentrification increasingly face (1996, 2002) encouraged me to look at the effects of gentrification and at working-class life experiences in gentrifying, or gentrification-harassed, areas.

Box 1 PORTO: A SHRINKING AND DUALISING CITY

Porto is a city in northern Portugal with around 220,000 inhabitants. It lies at the centre of a metropolitan region with over 1.5 million residents. It is the second largest metropolitan region in the country after the region containing the capital, Lisbon. Porto is what many urban geographers and urban sociologists refer to as a 'shrinking city': Since 1981, it has lost a third of its population due to suburbanisation and

(*Continued*)

decreasing demographic dynamism. It now has fewer residents than it did in 1930.

The city centre – once a densely populated area – was most affected by these social and demographic trends. Between 1981 and 2011, this old and picturesque 1.5 km^2 riverfront location lost over two-thirds of its inhabitants, and is now the place of residence for less than 10,000 people, about 4% of Porto's population (in 1981, the area accounted for nearly 10% of the city's inhabitants). Porto's historic centre is also a degraded location with an increasingly aged population; despite some recent signs of urban and social change (see following), residents over the age of 64 almost triple the number of under 15's, and 15% of the area's buildings are severely run down, with 40% of all dwellings being vacant (respectively, 7% and 20% in the city as a whole).

Porto's historic centre, declared a UNESCO World Heritage Site in 1996, is the city's main tourist attraction and has become a focal point for real estate, commerce and tourism. Fostered by both the private sector and local public authorities, the renewal of buildings and public spaces has intensified and real estate prices have risen sharply in recent years (Queirós 2015, Imoestatística 2016). While the territory's major social and demographic features are a population that is ageing and deciding to leave (both for voluntary reasons and resulting from eviction processes), there is evidence pointing to an increase in the number of upper-class residents in a once largely working class location. Although the percentage of upper-class residents remains below the city average, the growth rate of this social segment is proving to be faster in the historic centre in recent years (Queirós 2015). This favours the consolidation of Porto as the place of residence for an increasing number of high-profile individuals and families (the number of upper-class residents is clearly above the regional and national average), and helps supporting the lay observer's stance of a dualising urban and social fabric.

Drawing on some results of a sociological fieldwork-based approach to the consequences of urban rehabilitation programmes in inner city Porto, this paper also highlights this particular aspect of physical and social change in central city neighbourhoods, which is to say, change resulting from gentrification as lived and interpreted by locals. The analysis privileges the perspectives of long-term residents of Porto's historic centre – people still living in the area, as well as recently displaced individuals – and by doing so, it aims to answer Tom Slater's (2006, 2008) challenge for the reinstatement of critical perspectives in the study of gentrification, and the development of new answers concerning the vital question of the 'right to stay put' in the gentrifying city.

Considering Porto's urban renewal policies and inner city urban and social transformations through Neil Smith's theoretical lenses

While I am far from being able to thoroughly intellectually engage with Neil Smith's extensive and diverse work, I have encountered his stimulating studies since my undergraduate studies, which I pursued some fifteen years ago. At that time, I became interested in studying the nature and effects of urban renewal policies as seen in Porto's city centre. As a student in a rather theoretically eclectic Sociology Department, I relied on a vast array of classic references, ranging from Marx to Simmel, from Weber to Benjamin, from Wirth to Lefebvre, and from Castells to Ledrut. By that time, I was also being intensely introduced to the works of Pierre Bourdieu and to the analytical and empirical specifications of his theories put forward in European and North American urban settings by authors such as Loïc Wacquant. The works of already internationally acclaimed urban geographers such as David Harvey, Neil Smith or Jamie Peck, however, lacked the diffusion and notoriety they deserved, and still do. Not a single Neil Smith book has been translated and published in Portugal, and there are few researchers directly engaging with his work, which is something that I find truly surprising to say the least (see Mendes 2014 for an exception). I was able to get to know the works of such authors and to find among them the theoretical support for some of the firmer and more heuristic work hypotheses I've been pursuing in my research since then.

In 2004 and 2005, when I was looking at the characteristics and impending consequences of what seemed to be a relevant policy shift concerning the urban renewal of inner city Porto (Queirós and Rodrigues 2005, Queirós 2007a), Smith's work provided me with a general theoretical framework of the nature and evolution of the capitalist city and invited me to engage in the gentrification debate. In fact, what I tried to do then – in a somewhat impressionistic, undergraduate student-like manner – was to empirically test and to analytically specify Smith's (2002) general hypothesis on 'gentrification as global urban strategy'.

In the early 2000s, Porto was facing some pertinent legal, political and institutional changes, which suggested a clear transformation in the nature and goals of state action concerning city centre urban renewal: In 2004, after several months of political debate, a new general legal framework was introduced, defining 'urban rehabilitation' as a 'national imperative'. This legal framework cleared the way for a whole new set of institutional arrangements and laid the groundwork for the rise of several local urban rehabilitation programmes. Before the end of 2004, new state-held urban renewal agencies (*sociedades de reabilitação urbana*, literally 'urban rehabilitation societies') were created in Porto and other major Portuguese cities to pursue what was consecrated by public authorities as the proper instrument for 21st century urban renewal, the public-private partnership. In their master plans, these newly created urban rehabilitation societies were explicit in underlining their mission to attract and facilitate private investment. Through legal and financial encouragements, the forwarding of several small-scale plans pointing out priority intervention areas and the development of a number of pilot-projects (see Figures 17.1 and 17.2), the urban rehabilitation societies wanted to persuade the real estate sector of the benefits of the return to the city centre, by evidencing the new urban renewal model's agility, flexibility and potential return rates.

In Porto, the new urban renewal agency's 2005 masterplan (SRU 2005; see also Queirós 2007a, 2015 for a critical appraisal) wanted to favour a Back to Downtown Porto movement by younger, more qualified individuals and families. It also wiled to favour the attraction of tourists, businesses and affluent city users, through public-private partnerships channelling investments to three major intervention axes: tourism, retail and knowledge-based

Figure 17.1 The 'pilot-project' of Carlos Alberto, in the northern limit of Porto's historic centre, before (2005) the operation put forward by the 'Urban Rehabilitation Society'.

Source: Photo by João Queirós (2005).

Figure 17.2 The 'pilot-project' of Carlos Alberto, in the northern limit of Porto's historic centre, after (2013) the operation put forward by the 'Urban Rehabilitation Society'.

Source: Photo by João Queirós (2013).

activities. Table 17.1 synthetises the main proposals of Porto's Urban Rehabilitation Society in each of these axes (SRU 2005).

Meanwhile, Porto's city centre, and especially its historic centre, was being intensely depicted by local authorities as an 'aged', 'shrinking' and 'impoverished' – yet 'highly promising' – area where therefore public and private resources should converge, as to ensure its physical and social regeneration (SRU 2005, 2008, Queirós 2007a, 2015).

Although I soon understood that what I really should do was to learn more about Smith's rent gap theory in order to empirically test it in Porto (Smith 1979b; see also Slater 2015), I remained convinced that I was at least gathering relevant evidence of the local expression of a transnational trend pointed out by Smith, according to whom the dissemination of gentrification throughout the globe was the result of a new 'roll-out' neoliberal urbanism, moving into 'the vacuum left by the end of liberal urban policy', and fostering gentrification through 'a concerted and systematic partnership of public planning with public and private capital' (Smith 2002: 441; see also Hackworth and Smith 2001, and Peck and Tickell 2002 for a detailed account on current neoliberalisation processes).

Porto's city centre rehabilitation as state-led gentrification

According to Smith (2002: 441–443), the generalization of gentrification has various dimensions that can be understood in terms of five interrelated characteristics: (1) A transformed role of the state, meaning a more direct

Table 17.1 Porto's 'Urban Rehabilitation Society' strategic axes and measures, as seen in its 2005 master plan

'Intervention Axes'	*Main measures*
Tourism	Developing cultural and patrimonial tourism through the rehabilitation of the city centre's religious patrimony and local religious festivities. Livening up public spaces in the downtown area and on the city's riverfront through entertainment and cultural events (dance, theatre, music, exhibitions, circus, and street art). Upgrading the quality of the city's hotel offer and its capacity to meet the expectations of demanding tourists, by favouring the opening of charm hotels in important historical buildings in the city centre. Marketing the image of Porto as a tourist destination with many local singularities, all expressing the city's unique identity, history, culture, and festivities.
Retail and Shopping	Promoting new retail concepts and new forms of retail organization (green businesses, fair trade businesses, etc.), in order to welcome new residents, tourists and city users. Introducing thematic shopping (Port Wine, jewellery, regional and local products, antiques, books) and relating it to the city's new cultural life. Promoting luxury shopping, by attracting luxury brands to the city centre (such as Calvin Klein, Cartier, Chanel, Christian Dior, Cerruti 1881, Emporio Armani, Givenchy, Guerlain, Gucci, Hermes, Karl Lagerfeld, Kenzo, Louis Vuitton, Montblanc, Paco Rabanne, Prada, Thierry Mugler, Versace, and Yves-Saint-Laurent). Renovating Mercado do Bolhão, a traditional city centre market that should adopt a character and functional model 'similar to the one followed in London's Covent Garden'.
Knowledge-based Activities	Creating a 'technopole' (Innovation Park) in Porto's city centre, where creative and technology-based start-ups and established firms could be based. Developing a partnership between Porto Vivo, Porto's 'Urban Rehabilitation Society', and the University of Porto, in order to develop the activities of the new technopole. Creating a new infrastructure for R&D activities and cultural and scientific events (Centro do Futuro, the 'Centre of the Future'). Attracting national and international firms working to develop strong added-value activity sectors (environment, urban renewal, culture, technology).

Source: Adapted from SRU (2005).

and intense role of the state as an encourager and facilitator of gentrification; (2) penetration by global finance, meaning 'the reach of global capital down to the local neighbourhood scale' (Smith 2002: 441); (3) changing levels of political opposition with the emergence of the revanchist city and the growing repression and control over antigentrification movements, 'making the city safe for gentrification' (Smith 2002: 442); (4) geographical dispersal, meaning the outward diffusion of gentrification from the urban centre; and (5) the sectoral generalization of gentrification, which is to say, the evolution of the phenomenon into a 'vehicle for transforming whole areas into new landscape complexes that pioneer a comprehensive class-infected urban remake', as a result of the action of a 'new amalgam of corporate and state powers and practices' (Smith 2002: 443).

I knew the exploratory character of the research I conducted in Porto's city centre in 2004 and 2005 would not allow me to scrutinise thoroughly the empirical accuracy of each of the main characteristics of the latest stage of gentrification, as proposed and described by Smith, but the analysis of the configuration of the programme and initial projects put forward by local authorities to foster urban rehabilitation and the consequent physical and social revival and regeneration of the area (see above) suggested Porto could be envisaged as the place for yet another local expression of the new neoliberal urbanism that was and is disseminating gentrification worldwide. I was not able to collect solid evidence on the role of global finance in the gentrification-cum-rehabilitation projects that were being launched by that time, nor did I focus on how these projects were being perceived – and eventually opposed – by local actors and local organizations. The hypothesis of Porto's city centre rehabilitation as state-led, or municipally-led gentrification, however, seemed worthy of pursuit, taking into account the proactivity of public authorities regarding the planning, encouragement and launching of the public-private action that would ensure the supposed much-needed urban and social transformation of the city's central neighbourhoods.

I also noticed how the gentrification-cum-rehabilitation of Porto's city centre was being proposed by local public authorities in an increasingly intense fashion, with the direct assistance of the real estate sector's spokespersons and the compliance of some academics and several representatives of the city's cultural, nightlife and commercial activities as the 'inevitable', 'last chance' way of reversing the ageing population, and the degradation and impoverishment of an area that should in reality be the city's crown jewel (Queirós and Rodrigues 2005; Queirós 2007b, 2015).

The gradual establishment of a 'gentrification *doxa*', as Bourdieu would call it, meaning the taken-for-granted, undisputed belief of how 'things are' and 'should be' – in this case, how urban policy and action should inevitably be conceived and conducted (see Bourdieu 1997, 1998; see also Wacquant 2004) – seemed to be lubricating the urban rehabilitation apparatus, while functioning as a screen-discourse, in at least three ways: (1) by underlining the crucial role of the state in favouring private investment in gentrification-cum-rehabilitation, while reifying the area's existing demographic, social and physical conditions, namely through the obscuring of the decisive historical role the state played in the production and reproduction of those conditions; (2) by arbitrarily imposing the gentrification-cum-rehabilitation model as the single realistic, feasible city centre urban and housing renewal strategy; and (3) by diverting the focus from the existing social settings and centring attentions on what had to be done to attract private investment and the myriad of gentrifiers that would assure the success of urban rehabilitation projects.

In my PhD research, which I carried out in Porto between 2010 and 2014, I tried to look deeper into these matters. I wanted to examine in detail the major urban and social effects of the first decade of state-led gentrification in the historic centre through reading official statistics and gathering evidence of physical and social change in the area. I was also committed to the purpose of answering Slater's (2006) challenge for the reinstatement of critical perspectives in gentrification research, namely through the study of everyday life in gentrifying, or 'gentrification-harassed' areas and through the discussion of the meanings and personal and collective impacts of displacement. My goal was to further specify Smith's (2002) hypothesis on 'gentrification as global urban strategy', and simultaneously contribute to refocus academic concerns on the impact gentrification produces in the life of local communities, within a broader reflection on class struggles in the contemporary city.

In this paper I can present but a brief sketch of some of the findings of the research I conducted in Porto's historic centre in the first half of the 2010s. One of the most noticeable aspects of this area's recent evolution is that it has prolonged several urban and social processes endured in previous decades: The intense population decrease due to negative net growth and migration rates (between 2001 and 2011, the number of residents in Porto's historic centre fell by 30%), and the severe ageing of local residents proceeded, hence contributing to social devitalisation and institutional desertification of the area; the social conditions and prospects of several local families have declined, following welfare retrenchment and the disappearance of relevant locally-based public institutions and facilities. Abandonment and physical decay subsequently progressed, especially in inward, more recondite quarters (see Figures 17.3 and 17.4; see Queirós 2015 for details).

Meanwhile, several other pieces of evidence suggest the development of a different set of social processes, which are responsible for a number of relevant changes in Porto's historic urban centre and social fabric. In the last six to eight years, tourism in the city has observed a strong, steady growth, thus reinforcing the historic centre's function as the city's major tourist destination. As a consequence, many new hotels, restaurants and retail activities have been created, thus changing the economic profile of the area; local accommodation and a myriad of other tourist activities have also erupted, impacting on rents, daily exchanges and community life.

Since 2005, several pilot projects fostered by the Urban Rehabilitation Society have been concluded, and there has been a steady growth in the number of new private-led rehabilitation projects, leading to a rise in real estate prices (according to real estate sector's sources, prices rose by 17% in 2015; see Imoestatística 2016). Although people getting older and moving

Figure 17.3 Image of current ageing in Porto's historic centre.
Source: Photo by João Queirós (2011).

Figure 17.4 Image of current abandonment in Porto's historic centre.
Source: Photo by João Queirós (2011).

Figure 17.5 A rehabilitated quarter in the northern limit of Porto's historic centre.
Source: Photo by João Queirós (2011).

away are persistent social phenomena, and despite the fact that the majority of local residents are working class or have working class origins, official numbers point to an increased growth in the number of managers, high-level professionals and intellectuals residing in the historic centre. While still far under the city's average (40% in the city and 21% in the historic centre), the number of people in these professional groups more than tripled in the area under scrutiny in the last two decades (it doubled in the city as a whole), suggesting 'social filtering' and 'social upgrading' are underway here (see Figures 17.5 and 17.6; see Queirós 2015 for details).

What about Porto's historic centre long-term residents? How do they envisage and how do they relate to what seems to be the stepwise, systematic expansion, in a nonetheless shrinking and degrading area, what Smith (1996) called the 'new urban frontier'?

The ever-expanding 'frontier of gentrification' through the eyes of Porto's historic centre long-time residents

Nowadays, Porto's historic centre faces a contradiction resulting from its situation as both the city's main tourist attraction and one of the most degraded and impoverished areas within the city.

Not surprisingly, the discourse of local residents tends to express the splintered representation of the situation and prospects of their place of living. On the one side, the enduring phenomena of physical decay, social devitalisation, institutional devolution and welfare retrenchment, and the

Figure 17.6 An overview of Ribeira do Porto, the historic centre's riverfront and main tourist site.
Source: Photo by João Queirós (2011).

increasing reality – or threat – of eviction and displacement, whether immediate or perceived, seems inevitable for the members of the next generation ('They want to send us to the *bairro* [meaning the social housing estate in the city periphery]'; 'Our children and grandchildren won't be able to stay'), generating feelings of indignation and anguish. Additionally, the intensifying impression of living in a space progressively taken by 'outsiders' and 'outside interests', due to the mounting presence of tourists, tourist activities and real estate agents, erodes the long-time residents' affection for the area and even their will to stay, especially when leaving has taken away a relevant number of family members and long-term neighbours ('This is empty now'; 'There're people here who are not local people'; 'We only see our old acquaintances at weddings and funerals').

On the other side, there's a shared affirmative self-portrait of Porto's historic centre residents as 'genuine', 'self-determined' and 'generous' people, the 'true bearers' of Porto's 'characteristic way of life'. This self-portrait has recently become one of the main justifications put forward by long-time residents for a supposed right to stay put. The 'social mix' discourse, while functioning as the public authorities' rationale for 'urban rehabilitation' in an 'aged', 'underqualified', 'impoverished' area, has been converted, by many residents, in a kind of last resort ideological-discursive rationalisation against displacement ('The people that left are the ones that really should be here', 'We really love the place', 'What would tourists have to see if real Porto residents were not here anymore?') (Figure 17.7).

Figure 17.7 'Attention: Do not park in front of the door – residents live here. Thank you'.
Source: Photo by João Queirós (2012).

Considering the leading forces implicated in, and the course of, the ongoing urban and social transformations in Porto's historic centre, a rationalisation of this sort seems to be wholly ineffective. When confronted with the apparently irresistible expansion of the 'frontier of gentrification', the relevant question here concerns the conditions – both subjective and objective – of Porto's historic centre long-term residents who, in reality, have to guard their 'stronghold' and their 'world'. The necessary debate over resistance and alternatives means that we now need Neil Smith's inspiration more than ever.

References

Bourdieu, P. (1997). *Pascalian Meditations*. Stanford: Stanford University Press.

Bourdieu, P. (1998). *Acts of Resistance: Against the Tyranny of the Market*. New York: The New Press.

Imoestatística (2016). 'Centro histórico do Porto valoriza 17,1% em 2015', *Confidencial Imobiliário*, March.

Hackworth, J. and Smith, N. (2001). 'The changing state of gentrification', *Tijdschrift voor Economische en Sociale Geografie*, 92(4), pp. 464–477.

Mendes, L. (2014). 'Gentrificação e políticas de reabilitação urbana em Portugal: uma análise crítica à luz da tese *rent gap* de Neil Smith', *Cadernos Metrópole*, 16(32), pp. 487–511.

Peck, J. and Tickell, A. (2002). 'Neoliberalizing space', *Antipode*, 34(3), pp. 380–404.

Pereira, V. B. and Queirós, J. (2012). 'State, housing and the "social question" in the city of Porto (1956–2006): an analysis on the making of doxa, orthodoxy and

"alodoxia effects" in the (re)production of state housing policies', *Social Sciences – Annual Trilingual Review of Social Research*, 1, pp. 203–214.
Queirós, J. (2007a). 'Estratégias e discursos políticos em torno da reabilitação de centros urbanos. Considerações exploratórias a partir do caso do Porto', *Sociologia – Problemas e Práticas*, 55(3), pp. 91–116.
Queirós, J. (2007b). *O lugar da cultura nas políticas de reabilitação de centros urbanos: apontamentos a partir do caso do Porto.* Working Paper do Instituto de Sociologia da Faculdade de Letras da Universidade do Porto (Série 2000–2010), 17.
Queirós, J. (2014). *No centro, à margem. Estudo sociológico das intervenções habitacionais do Estado no centro do Porto e das suas consequências sociais.* PhD Thesis. Porto: Faculdade de Letras da Universidade do Porto.
Queirós, J. (2015). *No Centro, à Margem. Sociologia das intervenções urbanísticas e habitacionais do Estado no centro histórico do Porto.* Porto: Edições Afrontamento.
Queirós, J. and Rodrigues, V. (2005). *Recriar a Cidade. Dinâmicas Culturais Emergentes e Reabilitação Urbana da Baixa do Porto.* Undergraduate Thesis. Porto: Faculdade de Letras da Universidade do Porto.
Slater, T. (2006). 'The eviction of critical perspectives from gentrification research', *International Journal of Urban and Regional Research*, 30(4), pp. 737–757.
Slater, T. (2008). '"A literal necessity to be re-placed": a rejoinder to the gentrification debate', *International Journal of Urban and Regional Research*, 32(1), pp. 212–223.
Slater, T. (2015). 'Planetary rent gaps', *Antipode* Available at: doi:10.1111/anti.12185.
Smith, N. (1979a). 'Gentrification and capital: practice and ideology in Society Hill', *Antipode,* 11, pp. 24–35.
Smith, N. (1979b). 'Toward a theory of gentrification: a back to the city movement by capital, not people', *Journal of the American Planning Association,* 45(4), pp. 538–548.
Smith, N. (1996). *The New Urban Frontier. Gentrification and the Revanchist City.* London: Routledge.
Smith, N. (2002). 'New globalism, new urbanism: gentrification as global urban strategy', *Antipode*, 34(3), pp. 434–457.
SRU (2005). *Masterplan.* Porto: Porto Vivo, SRU (Porto's Urban Rehabilitation Society).
SRU (2008). *Plano de Gestão do Centro Histórico do Porto Património Mundial – Volumes I, II e III.* Porto: Porto Vivo, SRU (Porto's Urban Rehabilitation Society).
Wacquant, L. (2004). 'Critical thought as solvent of doxa', *Constellations*, 11(1), pp. 97–101.

18 Architecture of violence

'Anti-beggar architecture' as the 'eureka' of urban regeneration[1]

Sonia Maria Taddei Ferraz, Luis Gustavo Rosadas Campos, Mayra Duarte Lima, Paula Ramos C. Carvalho de Mendonça, and Leticia Lyra Acioly

Introduction

Based on Neil Smith's ideas (2000: 135) that both the real estate market and elite groups cause the homeless banishment from the urban landscape, and that 'their [the homeless'] visibility is constantly erased by the institutional efforts to move them to other places', we intend to develop an analysis of the homeless banishment from the upscale neighbourhoods of the main cities in Brazil.

We understand that the increase of public policies and private initiative actions regarding urban/human cleansing is related to the conflicting nature of the social relations that take place in cities. Such urban/human cleansing actions are the real reason for social and economic exclusion, which hastens the processes of urban gentrification. It is by means of these actions, which are revealed in physical interventions in the city spaces, that the elite groups and the public power make use of 'anti-beggar architecture' equipment. This movement intensifies the representation and action of repulsiveness and banishment of the poor from the best areas of cities. Consequently, this 'anti-beggar architecture' makes the homeless invisible; however, it disrespects the democratic right to the city, which strengthens and naturalises a particular model of sociability.

Martins (1993: 37), in her book *Livres Acampamentos da Miséria* [Free Camps of Misery], points out that in the 1980s, it was the use of hollow spaces and flyover spans as shelter that 'greatly marked the worsening of the housing problems' in the cities. It is important to highlight that the period observed by Martins was influenced by the economic downturn of the 1970s and by the emergence of global cities as the *loci* of realisation of the neoliberal economy.

'Anti-beggar[2] architecture' emerged in the media at the beginning of the 1990s, when devices for such architecture were developed, diversified and spread alarmingly quickly throughout squares and cities. In the beginning, these devices caused surprise and seemed not to be in harmony with the urban landscape. Today, however, they are something natural and appear to take on the aspect of the necessary generalization in favour of a desired urban regeneration. Thus, it is possible to say that 'anti-beggar architecture' is neither a local nor a national problem. It is a global issue, as the strategic planning that offers prior support to the real estate capital has been globally adopted.

The erasure of the undesirable

Violence, generally assigned to the poor and miserable, has caused a constant movement towards amplification and diversification of self-protection and asset protection strategies against any kind of delinquency. The real estate market capitalises this panic by turning it into calls and variation in the kind of materials, equipment and services provided. As a result, it systematically expands the billing of this business sector. According to the Mello (2015), the private security witnessed a 30% annual increase in Rio de Janeiro.

The amount of growth illustrated by Antônio Werneck (2016) in an article in the same newspaper is startling: 'Companies owned by Military Policemen besiege condominiums located in the south zone of the city presenting homeless removal services' [author' translation]. That is, 'through private security companies, policemen offer to remove homeless people'. This offer is a private service that frees any condominium from 'beggars or homeless on its sidewalks. Besides, with this service, the condominiums can rely on a 24-hour monitoring'.

Fear and aversion-based actions have been added to security movements against the 'rest' of the population. By exclusion and elimination of the *other*, the public space protection strategies are extended, exceeding the number of walls and iron gates, and taking away the collective right to the city. The consolidation of this perspective over the security issue undoubtedly produces an increasing and consensual wish to struggle against crime, in spite of the overlapping of security policies for the elites and the disrespect to poor people's rights. In this context, anti-beggar architecture grew quickly in mid-sized and large cities in the past few years in order to keep poverty under control and out of the sphere of action and the line of sight. Once more, we affirm that such architecture has contributed to the gentrification process on behalf of an urban regeneration initiative, as Neil Smith confirms (2000: 135) (Figure 18.1):

> Banished from the private spaces of the real estate market, the homeless occupy the public spaces, but their presence in the urban landscape is contested with fury. Their visibility is constantly erased by institutional efforts to move them to other places – to shelters, to places out of buildings and parks, to poor neighbourhoods, to places outside the city, and to other marginal spaces. Those who own a home and are engaged in desperate personal campaigns not to see the homeless, even when they

trip on the homeless bodies on the pavements, also erase the banished people. This ongoing deletion is strengthened by the media-produced stereotypes that blame the victims and, consequently, justify their invisibility.

Data from the Rio de Janeiro Municipal Secretariat of Social Development shows that, from 2010 to 2012, there was a 31.25% increase in people living on the streets of the city. The anger against the distressing evidence of the increasing misery can be seen both in the countless attacks against the homeless, which became natural over time, and in the interventions of the public power. Such interventions emerge to reach the total control over the streets, to displace or to erase the homeless population, which becomes a symbol of intolerance and class divide.

With sport mega-events in Brazil, such as the World Cup in 2014 and the Olympic Games in 2016, not only has the banishment of the poor been happening by means of 'anti-beggar architecture' but also through public urban/human cleansing policies that take the homeless to shelters.[3] A recent analysis of public urban/human cleansing policies points out that the homeless have been extensively banished from the upscale neighbourhoods of Rio de Janeiro. With this cleansing, the real estate market adds value to areas around the slums and/or to the ones that had the homeless removed, and contributes to the gentrification process. A good example of such urban/human cleansing public policy is the 'Choque de Ordem' ['Shock of Order'] operation. It promised to re-establish public order to the city by collecting the homeless and transferring them to municipal shelters, as registered in the

Figure 18.1 Homeless sleeping in front of an Olympic sign, which says: 'A new world', in Botafogo (Rio de Janeiro, Brazil).

Source: Photo by Felipe Barcelos (July 2016).

news website of the Rio de Janeiro Municipal Department of Finance,[4] as related in the two following excerpts, which were chosen among many from this web: '*Choque de Ordem* collects 109 street dwellers in the south zone' (23 June 2010), '*Choque de Ordem* collects 21 street dwellers in Copacabana' (14 April 2011).

The housing 'solution' came down to the transference of the 'sheltered', which was carried out with the support of the Social Assistance Service. The operation, in turn, became an apparent way for society and the public power to feel free from guilt. On the one hand, there is a resigned society sustained by the banishment. On the other hand, there is the government, who only invests in unscripted actions and does not play its constitutional role of ensuring dignified housing for everybody. With a charitable tone (as it was carried out by the Social Assistance Service), the action of collecting the homeless was announced as the 'sheltering of street dwellers'. The use of such tone blurred the truculence applied in the act of 'sheltering', as denounced by popular manifestations (Figure 18.2).

In the context of private practices, the intolerance and anger that Smith (2000: 135) discusses can be exemplified in materials published in various journals. They are speeches that justify themselves on behalf of the alleged necessary urban regeneration. The idea of this regeneration can be understood from its other meanings: rehabilitation, purification and correction. These connotations may lead us to suppose that the city faces a state of degeneration that needs to be corrected. Such thinking would permit a kind of repulsiveness towards the degenerate people, who start to be seen as the ones that 'dirty' the streets and parks with their presence.

Figure 18.2 Homeless sleeping on sidewalk in Niterói (Brazil).
Source: Photo by ArqViol team (May 2014).

Neil Smith (2006: 83) helps us to clarify the importance of this issue by bringing into discussion the euphemistic use of the language as the carrier of social change of meanings:

> To begin with, where does this language come from? A biomedical term applied to plant specimens, species or group of species – livers and forests can regenerate. That is why this metaphorical use is so revealing. Even though the language of regeneration has always been used to describe the city, this term only caught on in the neoliberal years of the 1990s. This means that the strategic gentrification of the city is really a natural process. Hence, the regeneration strategy supporters dissemble the subtle social origins and the purposes of the urban change. In addition, they erase winner and loser policies from where such lines of action emerge. This way, a second systematic silence comes from the losers.

The author (2006: 141) complements his thought about the language traps by saying: 'An object, event or situation understood in an incomplete manner is another object, event or situation. And this may reveal a truth or an insight'.

In this way, the journalistic statements bring out unquestionable examples of discursive strategies that produce new class consensuses. One example of this is of a retired lady who, in 2006, received a distinction for her courage to shoot a thief. In her speech, when receiving the honour, she declared: 'street dwellers should be left in the middle of the sea' (de Paiva 2012). Another example is one of an assemblywoman who, in 2013, asserted: 'the homeless do not have the same rights as citizens' (G1 Rio 2013).

The public policies associated with the journalistic articles end up feeding class anger, which frequently turns the cleansing of valuable areas into the annihilation of the homeless. In the language field, the media statements are also responsible for metaphorically describing the homeless population as 'street dwellers' and 'invaders'. By insisting on the usage of these words, the media naturalises and criminalizes their condition.[5] In addition, the language applied by the media encourages and legitimates the conflicts, repulsiveness, attacks and acts of anger performed by alleged vigilantes against the homeless. Many homeless people are executed with firearms or hit with stones while asleep. Others suffer attempts of being buried or burnt alive. In Brazil, a homeless person dies every two days (Agência Brasil 2012).

In different newspaper columns, the readers' opinions show repulsion against the presence of the homeless in parks and streets, alleging that they are dirty, dangerous and smelly. The fact that homeless people live in these areas brings profound class contradictions in cities to light. In this manner, as we can learn from Manuel Delgado (2011: 10), the public space becomes less and less public, because it is more and more at the service of the capitalist appropriation. At the same time, this space is occupied by an increasing poverty that cannot be hidden. Also according to Delgado (2011: 22), the public space is quintessentially the democratic space. The erasure of the poor through institutional efforts of removal corresponds to the human cleansing on behalf of selective use of the valorized spaces.

Data from the newspaper (Do Rio 2011) showed that one rich Brazilian earned as much as 40 poor Brazilians; this situation has not changed since then. As an illustration of it, there is the fact that the Brazilian cities that present the highest number of public human cleansing interventions and acts by vigilantes, such as Rio de Janeiro and São Paulo, are exactly the ones that have the greatest real estate values in the country. They are among the most elevated in the world. In this manner, the real estate market interests are advantageously met in silence, cooperating to the gentrification process, as we can learn from Neil Smith's words (2006: 83):

> It is historically and not accidentally that the gentrification is associated with the increase in the banishment of homeless people. Even so, this important regeneration effect went completely silent. In the 'urban regeneration discourse', there is no consideration about the destination of the people who, by virtue of a huge reconquest of the city, were displaced or thrown in the streets.

Smith's objection, taking into account the great number of homeless people that make do on the streets, goes beyond the removal of the bodies that do not fit in the urban requalification/regeneration model. The presence of the homeless in high-valued public spaces with potential for gentrification starts to be avoided by means of design elements projected in order to repress the stay of the poor in the affluent areas of the cities.

Smith (2000: 134)[6] shows the homeless vehicles in New York projected to evacuate the homeless, but enables their political reappropriation of public spaces. Pictures on the homeless vehicles can be found in Smith (2000: 162–167).

Differently from the homeless vehicles, the 'anti-beggar architecture' emerges as an effective measure of silent erasure and definite evacuation of the homeless from public spaces. As mentioned before, such architecture appeared in the big cities in the 1990s. Over time, it spread throughout, even becoming permanent and natural. In 1994, in an article written by Antonio Rocha Filho, the *Folha de São Paulo* (4th November) announced the creation of 'anti-beggar architecture'. One of the first devices of such architecture were sprinklers. They were installed under the thresholds and activated at random times in order to startle and chase away the homeless who slept on the pavements. Afterwards, buildings with no marquees appeared, and burnt oil used to be spilt on sidewalks. The journalist Selma Schmidt (2005) pointed out that 'as the public power cannot find a solution to prevent sheltered external areas and sidewalks from being occupied by the homeless, anti-beggar architecture becomes creative. Floodlights and iron gates surrounding buildings are not the only options anymore' (Figures 18.3–18.6).

Figure 18.3 Protesters denounce violence during shock order: 'Shock order kills poor'.

Source: Photo by Bruno Amadei (2009).

Figure 18.4 Iron rods on a residential wall in Salvador (Brazil).

Source: Photo by ArqViol team (April, 2001).

Figure 18.5 Spikes on a threshold in Paris (France).
Source: Photo by ArqViol team (2015).

Figure 18.6 Spikes as 'Fakir bed', São Paulo (Brazil).
Source: Photo by ArqViol team (2004).

The public power starts to avail facilities that intensify the representation of the repulsiveness of degenerate people, the anti-beggar architecture, materializing the disrespect to the democratic right to the city. The City Hall of Niterói, a city located in the metropolitan area of the state of Rio de Janeiro, decided to stick sharp stones into the pillar bases of the flyovers that give access to the Rio-Niterói bridge. People who drive around this area call this technology 'the fakir bed'. Another solution to avoid the homeless, used mostly by bank branches, was to place iron protectors on its front steps (Figures 18.7 and 18.8).

Also in 2005, the mayor of the city of São Paulo authorised the construction of anti-homeless ramps in order to try to block the concentration of the homeless under flyovers. Subsequently, anti-homeless benches were placed in squares and at bus stops so that the homeless could not lie across them (Figure 18.9).

The private power from homeowners in upper class areas has been successful in legitimising the establishment of anti-beggar architecture on the perimeters of their properties. This fact represents the metaphors of class repulsion as it definitely takes away spaces that used to shelter the banished.

Brief concluding observations

On one side, the public power contributes by clearly spreading obstacles to homeless settlement in public city spaces. It also plays the sanitizer role, making possible the reproduction of capital in the cities. On the other side, the private power of the noble area home owners has effectively legitimised

Figure 18.7 Homeless sleeping on the entranceway of a bank branch in Copacabana (Rio de Janeiro, Brazil).

Source: Photo by ArqViol team (September 2016).

Figure 18.8 Anti-beggar staples on a shop window (Spain).
Source: Photo by ArqViol team (2015).

Figure 18.9 Anti-beggar benches (tilted, rib bench, with dividing rods) in Rio de Janeiro (Brazil, 2015), São Paulo (Brazil, 2016), Eindhoven (The Netherlands, 2016), Niterói (Brazil, 2015).
Source: Photos by ArqViol team.

the application of anti-beggar architecture on the perimeters of their properties, unequivocally cancelling out the spaces that once sheltered the destitute. Thereby the anti-beggar architecture, as an array of devices, turned into the most efficient strategy to take away the dispossessed from the 'safe and beautiful landscape for tourists'.

While the homeless vehicles in New York City, shown by Smith (2000: 135), would be considered 'an impertinent invention that gives the banished the power to erase their own erasure of the spaces from where they were expelled', anti-beggar architecture is an absolutely silent measure of definite banishment. The homeless who are excluded from New York City have the option to come back through the political appropriation of the spaces, making their bodies present again. Anti-beggar architecture, however, casts out the homeless and makes them existent again only symbolically, by means of elements that testify their compulsory absence or erased existence from society.

The definite and silent deletion, with no chance of recovering the spaces, productively embodies a *eureka* as a metaphor of a given urban

Figure 18.10 Street dweller appropriates a bus stop shelter (Rio de Janeiro, Brazil, 2001).

Source: Photo by João Cerqueira/CPDoc JB.

requalification. According to Smith (2000: 140), 'the use of spatial metaphors, far from providing innocent and evocative images, goes straight to matters of social power'. That being the case, it will be only by actions of resistance against the public space closure and by the courageous re-occupation of these spaces that the homeless will be able to highlight their misery and the magnitude of the oppression of social power (Figure 18.10).

Notes

1 This work was carried out by the research group 'Architecture of Violence' linked to the Graduate Architecture and Urbanism Program of the Fluminense Federal University, Rio de Janeiro, Brazil. The group was created in 2000 with the financial support of FAPERJ-Fundação Carlos Chagas Filho de Amparo à Pesquisa do Estado do Rio de Janeiro and started its studies by investigating how the fear of violence had been designing new functional and formal patterns of architecture and cities. Its current interest is focused on the analyses that deal with urban interventions performed by either the private or public sectors, specifically the ones related to security policies that attack the slums and the homeless. This text was originally presented at the "Global capitalism and processes of urban regeneration. Hommage to Neil Smith, international conference" held in Barcelona (14th–16th September 2015). Afterwards, it was considerably expanded and fleshed out in order to be published in *MOVIMENTO-Revista eletrônica de educação*, in January 2016 [http://www.revistamovimento.uff.br/index.php/revistamovimento]. For the aims of the present work, it is again in its original version.

2 The alternate use of the expressions 'beggar' and 'homeless' was necessary because in Brazil the strategies to banish homeless people were insistently referred to as 'anti-beggar architecture' by the media. Consequently, this designation became commonplace. Such difference in the terms evokes new and important reflections, which will be dealt with in future work.

3 On the 21 January 2009, the newspaper *Jornal do Brasil* reported that despite the public authorities' efforts, 80% of the homeless carried to shelters returned to the streets.

4 Available at: http://www.rio.rj.gov.br/web/smf/listaconteudo?search-type=-assunto&assunto=ordem-publica. Accessed on 28 July 2012.

5 This issue has already been dealt with in Ferraz and Machado (2014).

6 Ironically, the little cars for the homeless were designed by the artist Wodiczcko in New York in the 1980s in order to put the mobility issue into discussion, since it figured as a central problem for the population that was thrown out from the public spaces and the private spaces of the real estate market.

References

Agência Brasil (2012). 'Ao menos 165 moradores de rua foram mortos no país em um ano', *Último Segundo – IG,* 15th March. Available at: http://ultimosegundo.ig.com.br/brasil/ao-menos-165-moradores-de-rua-foram-mortos-no-pais-em-um-ano-diz/n1597695622274.html [Accessed 25 July 2016].

Delgado, M. (2011). *El espacio público como ideologia.* Madrid: Catarata.

De Paiva, G. (2012). 'Idosa ganhou medalha da Câmara em 2006 após atirar em bandido, mas decisão causou polêmica', *O Globo,* 11th May. Available at: http://oglobo.globo.com/rio/idosa-ganhou-medalha-da-camara-em-2006-apos-atirar-em-bandido-mas-decisao-causou-polemica-4872374 [Accessed 25 July 2016].

Do Rio (2011). 'Rico brasileiro tem a soma da renda de 40 pobres', *Folha de São Paulo,* 17th November. Available at: www1.folha.uol.com.br/fsp/cotidian/9443-rico-brasileiro-tem-a-soma-da-renda-de-40-pobres.shtml [Accessed 28 July 2016].

Ferraz, S. and Machado, B. (2014). 'Eu não tenho onde morar, é por isso que eu moro na rua. Os 'sem-teto': moradores ou transgressores?', *Cadernos Metrópole*, 16(32), p. 609.

G1 Rio (2013). 'Vereadora do Rio diz que mendigo não tem o mesmo direito que cidadão', *G1-Globo*, 31st October. Available at: http://g1.globo.com/rio-de-janeiro/noticia/2013/10/vereadora-do-rio-diz-que-mendigo-nao-tem-o-mesmo-direito-que-cidadao.html [Accessed 25 July 2016].

Martins, A. L. L. (1993). *Livres acampamentos da miséria.* Rio de Janeiro: Obra Aberta.

Mello, I. (2015). 'Medo faz segurança privada crescer 30% em 1 ano', *Infoglobo – O Globo*, 26th July. Available at: http://oglobodigital.oglobo.globo.com/epaper/showarticle.aspx?article=ec2f58921b8f453084eed557b676c2e2&key=jPi4Vd5q9TnBehs7h4P8Kw%3D%3D%E2%80%A6 [Accessed 23 November 2015].

Schmidt, S. (2005). 'Chuveirinho vira arma para espantar mendigos', *O Globo*, 5th June, p. 31.

Smith, N. (2000). 'Contornos de uma política especializada: veículos dos sem-teto e produção de escala geográfica', in A. Arantes (org.). *O espaço da diferença.* Campinas: Papirus, pp. 132–159.

Smith, N. (2006). 'A gentrificação generalizada: de uma anomalia local à 'regeneração' urbana como estratégia urbana global', in C. Bidou-Zachariasen (org.). *De volta à cidade: dos processos de gentrificação às políticas de revitalização dos centros urbanos.* São Paulo: AnnaBlume, pp. 59–87.

Werneck, A. (2016). 'Por meio de empresas de segurança, policiais se oferecem para remover moradores de rua', *O Globo*, 12 April. Available at: http://oglobo.globo.com/rio/por-meio-de-empresas-de-seguranca-policiais-se-oferecem-para-remover-moradores-de-rua-19063604#ixzz4M8KHNwkI [Accessed 19 April 2016].

PART V

Activism and resistance

19 The urban frontier

Gentrification as ideology and class politics in the remaking of marginal urban space

Catharina Thörn and Katarina Despotovic

Figure 19.1 Kvillebäcken, Akbar Chark.

Remaking Gothenburg

In a time when gentrification has become the dominant form of urban transformation, it is easy to believe that the natural evolution of a city benefits everyone. Terms like renewal, vitalization, renaissance, inclusion, mixed-use city, attractiveness and sustainable urban development throw a mystifying veil over the process, obscuring the fact that gentrification benefits some groups at the expense of others. When Ruth Glass coined the term 'gentrification' in the 1960s, the tone was very different.

> One by one, many of the working class quarters of London have been *invaded* by the middle-classes – upper and lower. Shabby, modest mews and cottages – two rooms up and two down – *have been taken over*, when their leases have expired, and have become elegant, expensive residences [...] Once this process of 'gentrification' starts in a district it goes on rapidly until all or most of the original working-class occupiers are *displaced* and the whole social character of the district is *changed*.
>
> (Glass 1964: xviii–xix, our emphasis)

By using words such as 'invaded', 'displaced' and 'changed', we can see Glass's language was designed to point out that gentrification is ultimately a matter of power and resources: who are allowed space in the city and on what terms. Gentrification is neither natural nor inevitable; it is facilitated through political control of land use and capital flows. In *The New Urban Frontier* (1996), Neil Smith shows how gentrification is fostered and driven by unequal distribution of power and resources in the city. Instead of being a solution to urban decay, Smith argues that gentrification enables and maintains a dynamic between decline and capital investments. Smith used to term 'rent gap' in an attempt to capture this dynamic. The rent gap is found in a piece of land whose value ('ground rent') diverges widely from the value the land would have ('potential ground rent') if put to its 'highest and best' profit-driven use. In this way, new urban frontiers, ripe for development, are created in an unending cycle. When municipalities push gentrification under the banner of 'urban renewal' or the imperative 'to build the city', property values are driven up at these frontiers, with profound impact on the people who live and work in the neighbourhoods.

Kvillebäcken was marketed as a new city district in Gothenburg and a display window for sustainable development. The development is the first step in the radical transformation of the central parts of Gothenburg, in which the inner city will expand to the other side of the river and include parts of Hisingen Island. The site where the residential area of 'new' Kvillebäcken is emerging was once an insignificant industrial area with no name, although often called the Gustaf Dalén district. The area was built in the 1930s and 1940s, when it was mainly dominated by metalworking and paint industries which were then gradually replaced by sheet-metal companies, auto repair and service shops. They were joined later by small commercial businesses, voluntary associations and mosques. The central location, low rents and the opportunity to refurbish the interiors of commercial spaces as needed made the area a socially significant place for people of limited means. As late as 2002, the area was still being praised for its characteristic diversity of enterprise and activity in a municipal planning document that declared, 'The district thus serves a meaningful social function and is probably significant to the growth of new enterprises. Activities are thus highly significant to the community' (Gothenburg Urban Planning Department 2002: 39).

In spite of the geographically central location, Hisingen Island was regarded as peripheral for a long time. The situation changed when the map of the city was redrawn. Suddenly, in the political vision of 'River-City', with the city centre extending on both sides of the river, this peripheral land was regarded as prime urban land and potentially valuable;

hence, a rent gap was created. Demolition of the area began in 2006, before any official plans for the transformation had been presented; construction for the new Kvillebäcken began in 2010. It was planned as an exclusive residential area aimed at the eco-aware upper middle class. The demolition and subsequent gentrification of the Gustaf Dalén district required carefully orchestrated planning and manipulation to control the description of the problem and its solution. This meant that a few cash-flush players were given, with the city's help, almost total control of planning.

Gentrification as class politics

A new perspective on urban development, in which the 'city as brand' took on an increasingly prominent role, followed in the wake of the shipbuilding collapse of the 1970s and 1980s. Economic growth, once generated by heavy industry, would now be created through a combination of service, tourism and land development. This entails a demand for marketing in which an exciting and attractive inner city is expected to attract capital investments, business and tourists (Harvey 1989). The social infrastructure of the city, its political culture and spatial organization are thus understood as financial assets, where places are valued based on their potential to generate financial gains. In 1996, the Urban Development Department tasked the Trade and Industry Development Agency (TIDA) with answering the question of what character and content the area surrounding Gustaf Dalénsgatan [Gustaf Dalén Street] should have to 'contribute to positive development for Gothenburg and the Gothenburg Region' (Trade and Industry Development Agency 1996). TIDA then emphasized the importance of improving Hisingen's image. According to TIDA, it was partly due to Hisingen's character as an industrial, working-class district that it was not regarded as attractive to the more 'middle-class service enterprises' (Trade and Industry Development Agency 1996: 4). Instead, they held up as a role model the transformation of the southern shore of Hisingen, which had been the central shipyard district and was later converted into high-end residential neighbourhoods and business clusters. In many ways, the TIDA report is symptomatic of an increasingly dominant view of the urban landscape in which focus is on attractiveness and image.

Creating a new identity in terms of a 'strong brand' has been especially important for former industrial cities whose images were often associated with smoking chimneys or dockland cranes. These are now being removed or given new functions as aesthetic trimmings in an otherwise transformed urban space. The land around former port and industrial areas is often considered abandoned and thus regarded as developable to a much higher extent than other parts of the city. These areas are turned into loci of new investment, value appreciation and revaluation of the city. In the new places that are emerging – where knowledge, creativity and consumption, rather than production, are rewarded – older industrial buildings are seen as relics of the past that no longer fit in. These 'abandoned' industrial landscapes become dream landscapes of a sort for politicians, architects and building clients aiming to leave their mark on the city without a second's thought for the past. History will dissipate along with the ashes of buildings, leaving land open for new visions and narratives for the city. Former industrial areas, however, are rarely abandoned and empty even if no one lives there. On the contrary, as in the case of the area around Gustaf Dalénsgatan, they are often used by people whose economic circumstances make it hard for them to become established and enjoy the increasingly upgraded inner city.

When TIDA (1996) writes that it is essential that 'the image of all of Hisingen is improved', it is specifically Hisingen's character as a working-class district that they are identifying as a problem. Improving the image, they argue, is necessary to attract financially strong service businesses. From this perspective, attractiveness is linked to class where the middle class is regarded as creating attractiveness and urbanity. This notion, Neil Smith argues, is at the heart of the frontier mythology. According to this, places should be gentrified by being taken over and 'civilized' through building codes that fit the tastes and preferences of the middle class. In the contemporary Swedish conceptual framework, this is articulated in terms of 'upgrading', 'safety' or 'urbanity'. The belief that the middle class will save the city from decay is part of the mythology, and the first people to move into newly built neighbourhoods are often called homesteaders or pioneers. The term 'urban pioneer', Smith argues, is 'therefore as arrogant as the original notion of "pioneers" in that it suggests a city not yet socially inhabited' (Smith 1996: xiv).

Consolidation of elite power

Today, gentrification is embedded in new governance forms for urban development. It is all about who has power in the city, who can influence decisions and what narratives can be created about a place. Like many other port cities, Gothenburg was hit hard by the shipyard crisis of the 1970s and 1980s, and in response to the economic downturn, the city began adjusting to a globalized economy in which its competitiveness had to be strengthened. The so-called 'Gothenburg spirit' played a key role in this transformation (Holgersson et al. 2010). Deeply rooted in history, this spirit – built on a culture of consensus between the city's political and financial elites – has been instrumental in the ability to make fast decisions and create unity surrounding the city's development (Falkemark 2010). The nature of the Gothenburg spirit has changed over time and in recent decades can be seen as an expression of growing consensus among politicians (across party lines) and business, where the major property owners in particular have been given a central role (Thörn, Franzén, and Hertting 2016).

In Gothenburg, this has opened up opportunities for forming municipal urban development companies that have been given greater influence over urban development. Unlike government agencies, these city development companies act according to market principles. The profit imperative creates opportunities for *political* decisions concerning urban development and land use to be settled between the company and the private stakeholders *before* they formally come up on the political agenda. The burgeoning flora of cooperative forms of management – in partnerships, networks, collaborative projects and (semi) municipal companies – poses a risk of fragmented institutions and actors that are difficult to monitor. This also creates room for economically powerful actors to mobilise and impose demands on development. The risks of a democratic deficit and lack of opportunity to demand accountability are substantial when it is difficult to identify decision paths and who is representing what (Swyngedouw 2005).

The municipal company Älvstranden Utveckling (Riverbank Development), wholly owned and controlled by the municipality, has played a key role in the structural transformation of Hisingen. In 1996, the municipality bought the land next to the river previously owned and managed by the state during the shipbuilding era, and thus consolidated its landholdings along the river. In the 'Älvstrand Model', Älvstranden Utveckling and the building clients with whom they work have a joint 'vision responsibility', which for the latter entails significant influence over planning. One consequence of this, however, is that the lines of demarcation between Älvstranden Utveckling, private stakeholders, the Property Management Administration and the Urban Planning Department, and their individual roles, are blurred. Through the model, Älvstranden Utveckling is not required to comply with the land allocation policy upon which the work of the Property Management Administration is based, and when it comes to the actual planning, the Urban Planning Department has a subordinate role in relation to the company. As a result of this structure, the two major private landowners in the Gustaf Dalén district, NCC and Wallenstam, were given *direct influence* over the development of the area and could impose demands on the extent of development.

Even if the alteration of the Gustaf Dalén district is meant to be understood as the outcome of new forms of collaboration, it must also be viewed in the light of history. In his study of the slum clearance of the

Haga district of Gothenburg, sociologist Håkan Thörn emphasizes the active role played by the semi-municipal company *Göta Lejon* (Thörn 2013). The company was founded in the early 1960s as the 'joint city/private enterprise body for slum clearance in Gothenburg' (Thörn 2013: 165) and was the most hard line proponent of total demolition in Haga. 'Slum clearance' was one of the most central concepts in the art of Swedish urban engineering and referred to 'clearing' neighbourhoods such as Haga, which was described in terms of 'the worst slum in Sweden' (Thörn 2013: 164). The slum clearances of the 1960s and 1970s triggered widespread criticism, particularly with reference to their problematic cultural and social consequences.

Forty years later, the talk about slums returned in coverage of the Gustaf Dalén district in the local daily newspaper, *Göteborgs-Tidningen.* Reconnections to the conceptual framework of the 1960s were made with phrases like 'the slum must be torn down' and epithets like 'shanty town' (*Göteborgs-Tidningen* 2007a, b). This time, as well, there was talk about cleaning up a part of the city characterised by a vibrant social community, but now the spotlights were trained on a dilapidated industrial area instead of an inner city neighbourhood dominated by rundown housing. While the slum clearance – and consequential gentrification – of Haga took place within the confines of welfare state strategies, the teardowns in the Gustaf Dalén district occurred in an altered local context that was more focused on meeting the profit needs of private developers and creating attractive city neighbourhoods for an affluent middle class.

Power over time and space

Time, according to Neil Smith, is a central strategic component of gentrification. The rent gap is created over time where a systematic lack of maintenance and investment make it possible to devalue the buildings and social community of a neighbourhood even as land values may rise. It is all about the right timing for development. The ability to control time confers the power to strategically wait for the right opportunity or to accelerate the decision process when necessary. Control over time is also an important factor in undermining any organized opposition to gentrification (Sakizlioglu and Uitermark 2014). Time has played a central role in the transformation of the Gustaf Dalén district; both the municipality and property owners have made strategic use of time as a factor in controlling developments.

For a long time, the municipality's development plans were at a standstill in the Gustaf Dalén district. Located on Hisingen Island, the land was considered peripheral, despite its geographically central location, and there was little interest in investment on the part of the municipality or private stakeholders. As time went on, the district was put into a deadlock that was maintained through temporary building permits and short-term leases that were extended annually; this put the people doing business in the area in a vulnerable position. The businesses and other activities were never accorded any legitimacy by the property owners or the municipality. The poorly-maintained properties, short-term leases and temporary building permits were turned into arguments that the businesses and other activities were also transitory, even though many had been there for a long time (Olshammar 2002: 93). The fact that the district could be described as 'permanently temporary' was, in a nutshell, not a result of the nature of the activities but of the regulations that had for decades prohibited the district from changing. It should thus be understood instead as 'the result of that a certain type of business, or low status (ability to pay), is actually not accepted in the central districts of the city other than by virtue of its transitoriness' (Olshammar 2002: 93).

The long period when nothing happened and property owners left the district to deteriorate was followed by a speedy and accelerated development process. Tenants with temporary leases were given notice to move and temporary building permits were revoked. With an ambition to craft a plan that did not require a consultation process, Älvstranden Utveckling worked with the major property owners to create consensus among heavyweight authorities, *before* the plan was formally adopted, in order to avoid delay. Legitimizing a rapid and total transformation in the way that happened in the matter of the Gustaf Dalén district is a process with two requisite steps: First, to stigmatize the district in its current state and second, to present a solution and a vision for the future character of the district. The Gustaf Dalén district underwent several stages of stigmatisation, and the pace intensified when the central location was put forth as quality. At first, the district was described as disorderly, littered and dilapidated, only to be characterised a few years later as a slum, lawless and, repeatedly in both media and municipal descriptions, the Gaza strip of Gothenburg. Thus, total clearance emerged as not only necessary but entirely legitimate (see also Kallin and Slater 2014).

The need for swift change was also used to justify the rapid demolition of the buildings in the Gustaf Dalén district once they had been acquired, even though the detailed development plan had not been formally adopted. The aim was to bring about a *fait accompli* as fast as possible and stage the district as empty. The many temporary building permits and short-term leases did not only affect the status of the district, but also the opportunities of the people who were active in the area to have any influence over future plans. The balance of power between the municipality/building clients, on the one hand, and the people who were active in the district, on the other hand, thus became very uneven and opportunities to oppose the development were undermined.

The buildings that were razed to the ground and the people who used to work there had been written out of history, but they were also written out of the future of the district (see also Holgersson 2014a,b). As the plans proceeded, claims were staked on the denuded parcels by placing containers stacked on top of each other adorned with huge advertising banners proclaiming 'Kvillebäcken: A new city district in Gothenburg'. The banners faced outwards, away from the area and towards passersby, with texts like 'With a little imagination, you can see people having a crayfish party on the balcony'. In this manner, the vision project of a new future was begun on site before any new buildings were built. By means of a carefully orchestrated marketing campaign, the myth of a diamond in the rough was created and now discovered, cut and polished for the benefit of the whole city.

Gentrification as ideology

In his last lecture in Gothenburg (29th October 2010) Neil Smith stated: 'Gentrification is about a class-based restructuring of urban space. It is a rethinking of the city. There is a certain possessiveness involved in the process of gentrification and it is a statement of a set of social claims about who does and who does not deserve to live in the city'. Gentrification is now a dominant ideology in urban development. Embedded in the planning machinery, it acts as a strategy for renewal and is emotionally charged with positive images of a safe and peaceful future. In *Lectures on Ideology and Utopia,* Paul Ricoeur (1986) writes that ideology is fundamentally about legitimizing power. The ideology emerges in the tension between the claim to legitimacy by the ruling authority and the belief in the legitimacy granted by its subjects. The role of ideology is to create a view of social order rooted in necessity and naturalism. The marketing campaigns for the new, green, sustainable Kvillebäcken presented the future with the former industrial area as its implicit antithesis. The campaigns transmitted images of white upper middle class people jogging in parks or working with their laptops on their knees in public settings. Our photographic documentation of the place[1] should be read and understood as a direct intervention and an ideological criticism of the municipal planning machinery that is de-politicizing urban development. In contrast to the marketing pictures aimed at instilling a sense of peace and harmony, the photographic documentation seeks to politicize urban development and concretize the consequences of gentrification. The people in the photographs have been directly, and in several cases brutally, impacted by the gentrification of the area while their history has, time and again, been erased by the municipality. Unlike the abstract visions of everyday life offered up by the marketing, the photographs seek to concretise the ordinary lives and work of the people who used to be there.

Figure 19.2 The sign says: "With a little imagination you can smell the freshly brewed coffee in the kitchen of your new three-room flat."

Figure 19.3 Kvillebäcken.

Figure 19.4 Kvillebäcken, Turkish Mosque.

Figure 19.5 Kvillebäcken, Gambian Mosque.

Figure 19.6 Kvillebäcken. The demolition has started.

Figure 19.7 Kvillebäcken, Shop AfroMix.

Figure 19.8 Kvillebäcken.

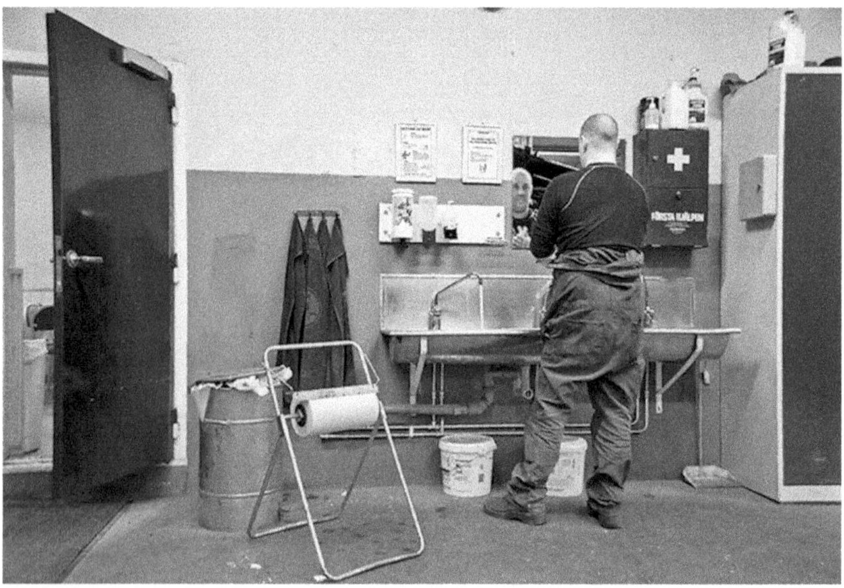

Figure 19.9 Kvillebäcken.

Figure 19.10 Kvillebäcken.

Figure 19.11 Kvillebäcken, MMA Fighting Club.

Note

1 All photographs taken by Katarina Despotovic.

References

Falkemark, G. (2010). 'The Gothenburg sprit: from donations to networking', in H. Holgersson, C. Thörn, H. Thörn and M. Wahlström (eds.). *(Re)searching Gothenburg: Essays on a Changing City*. Gothenburg: Glänta Produktion, pp. 87–93.

Glass, R. (1964). 'Introduction' in Centre for Urban Studies (ed.). *London, Aspects of Change*. London: Macgibbon & Kee, pp. XII–XLI.

Göteborgs Tidning (2007a). 'Här skall 3000 bostäder byggas', 12th April 2007.

Göteborgs Tidning (2007b). 'Nu ska slummen rivas', 31st August 2007.

Gothenburg Urban Planning Department (2002). *Program inför förnyelse av Östra Kvillebäcken*. Dnr 1326/01.

Harvey, D. (1989). 'From managerialism to entrepreneurialism: the transformation in urban governance in late capitalism', *Geografiska Annaler B*, 71(1), pp. 3–17.

Holgersson, H. (2014a). 'Challenging the hegemonic gaze on foot: walk-along as a useful method in gentrification research', in T. Shortell and E. Brown (eds.). *Walking in the European City*. London: Routledge, pp. 207–224.

Holgersson, H. (2014b). 'Post-political narratives and emotions: dealing with discursive displacement in everyday life', in H. Jones and E. Jackson (eds.). *Stories of Cosmopolitan Belonging: Emotion and Location*. London: Routledge, pp. 115–126.

Holgersson, H., Thörn C., Thörn, H., Wahlström M. (2010). 'A critical view of Gothenburg' in H. Holgersson, C. Thörn, H. Thörn and M. Wahlström (eds.). *(Re)searching Gothenburg: Essays on a Changing City*. Gothenburg: Glänta Produktion, pp. 7–26.

Kallin, H. and Slater, T. (2014). 'Activating territorial stigma: gentrifying marginality on Edinburgh's periphery', *Environment and Planning A*, 46(6), pp. 1351–1368.

Olshammar, G. (2002). *Det permanentade provisoriet: Ett återanvänt industriområde i väntan på rivning eller erkännande*. Göteborg: Chalmers Universitet, Sektionen för Arkitektur.

Ricoeur, P. (1986). *Lectures on Ideology and Utopia*. New York: Columbia University Press.

Sakizlioglu, N. B. and Uitermark, J. (2014). 'The symbolic politics of gentrification: The restructuring of stigmatized neighborhoods in Amsterdam and Istanbul', *Environment and Planning A*, 46(6), pp. 1369–1385.

Smith, N. (1996). *The New Urban Frontier: Gentrification and the Revanchist City*. London: Routledge.

Swyngedouw, E. (2005). 'Governance innovation and the citizen: the Janus face of governance-beyond-the-state', *Urban Studies*, 42(11), pp. 1991–2006.

Thörn, C., Franzén, M. and Hertting, N. (2016). *Stad till Salu. Entreprenörsurbanism och det offentliga rummets värde*. Göteborg: Daidalos.

Thörn, H. (2013). *Stad i rörelse: Stadsomvandling och striderna kring Haga och Christiania*. Stockholm: Arena.

Trade and Industry Development Agency [Näringslivssekretariatet] (1996) *Området vid Gustaf Dalénsgatan*, 4th March 1996.

20 Alternative geographies for social action in Medellín

Eryka Y. Torrejón Cardona

Introduction

According to the last National Population and Housing Census (National Department of Statistics, *DANE* for its Spanish acronym: 2005), the population of Medellín is 2,216,830. It is located in a long and narrow 310 km² valley area. For more than five decades, over half its population has been concentrated along the slopes of its northern region. This trend has caused differences in the urban lives of neighbourhoods in the northern and southern zones of the city, which play a determining role in the divisions and social distances in the city today.

Between 1980 and 1990, the city experienced the establishment of an economy of drug trafficking and with this, the local and international effects of a rigorous and ongoing drug policy between Colombia and the United States. Urban life recovered slowly during this period and required a wide range of social programs that were funded with local, national and international resources. According to experts, the drug trafficking economy became established in Medellín due to the unequal distribution of wealth, city dwellers' access to social services (education, employment, healthcare, housing and recreation) and a standing and close relationship of local governments with the city's business elite.

According to James Parsons (1990), this mix of factors was instrumental for how the industrial development of Medellín at the beginning of the 20th century would take its toll decades later. The ethos of *Paisas*[1] as promoters of values rooted in the Catholic religion, rooted in a capitalist economy and in the ideal of a nuclear family, took on a regionalist sense of supremacy that has stood out. *Paisas* are the equivalent of 'South American Yankees', according to the term used by Parsons.[2]

Between 2000 and 2015, different local government administrations have taken efforts towards recovering the city zones[3] with public infrastructure works and social investments. They have done this in response to the consistent demands made by grassroots social organizations, by the community in general and due to the social crisis that has taken the lives of many people throughout previous decades due to the confrontations of legal and illegal armed groups. Among these efforts, the following stand out: the social programs and infrastructure of the city's northern zone, large-scale road infrastructure projects, the city's promotional campaigns and the regular hosting of international events: the 8th Congress of Academies of the Spanish Language (2007), the 11th Medellín South American Games (2010), the 7th Social Urban Forum (2014) and the 7th Latin American Council of Social Sciences – CLACSO – Conference (2015).

These efforts towards capital recovery led by local government administrations, especially between 2005 and 2015, concur with the characteristics and phases of the neoliberal city as presented by Neil Smith throughout the majority of his works. For this reason, this chapter will present a case study on the city of Medellín, Colombia, which will serve as an example of the gentrification processes that occur in Latin American cities.

This chapter is divided into three parts. The first section introduces the characteristics of Medellín as a neoliberal city. It begins with describing local and international capital recovery efforts of large-scale construction projects conducted under the authorization and joint financing of local government and private and public agencies. It then details the developments and scalar changes in the city's uneven geographic developments and ends with how the local culture is marketed (landscapes, identities and images) through the internal and external promotion of the city in the form of city marketing, in an effort to generate urban income. Due to the results of neoliberal processes, the second section highlights the effects of these processes on social organizations and remarks the role of grassroots organizations in the criticism of neoliberal cities. The third section provides conclusions and summaries on: (1) the constant relationship between the local government and financial capital in the urban development of Medellín; (2) the construction of different and varied scales within the city; and (3) the generation of urban income through the marketing of the local culture in the form of city marketing and its counter stories as emerging expressions of grassroots organizations, which are fuelled on unique and attention-attracting social actions where a global sense of place is a constant need.

Andean neoliberalism

In the first half of the 20th century, Medellín was characterized by the development of its industrialization process. It also established and furthered its urban growth, which experienced great social inequalities throughout the 1970s. Medellín went from a rapid industrial growth process and its eventual crisis to a post-industrial period characterized by the breaking down of worker and social rights in general, the impoverishment of its most vulnerable residents, a rising humanitarian crisis and a service-based and capital flow-based economy. Decades later, in Central and South America, a heavy presence of locally based companies that offer financial, power and mining services has risen (Bancolombia, Empresas Públicas de Medellín and the cement company, ARGOS S.A, respectively).

Between 1990 and 2010, a surge of road infrastructure, education and technology projects were carried out. For example, local government administrations established the conditions for public and private local capital to finance large-scale transportation projects in the city. This development of large-scale projects began with the Metro of Medellín (1995), which, together with the following, consist of Medellín's current public transit system: the Santo Domingo Metro Cable (2004), the San Javier Metro Cable (2008), the Sierra Metro Cable and the Pan de Azúcar Metro Cable (under construction), the Metroplus (a Bus Rapid Transit system, 2011) and the Ayacucho Trolley (2015). Other infrastructure projects such as the Tunnel of the West (a route that leads to the Caribbean Sea), the Library Parks located throughout the different zones of the city and Arví Park are examples of the capacity for local funding in the construction of large-scale public works and, therefore, according to the contributions of Neil Smith (2005) and David Harvey (2005), are examples of the capacity for urban income.

The first characteristic of the neoliberal city is the continual transformation of space in order to generate greater levels of urban income through local and international investment flows. This is evident in the case of Medellín due to the investment of public resources in local works and to the ability of local capital to reach foreign markets. At present, however, the change occurring in the creation of urban income is characterized by the reproduction of capital through the constant transformation of the use of urban grounds. The city's zones where these works have taken place have also fostered change in the different forms and styles of urban life.

The second characteristic of the neoliberal city refers to the forms of urban life that have been fostered throughout the different times and scales of Medellín's recent history. From the perspective of dynamic socio-spatial organization, these ways of urban life are scales that are under development. The administrative organization of Medellín in the 1990s is therefore not just the result of processes of social exclusion and gentrification in the city's northern and southern zones, but also the result of divisions and social distances to the scale of neighbourhoods in slope or flat zones.

The slope zone as a way of urban life is to become a space of scalar relationships from within the city with other neighbourhoods and from outside the city with the rest of the world. This scalar relationship of the slope zone is characterized by the prominence of the social actions of youth organisations (anti-military groups, musicians, cultural groups and sports organisations) that take on a spatial identification from the margins of the city and that are in constant reconstruction.

Scalar relationships, and specifically the continual reproduction of new social scales within the city, are one of the characteristics that identify several of the alternative geographies that persist in Medellín, in contrast to an organisation whose administration is centralised, in the planning of local governments and sources of private capital. These types of organization can be a socio-spatial organisation that takes social action in the city's zones and neighbourhoods, a socio-spatial organisation that is based on civic participation networks where space provides an opportunity for identification and social branding in the city and in the world, and a socio-spatial organisation that focuses on the urban border conditions of the city's peripheral zones and border. In some cases, in Latin American cities such as Caracas, Lima and La Paz, there are a variety of citizen actions, where to the exclusion and marginalisation of other organisational models in the city, the slope zone provides an opportunity for urban life. While the slope zone is a social universe, it can also become a social prison. This is the difference between a gentrification process of capital investments in public works to divide and separate the city's zones or in projects and local processes that make it possible for the slope zone to be a place, in relation to others, in the exchange of identifying elements that are experiencing ongoing reconstruction.

The neoliberal city is characterized by the reconstruction of social scales. In the case of Medellín, micro-social scales as forms of urban life stand out as alternatives to Cartesian and official organizations. Micro-spatial scales correspond to the spatiality where daily life occurs and refer to a sense of what is locally conceived and understood through spatial perception. Micro-spatial scales are found in the spaces connected to the daily lives of people such as the stress, the homes, the neighbourhoods and the communes.

The third characteristic of the neoliberal city, and what has become an ongoing practice among local government administrations in Medellín, is the generation of urban income through promotional campaigns for the city, or what has been called 'city marketing.' This third characteristic was common during the decade following the 1990s, and again between 2005 and 2015, when advertising efforts clearly promoted the city as a space of urban life for business, entertainment and education among other urban uses and services. Neil Smith (2002, 2005) refers to this as the marketing of local culture, and, in the case of Medellín, this is reflected in the use of a series of cultural elements such as mountainous landscapes, the image of a modern city, and religious and economic symbols. This level of specialized publicity, which focuses on an interpretation of the supply and demand of cities, is one of the indicators in which urban life becomes the expression of a context, marked by both the demands of the global economy and the local demands of social inclusion. Medellín is not an exception and, in recent years, has displayed great interest in protecting its image as a city that subscribes to international needs. In exploring this field of publicity and of local policy that uses publicity as a spatial practice, its most current promotional efforts focus on selling the city as a space for capital investment, urban renewal, constructing spaces for large-scale events and the provision of urban infrastructure (such as libraries, bridges and road infrastructure).

Drawing on these three characteristics of a neoliberal city, there appear to be close ties between the zones in Medellín where large-scale infrastructure projects took place and between the reconstruction of micro-social scales and the city's promotional content. These promotional spaces correspond to the neighbourhoods of the northern zone, the ones along the border of the urban perimeter of the city and to some elements of the local culture.

The different advertising campaigns that have taken place throughout the last three decades have stood out due to both their international scope and to the counter-stories that arise from within the city; due to a combination of scalar relationships, these campaigns express both the demand for local efforts to take place due to the international demands of globalization and a global sense of place, or in other words, spatial justice (Soja 2010).

Social action and socio-spatial organization in Medellín

The social actions of grassroots organizations in Medellín provide an alternative geography to the administrative methods by which the territory is structured. The social space that is the basis of these actions are the neighbourhoods and communes of the northern zones of Medellín. An example is *La Alianza para el Desarrollo de la Zona Norte* [the Alliance for the Development of the Northern Zone], which is made up of several organizations that have provided service to the community for more than two decades, among which include *La Corporación Cultural Nuestra Gente, Fundación Sumapaz, Corporación Con-Vivamos* and *Corporación Mi Comuna Dos*, among others.

For these social organizations, the neighbourhood has served as a base for promoting scalar relationships with other neighbourhoods and with the communes from other zones of the city. Neighbourhood grassroots organizations are therefore those that give a new spatial construction to the city, which is now promoted through local and global geographic consciousness (Albet and Benach 2012). The global sense of place of neighbourhood grassroots organizations is characterized for preserving the urban life of the neighbourhood; it prioritizes the non-military citizen participation of youth and children through arts, culture, sports and the continual advocacy for the local identities of marginal and peripheral neighbourhoods. A clear sense of localisation and geographic consciousness, local neighbourhood identities in constant reconstruction and new social, spatial and time relationships with the city and with the world are what stand out in the social actions of grassroots organizations.

Some neighbourhoods that serve as examples of these characteristics of social action include Barrio El Picacho (in the north-western zone of Medellín), home to *La Corporación para el Desarrollo Picacho con Futuro,* and Barrio Santo Domingo Savio (north-eastern zone of Medellín), home to *Corporación Con-vivamos.* With a history of over two decades of community service, these organizations have advocated for alliances with other local and international organizations. They have had experience in the community planning of their territories during crucial times of intervention in the construction of mega projects led by local government administrations. Their efforts focus on protecting the collective rights of neighbourhood residents against the effects of physical interventions in their territory.

Conclusions

Together with the contributions of Neil Smith (2002, 2005), the basic characteristics of a neoliberal city have been examined: in this case, the characteristics of Medellín in Latin America. This chapter has demonstrated how, throughout the last three decades, the investment of local capital in large-scale projects has been a constant factor in the urban development of the city. Along with this trend, through the use of elements such as mountainous landscapes, infrastructure projects for transportation and large-scale events, the city shows how the marketing of local culture through 'city marketing' generates the reproduction of urban income. Lastly, the grassroots neighbourhood social organizations of Medellín reconstruct its neighbourhoods, communes and zones.

Other forms of socio-spatial organization emphasize the localization and social consciousness of marginal and central positions (Lefebvre 1983), local identities that are being constructed based on the anti-military practice of art, culture and sports, and new social, spatial and time relationships in which the subjects experience the turmoil of what is considered local and yet the sense of place inevitably requires them to connect with places that are located outside of the neighbourhood and of the commune.

The uneven geographical development between the city's zones is increasing, and the conditions and the (geographical) peripheral consciousness of grassroots organizations are creating a new social geography of the city as a counter story to the global development of the neoliberal city.

Notes

1 *Paisa* is the demonym in Spanish for the inhabitants of Medellín as well as for those from other surrounding departments and cities in Colombia.
2 Parsons describes the key elements of the *Paisa* ethos. For him, it is analogous to the mountainous geography that characterizes the region of Antioquia. Paradoxically, it is a growing capitalist economy, yet it has a collective mindset that is closed in on itself.
3 A zone is one of the administrative units that the city is organised into. Each zone consists of several communes, and each commune consists of several neighbourhoods. Medellín has six zones: northeast, northwest, mid-east, mid-west, southeast and southwest. This chapter uses this unit to illustrate the geographic location of the northern, central and southern zones.

References

Albet, A. and Benach, N. (2012). *Doreen Massey. Un sentido global del lugar.* Barcelona: Icaria.

Departamento Nacional de Estadística (2005). *Censo Nacional de Población y Vivienda.* Available at: www.dane.gov.co.

Harvey, D. (2005). 'El arte de la renta: la globalización y la mercantilización de la cultura' in D. Harvey and N. Smith (2005). *Capital financiero, propiedad inmobiliaria y cultura.* Barcelona: Museu d'Art Contemporani de Barcelona & Servei de Publicacions de la Universitat Autònoma de Barcelona, pp. 29–58.

Lefebvre, H. (1983). *La presencia y la ausencia. Contribución a la teoría de las representaciones*. Mexico: Fondo de Cultura Económica.

Parsons, J. (1990). *La colonización del occidente colombiano*. Medellín: Universidad de Antioquia.

Smith, N. (2002). 'New globalism, new urbanism: gentrification as global urban strategy', *Antipode*, 34(3), pp. 427–450.

Smith, N. (2005). 'El redimensionamiento de las ciudades: la globalización y el urbanismo neoliberal' in D. Harvey and N. Smith (ed.). *Capital financiero, propiedad inmobiliaria y cultura*. Barcelona: Museu d'Art Contemporani de Barcelona & Servei de Publicacions de la Universitat Autònoma de Barcelona, pp. 59–78.

Soja, E. (2010). *Seeking Spatial Justice*. Minneapolis: Minnesota University Press.

21 Alternative narratives from an invisible city

Gentrification, counter-proposals and women activism

María del Socorro Pérez Rincón Fernández

Women in disadvantaged areas that are temporarily forgotten by the capital are silently involved in everyday innovative practices that transgress hegemonic urban policies, participating in creative strategies that focus on the ways spaces are used and the existing urban structure. The aim of this chapter is to show the involvement of women fighting against the renovation works that drive gentrification, while defending their rights in the places they live.

The analysis was carried out based on urban projects (2005–2009) that have transformed the landscape of the Ciutat Vella, in the context of large-scale public investment in partnership with private capital in Barcelona (Spain). The ageing of the local population and the cultural diversity of the immigrant population in central districts are an uncomfortable social reality for property developers. These arguments are used by the media to show the 'ghetto' conflict in the old city centre and the need to domesticate the landscape using small-scale urban measures aimed at an intensive transformation of the urban environment.

This text is divided into four sections. The first section provides context for the case study and describes the urban intervention projects in Barcelona's city centre. The second part shows the close relationship between transforming the landscape for the tourist economy, urban protest movements and the role of women in underprivileged neighbourhoods. Finally, the paper shows two cases of the revolutionary role of women in the Casc Antic and Barceloneta districts.

The Ciutat Vella, urban intervention in 'underprivileged neighbourhoods'

The category of 'underprivileged neighbourhoods' is used by local governments to identify areas that are considered socially conflictive and that are a priority for local and European funds aimed at urban regeneration. These neighbourhoods feature contradictory economic, cultural and social dynamics that mean they are generally located in central areas that are first stigmatised before being subject to urban regeneration to attract the tourist industry.

This is the case for the historical city centre of Barcelona, as from the mid-19th century it has been subject to several renovation projects that represent conflicting desires in terms of the area's cultural heritage, providing a living space for the working-class (López 1986), and acting as symbolic centre of power. On several occasions, the Ciutat Vella, the heart of the city of Barcelona, has been subject to large-scale remodelling to transform its image and to show it as the financial centre of a modern city with new functions (Benach 2009).

The significant urban transformation in the Ciutat Vella has always seen the poorest residents move, and this trend has continued to the present day. The maximum population density of the Ciutat Vella reached its peak in 1950, with 581.4 inhabitants per hectare: an important feature was the population of over 65-year-olds, who made up over 26% of inhabitants. Social problems such as delinquency, prostitution and, later, drug use extended throughout different sectors of the central neighbourhood, which deteriorated due to a lack of investment by building owners, who used to own factory property, and due to the concentration of a poor population renting their living space. Rooms were also sublet, housing was broken up, and Barceloneta was divided (Tatjer 1973). After years of degradation and lack of urban investment, the restoration of democratic local councils in 1980 saw processes begin to carry out the regeneration of the historic city centre, in association with the 1992 Olympics and building a new image for the city (Benach 1993).

We must not forget that the old housing estate in the Ciutat Vella was the leading provider of affordable rental sites, both for permanent residences with frozen rent prices and former tenants who came to live in the mid-1960s, and as housing for immigrants. On separate occasions, this has generated great contrasts in the same neighbourhood: On the one hand, marked pockets of poverty have resulted in labelling certain areas as 'ghettos' and, on the other hand, the permanent attempt to renovate the estates and remove the poor people to take advantage of the higher land value and stimulate the potential property market. Ciutat Vella has undergone gentrification as developers identify an economic opportunity on which to capitalize its landscape potential value (Smith 1979).

Since the 1980s, the Plans Especials de Reforma Interior [Special Plans for Internal Reform] and the Àrea de Rehabilitació Integrada [Integrated Rehabilitation Area] have been used to rebuild the districts and modify the spontaneous population of a high concentration of immigrants and disadvantaged populations (García and Vilanova 1989). In each operation, an insufficient percentage of protected property was added in comparison to the property destroyed. At the same time, privately owned land was freed up to create public spaces and increase the value of the property that, in addition to being profitable, was justified through a political discourse praising reurbanisation as a requirement to improve living conditions and create a safer environment.

Areas with urban decay are represented as obsolete, conflicted spaces in order to promote new centralities (Benach and Tello 2004). In 1996, Mikel Aramburu (Aramburu 2000) and Nadja Monnet (Monnet 2000) carried out anthropological studies in the Ciutat Vella to analyse the popular discourse between residents, showing how the discourse labelled urban spaces as ghettos, stigmatising them as dangerous and unhealthy. This discourse does not correspond to residents, but rather to those who pass through these spaces, who feel the stigma of considering them conflictive spaces due to the number of immigrants. The institutional discourse and that of many residents is handled so that immigrants are made responsible for the urban decay of the area they are living in or frequenting. Mikel Aramburu (2000) described a resident protest in 1997 in the Ciutat Vella with the slogan 'the neighbourhood is a ghetto because of the city council', and with this phrase, the protesters linked the physical deterioration of the district to the high concentration of immigrants. If we update this slogan to 2009, we can also quote the campaign in Raval and Poble Sec, which has the slogan 'Volem un barri digne' [We want a dignified neighbourhood]. Again, immigrants are blamed for urban decay and uncivil behaviour. At the same time as the planned urban transformation, a 'social cleansing' is carried out of the groups in the most precarious housing situations, reinforcing the process of gentrification.

Gender representation in resistance from residents and the transformation of the urban landscape for tourism

The conflict between public and private is shown openly when urban intervention proposes reforms that will radically transform the urban landscape, organising it for tourist use and transforming daily life and relations between the inhabitants. The new landscape often leaves the old structure as an urban layout but changes the filled and empty urban spaces. The process of reurbanisation temporarily abandons spaces that become 'temporarily abandoned' while its functions and public and private spaces are redistributed to attract a new ideal population. Who wins and who loses with each new reorganisation of space? How does social mobilisation occur in urban spaces that experience an intense urban transformation? Who mobilises and why? What role do women play? In order to respond to these questions, we should consider that the political actions of women in a place goes through a symbolic conflict between the subjective perception of personal needs and the participation of women in public matters (Feldman and Stall 2001).

In urban resistance movements, women make a very close link between the public and private spheres (Wekerle 1984, Naples 1992). There is a functional continuity that is very sensitive to any changes in daily life, such as an extension to family reproduction and to identity. This feminine sensitivity and capacity for resistance in order to protect the infrastructure of daily life, which challenges the feminine mystique (Friedan 1974), is almost invisible in institutional discourse due to the discomfort it causes.

In the discourse of transformation in the Ciutat Vella, there has always been an idea of changing the social profile of the inhabitants as a solution to urban decay. From 2004 and the start of the government funding programme for deprived neighbourhoods known as the Llei de Barris [Neighbourhood Law], the actions carried out in the Ciutat Vella mean that this public intervention takes a different shape in the neighbourhoods. In the Casc Antic, it further consolidates the tendencies for property speculation around the Santa Caterina market and the streets nearby with historical buildings such as the Palau de la Música, and in Barceloneta, there is a high level of pressure on working-class housing (the *quart de casa*).

All of these plans aim to find a structure in a new perception of public space for a project aimed at temporary 'tourist' inhabitants. In this new framework, there are singular goals designed by international architects, such as the Santa Caterina market, designed by Enric Miralles and Benedetta Tagliabue, and built between 1997 and 2005. These architectural elements reuse the traditional landscape, converting facilities that are for everyday use, such as markets, into contemporary tourist icons and strategic urban weapons that reconfigure the relationship between public, semipublic (establishments that attract the public) and private spaces, in order to create a dynamic that magnifies a type of specialised commercial offer in key commercial areas.

The massive destruction within the historical urban fabric is conceived as a strategy to give greater visibility to architectural works. It gives a new reading to the symbolic power of the markets: from a public space for everyday use and social encounters, to a unique architectural, multi-function building, replacing a large part of its traditional purpose with a specialised culinary offer. This reform of the markets is part of a design for a new strategy aimed at tourists and a reformulation of the space, in order to offer tourists something that unites the city and gastronomy, as a planning 'recipe' that focuses on a creative tourism niche, in which the main resource is the intangible heritage of the markets, adapting tradition to a new tourist offer (Figure 21.1).

This strategy for defined routes predicts where tourists will go and what they will consume in a city with medieval streets, so they move around in a safe way and without getting lost in the historical labyrinth by simulating these paths inside a shopping centre. Women, who have historically been the users of these markets, are the first to notice the transformation of the landscape and the effects of this type of remodelling on daily consumption patterns: the death of small businesses and products on the market stalls getting more expensive. In the case of the Santa Caterina market, the market's surroundings have divided the everyday experience of a simple walk around. However, thanks to the movements of the inhabitants of the Casc Antic there is still a social fabric that resists the attacks on this predatory model of reurbanisation.

Figure 21.1 Santa Caterina market.
Source: Photo by Socorro Pérez Rincón.

By incorporating women's activism as a vector of analysis, it is possible to explore the versatility of the space and help to question the projects for urban transformation from the perspective of the everyday urban landscape. In every one of the neighbourhoods in the Ciutat Vella, reclaiming the space has taken on different nuances: the relationship between reorganising the public space, facilities and housing are dynamic factors that, when losing the focus on the common good, stimulate resistance to prevent the landscape becoming trivialised and turning it into a consumer brand.

In the two cases we will analyse in this paper, we will show dedicated women activism defending public spaces, public and private assets, which are fights related to basic rights and a strong resistance to imposing the model of a tourist city.

From being overrun to creating a park: El Forat de la Vergonya

In a progressive history of demolitions, at the end of the 1990s, one of the most famous counter-projects in Barcelona, and throughout Europe, took root. Its origins lie in the deterioration of society for the inhabitants in the Santa Caterina neighbourhood, caused by demolition work. The locals organised themselves after finding out that the city council was planning on building a car park due to a surplus of underground space and the proximity to the Santa Caterina market, in a space left after the demolition of over 1,000 homes.

El Forat de la Vergonya [The Hole of Shame] is located in the centre of the Casc Antic. This urban space represents one of the most important recent urban resistance fights in Barcelona. The protest started, countering the institutional participation process, to rescue the space created by the destruction of 1,078 properties in the district (70,000 m^2 of floor space), with a meeting held by local residents. Women were present throughout the protest process to convert the empty space into a park: as leaders of the neighbourhood organisations (Mas and Verger 2004), and as users of the same space.

The aim of the neighbourhood platform was to rescue the large empty space and to turn it into a community park to create a meeting point for the locals who had resisted the large-scale demolition works, and this site was given the name El Forat de la Vergonya. This counter-project brought together the contradictions of two different viewpoints in the same place: the conflict provoked by the city council's functional perspective on the neighbourhood and the social expectations of the locals to turn it into a communal park.

In addition, this empty urban space represented proof of structural violence in the urban renovation process, and the intense social unrest concentrated on this scar on the urban landscape managed to bring many different groups together in a coalition of movements called 'L'Espai d'Entesa' [The Space of Understanding]. The local inhabitants occupying the empty space to turn it into a park represented a powerful symbol of social protest against neoliberal urbanism. In 2002, local residents, tired of speculation and the fragmentation of the daily life caused by the demolitions, planted a symbolic pine tree in the empty space to represent the founding of the park. As part of a decision taken by the residents, this was the moment the park collective for the Forat de la Vergonya was born.

This neighbourhood counter-proposal was organised in different areas. It was made up of a child's play area with a 'pirate ship' made of recycled objects, an area with fruit trees planted by the locals, football and basketball pitches, a stage and a community garden. All of these uses represented community values and the reappropriation of the abandoned landscape left by the demolitions. The edge of the park was surrounded by self-built street furniture, with flower boxes made of recycled bicycles and the benches around the edge marking the new intensive use of the square as a living space. The Casc Antic, with its sinuous street layout and high density of housing, is not conducive to shared spaces, and the new configuration of the Forat de la Vergonya space was an opportunity to create a shared leisure area and a space for neighbourhood organisation. In addition, the use and appropriation of the space was a powerful political act.

As a result of years of fighting from the local inhabitants, an agreement was reached for the Forat de la Vergonya to be officially constituted as a neighbourhood park. The pressure of the neighbourhood movement managed to start a debate and prioritise social projects such as facilities for young people, an old people's home and the Forat de la Vergonya square. As

Figure 21.2 Community and protest activities in the Forat de la Vergonya, 2004.
Source: Photo by Socorro Pérez Rincón.

a condition to allow for formal urbanisation, Espai d'Entesa asked the city council to respect the following premises for use and design:[1] not building a square with a hard surface, consolidating a leisure area for locals, respecting the garden and the pitches, keeping a play area for children, building multi-purpose facilities and avoiding the park becoming privatised with bars and restaurants. Even though the park facilities were temporary, the basic design started by the locals formalised through public investment. The city council renamed the square 'Pou de la Figuera' [Figtree's Well] in order to erase the social roots of the project, as the name itself reflected two things: on the one hand, empty urban spaces as a result of urban violence and, on the other hand, social reconquest from the margins of a neoliberal city (Figure 21.2).

Barceloneta and the 'Elevator Plan': property speculation

Barceloneta is a district that is part of the Ciutat Vella and that was founded gradually in the 18th century on a sandbank reclaimed from the sea. The district was founded initially as compensation for the houses demolished in the La Ribera neighbourhood to build the Ciutadella park. It was to be inhabited by military engineers, and later by artisans and fishermen. This population had fewer resources, and resulted in the properties being divided several times. This resulted in the typical type of property in the Barceloneta district, known as *quart de casa* (quarter house, measuring $35m^2$).

The relation between women and public spaces in the Barceloneta district is so close thanks to the small size of the liveable domestic area, which means that practically speaking, the street becomes a functional extension of the home and work area. In the past, men and women could be seen repairing fishing nets in the street. The tradition of spending time on the streets has still been conserved, but with other activities. During public festivals, sections of the grid layout of the streets in Barceloneta are traditionally used as a temporary public space: The streets are decorated, and dances and communal dinners are organised during the annual neighbourhood festival *Festa Major*. This structure of organisation and communication, which has been put in place over the years, is mainly made up of women who direct and carry out a large part of the activities associated with the parties, such as the street decorations and the organisation of communal dinners.

The neighbourhood association l'Òstia was founded from the street parties organised in carrer dels Pescadors, a group of women who organise the activities for the annual local neighbourhood celebration, or Festa Major, and are in touch with the problems in the neighbourhood. These women were aware of the danger posed to the district as a result of property speculation caused by tourism and changes to both the Urban Plans approved in the 80s and to the 'Pla d'ascensors' [Elevator Plan] (2007) to radically transform the urban landscape in favour of tourism.

The Pla d'ascensors is the popular name given to the changes to the Pla General Metropolità (PGM) [General Metropolitan Plan] for Barceloneta in 2007. Approving this modification opened significant pressure on property, as it allowed property owners to carry out small reurbanisation projects to join buildings together and install elevators, and the rehousing the resulted from joining property together would be the responsibility of those managing each project. The Pla d'ascensors was a modification to the PGM that marked Barceloneta as an area for urban action that would mean the mass destruction of traditional property.

The Barceloneta Alerta association is one of the first organisations that started an analysis to identify the problem that many older people in the district found it difficult, or simply could not, leave their flats due to the tight staircases in the *quart de casa*. This study on social accessibility was looking for support for local residents and ended up as the pretext and preamble to the proposal for improving vertical access with lifts in all the properties at any urban or social cost. In addition, a year after the modification to the PGM, it was decided that Barceloneta would benefit from the Llei de Barris. The organisations l'Òstia and the Plataforma d'Afectats de la Barceloneta immediately reported the danger of the large direct and indirect investment needed for the transformation.

Figure 21.3 'Another plan is possible'. L'Ostia and the Plataforma d'Afectats de la Barceloneta meet every Saturday in the Barceloneta market.
Source: Photo by Socorro Pérez Rincón.

In 2009, the Barceloneta district was chosen to receive the Llei de Barris investment. This investment, as mentioned above, ran the risk of indirectly cofinancing the objectives of the Elevator Plan. The fight of the women in l'Òstia was invigorated as a list of non-agreed projects appeared under the promise made by the City Council to create a large participative workshop in Barceloneta, which, little by little, would redefine the future of the Barceloneta district.

This district is probably the only one that has been against receiving investment from the Llei de Barris due to the risk of indirect financing as indicated in the modification to the Pla General Metropolità, the Pla d'ascensors: a further example of plans that were not socially agreed and of local public action that prioritises private interests over public ones. Given the exclusion of residents from improving their district, they joined together to form a protest movement including the neighbourhood association l'Òstia and the Plataforma d'Afectats de la Barceloneta. These two organisations were founded first and foremost by women who saw the risk of this plan associated with expelling the most vulnerable residents (Figure 21.3).

The impact of removing residents and the proliferation of tourist apartments is a trend that continues to rise, and in 2014, it spurred a mass demonstration in Barceloneta against tourist apartments (elPeriódico 2014).

The social decoding of the Pla d'ascensors passed through several stages. First, it went through a community exercise to analyse how the urban measures would be applied, and then to what extent they would have an impact on society in different real-life scenarios. The market square known as 'La Repla' and the carrer dels Pescadors were the main points to share information to residents, after finding ways to share information, detect at-risk residents and see how properties were being turned into tourist apartments. In this way, it was clarified that the real objectives of this 'abstract' plan were to place urban planning in private hands and this meant that residents

Figure 21.4 No neighbor outside the neighborhood. L'Ostia and the Plataforma d'Afectats de la Barceloneta protest march.

Source: Photo by Socorro Pérez Rincón.

needed to challenge the changes to the PGM.[2] In the comic strip shown in Figure 21.4, there is a visual metaphor for how the city council came up with the Pla d'ascensors (Figure 21.4).

Showing the scope of the damage was the result of a significant community effort carried out by the women of l'Òstia and Plataforma d'Afectats de la Barceloneta, led by Emília Llorca, using resources such as models to show residents the practical dimensions of the plan, and organising debates with specialists, mass demonstrations and even performances in the public space to show the process of property speculation.

The residential counter-proposal consisted of putting different strategies into practice to show and symbolically represent the effects of the plan in terms of property speculation. Among the resources used to show residents the effects of the plan, a comic was designed, titled 'the elevator plan under a magnifying glass' (Velázquez, Pérez-Rincón and Barrechina 2008), which can be seen in Figure 21.5. The comic showed the social impact of the plan and covered different scenarios such as 'future landscapes' created by the elevator plan, to show how the politically correct discourse of 'accessibility' was twisted to put the social life of the working-class Barceloneta neighbourhood in danger.

Conclusion: women resistance to neoliberal urbanism

We have identified different protests to avoid the symbolic expropriation of public spaces in the Casc Antic and private spaces in Barceloneta. These interventions are carried out with a camouflaged discourse that argues for

Figure 21.5 The elevator plan under a magnifying glass.

Source: Illustration by Marta Barrechina, storyline by Socorro Pérez Rincón and Susana Velázquez.

improved access, participation and gender equality, when, in reality, they were encouraging urban property speculation.

Women participate in everyday life through different types of collective action, without the fact they are 'women' being verbalised, although they insist on sensitivity with regard to the destruction of the space around them. These women dare to publicly question urban policies and the simulated participate processes that back them up.

Despite advances made in terms of equality and urbanism from the perspective of gender, women are yet to play a significant role in the participative processes for rehabilitation, and in the best-case scenario, their role is merely a consultative one and not linked to the decisions that really build the city. However, women's capacity to fight in a landscape that is constantly being redefined through urban intervention for tourist use is clear. In this landscape, despite there being sufficient resources available to improve the community environment, a large part of the public budget is invested indirectly in gradually emptying the city. This situation leads us to question the essence of the ways in which so-called marginalised neighbourhoods are dealt with, emphasising the role played by women as an invisible force that does not receive recognition in the public sphere despite urban policies that highlight the need to include gender as a priority.

Notes

1 Agreement signed between the different organisations involved in Espai d'Entesa in 2005.

2 See Masala (2008) for an interview with Pere Comes, lawyer for the neighbourhood association l'Òstia, about the appeal against the modifications to the PGM.

References

Aramburu, M. (2000). *Bajo el signo del Gheto: imágenes del inmigrante en Ciutat Vella*. PhD Thesis. Bellaterra: Universitat Autònoma de Barcelona.

Benach, N. (1993). 'Producción de imagen en la Barcelona del 92', *Estudios Geográficos*, LIV(212), pp. 483–505.

Benach, N. (2009). 'El centro histórico de Barcelona bajo presión: flujos globales y derecho a la ciudad', in R. Tello and H. Quiroz (eds.). *Ciudad y diferencia: género, cotidianeidad y alternativas*. Barcelona: Bellaterra.

Benach, N. and Tello, R. (2004). 'En los intersticios de la renovación. Estrategias de transformación del espacio y flujos de población en Barcelona', *Revista de Geografia*, 3, pp. 93–113.

elPeriódico. (2014). 'Un millar de personas se manifiestan en la Barceloneta contra el turismo "low cost"', *elPeriódico*, 23 August. Available at: www.elperiodico.com/es/noticias/barcelona/millar-personas-manifiestan-barceloneta-sabado-tardecontra-turismo-incivico-3467735.

Feldman, R. and Stall, S. (2001). *The Dignity of Resistance: Women Residents' Activism in Chicago Public Housing*. Cambridge: Cambridge University Press.

Friedan, B. (1974). *La mística de la feminidad*. Madrid: Júcar.

García, E. and Vilanova, J. (1989). 'Distribució territorial de la població i de les activitats econòmiques', in Promoció de Ciutat Vella (Ed.), *Primeres Jornades Ciutat Vella: revitalització urbana, econòmica i social*. Barcelona: Ajuntament de Barcelona.

López, P. (1986). *El centro histórico, un lugar para el conflicto: estrategias del capital para la expulsión del proletario del centro de Barcelona, el caso de Santa Caterina y El Portal Nou*. Barcelona: Edicions Universitat de Barcelona.

Mas, M. and Verger, T. (2004). 'Un forat de la vergonya al Casc Antic de Barcelona', in AA.VV (ed.). *Barcelona marca registrada. Un modelo para desarmar*. Barcelona: Virus.

Masala. (2008). 'Cualquier defensa ante el plan de los ascensores pasa por impugnar el PGM', *Masala*, 42, p. 9.

Monnet, N. (2000). 'El uso del espacio público por parte de los nuevos habitantes del Casc Antic de Barcelona. Continuación e innovaciones' *Scripta Nova-Revista electrónica de geografía y ciencias sociales*, 69–48 Available at: www.ub.es/geocrit/sn-69-48.htm.

Naples, N. (1992). 'Activist mothering: Cross-generational continuity in the community work of women from low-income neighborhoods', *Gender & Society*, 6(3), pp. 441–463.

Smith, N. (1979). 'Toward a theory of gentrification: a back to the city movement by capital, not by people', *Journal of the American Planning Association*, 45, 538–548.

Tatjer, M. (1973). *La Barceloneta. Del siglo XVIII al Plan de la Ribera*. Barcelona: Los Libros de la Frontera.

Velázquez, S., Pérez-Rincón, S. and Barrechina, M. (2008). 'El plan de los ascensores a vista de lupa', *Masala*, 42, pp. 11–12.

Wekerle, G. (1984). 'A woman's place is in the city', *Antipode*, 16(3), pp. 11–19.

22 The onslaught against the Greek squatting movement and the value that it produced

Myrto Dagkouli-Kyriakoglou

Introduction

In 2012, the Greek right-wing government announced its attack against the political squats. This attack aimed to eliminate an urban movement that, in comparison to that of Northern European countries, was developed more as a political solution rather than a housing solution. Nonetheless, anti-authoritarians and anarchists formed the majority of the activists involved and created places that promoted self-organization in the fields of education, health, childcare, culture, creativity, entertainment and even work, food and shelter. They therefore produced social, economic and political value in neighbourhoods where they were developed, as squatting is always a political action regardless the individual motives (Pruijt 2011) and the impact in space in all cases is significant. Accordingly, because of their political orientation, squatters are a priori against authority and planning; thus, they tend to resist against gentrification and other functions of city planning and commodification. Moreover, they made up a threat to the Greek government, as they struggled against the cuts in public services that were initiated by the state with the excuse of the fiscal crisis.

Therefore, the state initiated an attack against them. The social acceptance of the assault against the squats, however, was strong and found support from a large portion of society. The Greek people became afraid and tired of the crimes, violence and chaos that existed in urban areas, and thus, they were easily controlled and persuaded. Specifically, this control was realised continuously through the majority of the local mass media, which claimed that the squats were local cells of violent and illegitimate activity managed by careless, irresponsible and dangerous people. Accordingly, the only way to weaken this wave of eviction was to make it unpopular to the public. The communication of the squats' actions to the public was the biggest challenge for the movement as some of them still remained strict and elitist with the society from the previous years.

The methodology used for this chapter is a combination of qualitative methods. With regards to the bibliography, it was crucial to use a lot of local information sources of both a radical and conventional nature in order to determine the way the media communicated the onslaught and how the people who were involved in the movement presented it accordingly. Moreover, public opinion was formed mainly from the regime media, and this is of great significance, as their view acted as consent to governmental actions. Therefore, in this piece, conventional media articles are gathered and related in order to amplify the next state decisions. In addition, as some aspects of these events are suppressed, it is important to also follow the radical media in their announcements. Nonetheless, as the media always presents even a simmering interest, it is important to communicate results of participant observation as well as comments from interviews with people who were, or are, involved in these events. Last but not least, the personal participation in the movement over the last five years will be used here in an ethical and careful way to reveal some insight into the situation.

In particular, this chapter is showing the initiation of political squatting in Greece and its primary concentration in the urban context, and it will express arguments about the alliance between state tolerance, real estate strategies and fascist activity in the Athenian centre. What is more, the impact of the economic crisis on urban life, politics and citizens' behaviour will be analysed in order to explain the success of state attack against the squats. Finally, the aftermath of the attack will be questioned in an obscure future because of the crisis and socio-political discomfort.

Presentation of squats

Squats can be seen throughout human history since the time that property was invented. The European squatting movement, as it is known today, however, has existed since the beginning of the 20th century. The first so-called squats were especially for shelter and occurred in Northern European countries. Nowadays, both squats and political occupations can be observed there. Nevertheless, we have to mention that in these countries a lot of the political squats are attacked, and the groups that manage to occupy buildings are now obliged to pay a normal or symbolic rent into banks and other capitalist institutions. Even though this new situation is not seen as traditional squatting, the inhabitants are still struggling to keep the buildings in their organization and most of them refer to themselves as 'anarchists' and 'communists' and resist repression, state attacks and observation as they attempt to self-organize and carry out their political actions.

It is interesting to note the way this culture arrived in Greece. Kalamaras (2013) explains that Interail and Magic Buses were the most common and cheapest ways for young Greek adults to travel to Western Europe in the 1980s. It was during these trips that people first experienced squatting and the culture and knowledge that came with it, and subsequently decided to take it back to Greece. We can therefore observe that the Antagonist movement greatly impacted Greek squats (Souzas 2015) as well as their prototypes.

According to Pruijt's (2011) categorization of urban squatting, Greek case refers to the categories of 'Squatting as an alternative housing strategy' and 'Political squatting'.[1] In most of the cases, the anarchist and anti-authoritarian movement supports the squats that are active in the political and social life of the cities through anti-systemic actions. Indeed, the Greek squatting movement, even though is not homogeneous, has as its main aim the cultural and political expression and as its secondary the housing issues. This can be attributed to the fact that housing practices in Greece are mainly a family issue (Siatitsa 2014). Furthermore, it is a member of the 'family of New Social Movement' that refers to a wide category of movements that share similar characteristics like informality, enthusiastic model of organization, flexibility in commitment and expressly a high percentage of middle-class participants (Pruijt 2011: 23). Furthermore, Della Porta and Rucht (1995: 121–123) state that squatters' movements overlap with less radical political people as part of a wider 'left-libertarian social movement family'. This was proven greatly after the uprising of 2008 in Greece, with squats that initiated from leftists and/or apolitical people as well as the support they demonstrated during the attacks. All in all, here the political squatting will be analysed that renders the visible anti-authoritarian squatting movement, which creates social centres, places of work and even accommodation opportunities.

The first political squatting action occurred in 1981 in the centre of Athens. The 'Valtetsiou Squat' began on Valtetsiou street, as its name suggests.[2] Furthermore, some of the most important and long-lasting squats are located in the centre. Therefore, they were often a shelter for demonstrators and a place for important celebrations and resistance. Among the most important and central squats (see Figure 22.2) are Villa Amalias (since 1993), Lelas Karagianni (since 1988), Skaramagka (since 2012).

Over the past twenty-five years, the anarchist movement primarily occupied buildings and, in many cases, rebuilt them in order to simultaneously secure homes and political space. Moreover, according to the records of formal and informal reports, a significant number of squats used buildings of the School Building Organization of Greece, University and also some of the Navy Veteran Fund.

Exarcheia, the 'Anarchist neighbourhood', lies in the heart of Athens, and logically, over the past decades, it has developed radical views and fierce politics. Moreover, it constitutes an urban frontier between the deteriorated and the upper class center, distinguishing class norms to each side. *Furthermore, in this so-called 'Anarchist neighbourhood', the radical or 'lifestyle radical' people tend to gather in order to socialize, eat, have assemblies and familiarize themselves with public spaces.* This, and other central areas of the city, could not be controlled by the police efficiently enough, and for decades, it was seen as a political issue, ever present on the government agenda (Euagrios 2013). Exarcheia, in particular, was inaccessible to the police until around December 2008, when an uprising caused by anger at the economic crisis, intense public dissatisfaction and government control took place. Thus, in the years that followed, police presence became commonplace. It also led to the introduction of new police groups holding greater power and desire to dominate as supported by the government. As Smith (1984: 156) points out, 'as the crisis progresses it adopts an increasingly spatial dimension'; likewise, the Greek authorities could not afford to have the centre out of their control during this transitional period.

Fascist real estate

In order to initiate squat geopolitics, it is important to refer, first of all, to the political and economic situation of the Athenian Metropolis at the time of the attack. Capitalism is a fundamentally geographical project, even during a time of economic crisis, as Smith (1984: 178) points out. Specifically, the real estate interest of the city centre became increasingly important as land value fell due to the effects of the economic crisis, which produced economic instability and insecurity for the future of properties and their owners. This was not helped by the state's unwillingness to offer support (Siatitsa 2014). Simultaneously, the restructure of the urban economy was attempted as a response to a series of wider global transformations where 'inner city is valuable again, profitable' (Smith 1996: 6). Therefore, real estate agents were able to purchase plots in central areas in order to transform the centre of Athens into a more profitable district. Understandably, during the economic crisis, the centre of Athens had to change in favour of the interests of certain politicians, entrepreneurs and urban planners (Makrygianni and Tsavdaroglou 2010).

Gentrification is a procedure that demands 'civilizing' the affected citizens and their obedience in the social capitalistic norms. Accordingly, during the gentrification procedure, the area is supposed to be transformed following the developers' strategy, aiming to the city commodification (Smith 1996). Therefore, along with state tolerance and activities in favour of central gentrification, another key factor that formed the social 'balances' of central land use appeared to be the 'fascist real estate activities'. Nonetheless, in order to analyse the fascist contribution of central geopolitics, we have to refer prior to the presence of immigrants in the Metropolitan centre.

It is only since the late 1980s that Athens has begun to welcome immigrants, and its destiny was bound up with this event. Despite the lack of a formal state strategy (Vaiou et al. 2007), they were placed in the centre of the city both for governmental interests and their proximity to basic services. Expressly, they settled in areas with good transport links in order to move around the city and find employment opportunities. Moreover, they found accommodation in abandoned factories and offices as well as in the basements of poorly-maintained central blocks of flats and shabby refugee shelters (Makrygianni and Tsavdaroglou 2010). Given that they stayed in the city centre, they managed not to be excluded from society and thus did not initially form ghettos (Vaiou et al. 2007).

After 2008, when the crisis grew worse, a group of conservative citizens (almost exclusively members of Golden Dawn, the Nationalist party) started the so-called 'committees of citizens', especially in central areas, and began protesting against immigrants, claiming they were criminals, sub-humans, dangerous and other terrible labels. These same activists today form Golden Dawn, a self-proclaimed Popular and Nationalist Movement with neo-Nazi views that gained popularity in Parliament while they proceed to kill, torture, stealing from and terrorize immigrants. The years before the 2012 elections[3] brought a right-wing government featuring a strong Golden Dawn presence, their members began a campaign to 'assist' older people in the centre. It is essential to mention that the elderly are a large percentage of the Greek population in the centre. Therefore, the party gained a lot of power by helping them carry shopping bags and evicting immigrants/tenants who could not pay rent for their apartments. This effectively gave state tolerance to vigilantes. They even started to spread 'business cards' using social networks to promote their services. In other words, they 'cleaned up' those central neighbourhoods they wished to gentrify, which in turn would lead to new economic interest, real estate and commerce.

In this profitable plan, resistance and uprisings were an obstacle. Thus, at the same time that the members and supporters of Golden Dawn were carrying out immigrant pogroms in the centre, the state had to clean out the rebellious, struggling radical people from the Exarcheia, the approximate squats or even the local universities. 'The effort to recolonize the city involves systematic eviction', as Smith (1996: 25) points out. This work was assigned to a Minister named Dendias, who managed to evict people from many squats throughout Greece but, importantly, some of the major squats in Athens (and Thessaloniki). This helped to extinguish the radical community who fought against city commodification and at the same time helped to give a new image of a 'gentrified' centre with the biggest land planning value in Greece. It has been always the developers' aim to disappear the past habits and activities in order to introduce new urban imageries (Smith 1996).

Crisis in the city

The fiscal crisis initiated in 2008 served as a catalyst for attacks on squats. We will therefore analyse the socio-economic and political situation of Greece during this period. According to the characteristics of the crisis as Saraceno describes (2013) it, there are not only the economic implications that people have to face but also the 'crisis of the welfare state'. In Greece, this kind of assistance was almost always lacking, but after the application of the first memorandum (2011), the situation only worsened (Dalakoglou 2012). First, crisis brought about changes to the labour market, social protection, demography, political cultures and balances, family and gender arrangements, household consumption and quality of life (Hadjimichalis 2011, Saraceno 2013). Furthermore, the application of austerity measures engendered salary shrinkages, unemployment (often long term) and weaker basic social protection (Saraceno 2013, Arapoglou, Gounis and Siatitsa 2015).

At the same time, the government, after the double elections in June 2012, was formed from a coalition of one right-wing, one central and one left-wing party (Nea Dimokratia, PASOK, DHMAR). This government proved from the very start to be conservative and decisive in what concerned obedience to the IMF's proposals and consequently repression of incidents or resistance. Minister Nikos Dendias took charge of the Ministry of Public Order and Citizen Protection and repeatedly stated after the eviction of Delta squat (the first to be evacuated) that

> The message that the Coordinate state sends today to every direction is clear: Zero tolerance against anomy and legitimacy's catalysis. Greek Democracy will defend the social demand for the full development of the public property in favour of Greek people, everywhere in every case.
> (Press release from the Minister of Public Order and Citizen Protection 12 September 2012)

Therefore, it is rather logical to regard as a starting point for the attack against the squats, the evacuation of Delta squat[4] in Thessaloniki (the second biggest city in Greece). Principally, Delta squat was a Student house that had been occupied since 2007 and belongs to the Technological Educational Institute of Thessaloniki. Its location on one of the most central roads in Thessaloniki, Egnatia, makes it an important position in need of being 'cleaned out' so that the new government can put its words into action. When people were evicted from this squat, they were also arrested, the rooms were damaged, and squatters' personal property was either destroyed or confiscated. They were accused of disturbing the peace, gun possession, perjury, forgery, disobedience and violation of the firecrackers' law. Thus, they collectively had to pay a fine of €7,950 (see http://villa-amalias.blogspot.gr) (Figure 22.1).

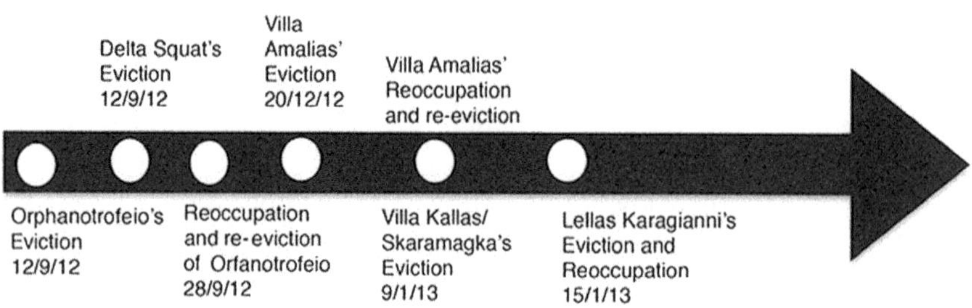

Figure 22.1 Timeline of the major squat evictions.

Following the attacks in Athens against occupied social centres, it is interesting to analyse the geography of the specific central area of Patision to which Villa Amalias and Skaramagka (the first of the evacuated squats) belonged. This area was crystalised as a space of resistance during the uprising of 2008 and the squats were involved or even initiated (e.g. Skaramagka) at that period. The map in Figure 22.2 shows the position of the two spaces, their proximity to two of the oldest and most noted public universities (Athens University of Economics and Business and the Architecture Department of the National Technical University of Athens) with both strong political history and symbolism,[5] the square of Exarcheia (with an anarchist icon), and even a building occupied by anarchist architecture students inside the school used as haunt.[6] Specifically, before the attacks, there was a loose connection between the main two squats, but this was mainly because of the common political reference, not the same political ideology. Afterwards, during the attack, all of them were organized locally and citywide, while at the same period, campaigns initiated from them as a reaction to collective threats.

First, Villa Amalias was evacuated on the morning of 20 December 2012. The fact was communicated immediately, and a lot of people gathered together outside in solidarity. Simultaneously, another group of people

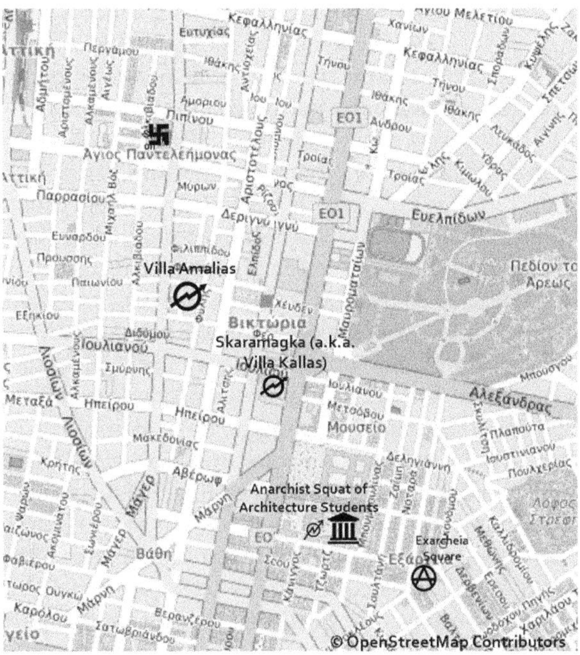

Figure 22.2 Map of the extended area of Patission street in the center of Athens, showing the referred squats, the anarchist area of Exarcheia Square and the localisation of the National Technical University of Athens.

gathered in the police department where the squatters were held. After the attack in Villa Amalias, some newspapers, working in the contemporary governmental interest, published long articles mourning the 'lost public buildings' to radical political squatters (Koumoukakas 2012, Santarakos 2013). They also published maps of Athenian squats and lists of all the buildings in Greece occupied by Anarchists. At this point, the public was misled as people realized some of the last squats in Athens and other Greek cities were also left-wing and public assemblies with a radical but not specific political identity (like the Strougka squat in New Filadelfia, Athens). One month later, there was an attempt to reoccupy Villa Amalias that had only a few hours success before a second eviction was carried out. At this time, a lot of people gathered in the area of Patision to support the squat. Simultaneously, the second central squat, Skaramagka, which was empty from squatters already for precaution, was attacked as well and closed. Patision can be seen as contested space for both government, capitalist forces and radical community.

Conclusion

As the fiscal and economic crisis expanded, it hit the gentrification process of Athens and impacted upon citizens' lives as well as urban quality. It can be seen that as the crisis 'progresses, it adopts an increasingly spatial dimension' (Smith 1984: 156). The neoliberal government tried to gentrify the centre of Athens and especially the area of Patision by getting rid of the radical presence and allowing the neo-fascist movement to act as a tool for the capitalistic ambitions of local real estate. Moreover, the geography of urban squatting in Athens changed forever as the state directly and indirectly 'destructed the everyday life' (Smith 1984: 164) by depopulating the area from people that were active and visible in the public life. Thus, instead of making it appealing, they increased the prospective insecurity as there were no people using the streets late at night (Jacobs 1961), organizing events and even showing urban interest. Concerning the urban squatting movement in Greece, even if it did not demonstrate clear anti-gentrification strategy, the squatters always played a major role in urban protests, and thus, they had to be eliminated from areas that were about to be gentrified.

Furthermore, as the Greek urban squatting movement had shelter's pursuit as a secondary aim, demonstrations against evictions wave slowly and some of the participants moved to other structures of this 'family of the New Social movements' (Kriesi 1995) or occupied, much later, other buildings. One year after the attacks, local elections took place in Greece and, surprisingly enough, hope arose for regaining the lost territory from a part of the squatting movement. This hope was placed on the candidacy of a young left-wing politician for the city of Athens who vanished in the second round of elections. Mayor Kaminis was re-elected: He supported evictions in the previous year. Both the hopes of some members of the movement and the political rise of the same people that assisted the initiation of gentrification of the area can be regarded as an evidence of the local geopolitical balances in the Greek Metropolis.

Notes

1 Nevertheless, after 2015, because of the great influx of immigrants, members of the antiauthoritarian and leftist movement also initiated the deprivation based squatting in order to shelter the immigrants.
2 It may be interesting to mention that this same street during recent years has become the 'centre' of the Anarchist neighbourhood (Exarcheia) and lifestyle of Exarcheia. Moreover, new bars and collectives, which are opening there, demonstrate a new attitude for the movement's habits.
3 A year before the assassination of the antifascist artist Paulos Fussas, which finally put Golden Dawn's members on trial for this and many other crimes.
4 Many years before Delta became a squat it was a hotel in this main street in Thessalonian, but because of financial difficulties, it was sold to the Technological Educational Institute of Thessaloniki. Thereupon, it was used as a student house until 2005 when it was closed due to governmental financial cuts. In 2007, two years after this huge building was abandoned, and in the middle of student demonstrations against the new educational law, it was used as a squat by people from the Anarchist/Anti-authoritative movement. Nonetheless, police never evaluated Delta squatters as dangerous but rather as moderate. Furthermore, they did not even participate in the fights between the Thessalonian squats the final years that led to 25 arrests (2 July 2012).
5 Many demonstrations and occupations took place here in the last decades like during the dictatorship or the uprising of 2008. They are regarded as Asylum so the police, in theory, cannot enter or arrest people inside their territories.
6 Figures 22.1 and 22.2 are based on the information taken from the following websites/blogspots: Former squat Villa Amalias blogspot (http://villa-amalias.blogspot.gr); Map of Social Resistance, Observation Post of Free Spaces (www.asda.gr/elxoroi/katalBantist.htm); Former squat Delta Blogspot (http://delta.squat.gr/); Athens Indymedia (https://athens.indymedia.org/); Ministry of Public Order and Citizen Protection (www.mopocp.gov.gr/main.php).

References

Arapoglou, V., Gounis, K. and Siatitsa, D. (2015). 'Revisiting the concept of shelterisation: insights from Athens, Greece', *European Journal of Homelessness*, 9(2), pp. 137–157.

Dalakoglou, D. (2012). 'Beyond spontaneity. Crisis, violence and collective action in Athens', *City: analysis of urban trends, culture, theory, policy, action*, 16(5), pp. 535–545.

Della Porta, D. and Rucht, D. (1995). 'Left-libertarian movements in context: a comparison of Italy and West Germany, 1965–1990', in J. C. Jenkins and B. Klandermans (eds.). *The Politics of Social Protest: Comparative Perspectives on States and Social Movements*. Minneapolis: University of Minnesota Press, pp. 229–272.

Euagrios, A. (2013). 'The short-long history of the squats in Greece', *Apatris Road Newspaper*, 21(April), p. 6.

Hadjimichalis, K. (2011). 'The pedestranalization of Panepistimiou and other planning fantasies for the city center', *Augi*, 15 May, p. 27 [original in Greek].

Jacobs, J. (1961). *The Death and Life of Great American Cities*. New York: Vintage.

Kalamaras, P. (2013). *Patision and Stournari Corner*. Athens: Eleutheriaki Koultoura.

Koumoukakas, K. (2012). 'End of the series "Delta Squat"', *Ta Nea*, 13 September, p. 4.

Kriesi, H. (ed.) (1995). *New Social Movements in Western Europe: A Comparative Analysis*. Minneapolis: University of Minnesota Press.

Makrygianni, V. and Tsavdaroglou, H. (2010). *Athens, Unfortified City*. Thessaloniki: Urban Anarchy. Available at: www.urbananarchy.gr.

Ministry of Public Order and Citizen Protection. (2012). *Report with the Declaration of the Public Order and Citizen Protection Minister, Mr Nikolaos Dendias Relevant to the Police Action in Thessaloniki*. 12 Sept. Available at: www.minocp.gov.gr.

Pruijt, H. (2011). 'The logic of urban squatting', *International Journal of Urban and Regional Research*, 37(1), pp. 19–45.

Santarakos, G. (2013). 'The circle of Vi(l)a', *Ethnos*, p. 8–9, 12 January [original in Greek].

Saraceno, C. (2013). 'Three concurrent crises in welfare states in an increasingly asymmetrical European Union', *Stato e mercato*, 33(3), pp. 339–358.

Siatitsa, D. (2014). *Demands for the Housing Right in Southern European Cities: The Speech and the Role of Social Movements*. PhD Dissertation. Available at: http://thesis.ekt.gr/thesisBookReader/id/35495#page/1/mode/2up [original in Greek].

Smith, N. (1984). *Uneven Development. Nature, Capital and the Production of Space*. Oxford: Basil Blackwell, 2nd edition, 1990. Athens: University of Georgia Press, 3rd edition, 2008.

Smith, N. (1996). *The New Urban Frontier: Gentrification and the Revanchist City*. New York: Routledge.

Souzas, N. (2015). *Stop Talking about Death, My Baby*. Thessaloniki: Nautilos [original in Greek].

Vaiou, D., Bacharopoulou, A., Fotiou, T., Hatzivasileiou, S., Kalandides, A., Karali, M., Kefalea, R., Lafazani, O., Lykogianni, R., Marnelakis, G., Monemvasitou, A., Papasimaki, K., Tounta, F. (2007). *Intersecting Patterns of Everyday Life and Socio-Spatial Transformations in the City. Migrant and Local Women in the Neighbourhoods of Athens*. Athens: L-Press and NTUA [original in Greek].

23 Revanchism and the racial state

Ferguson as 'internal colony'

Deborah Cowen and Nemoy Lewis

The Black Lives Matter movement has declared an outright crisis of domestic life within the US Empire. Initiated through social media by Black queer women in response to a string of police murders of Black people on the streets of American cities or in their back offices – American prisons – the movement has grown into a powerful and more organised form that faces extraordinary challenges but holds equally extraordinary potential. Black Lives Matter insists that racial violence is not an accident of the police acting badly, but the source of its genesis. Indeed, there is nothing new in the racial violence of the US police force; the formation of public and private police is rooted in racial and colonial violence that dates back to the capture of escaped slaves and the protection of colonial infrastructures. We suggest that there *is newness* in the social order that policing actively supports, however, and that this newness is assembled through the geopolitical economies of urban space and life.

In this chapter, we elaborate on some of the invaluable insights that Neil Smith's urban work offers on the theft of Black space and wealth through decades of gentrification and the more acute dispossession of the subprime crisis. We suggest that Smith's work on gentrification and revanchism are invaluable resources for understanding struggles in places such as Ferguson, precisely because of his emphasis on urban geopolitical economy as not only a space-time of capitalist transformation and anti-capitalist struggle, but one of imperial renovation and anti-imperial insurgence. In what follows, we investigate aspects of the changing nature of the US Empire and especially its domestic cartographies of urban accumulation. Looking to the changing geography of the American city and the recent rounds of dispossession and displacement of Black communities brings into focus not only the everyday violence and securitisation of the US racial state but also significant shifts in the workings of empire at home. In conjunction with historical debates about the space of 'internal colonial administration', we ask what is at stake in the new suburban geography of 'the Black ghetto'.

We see the police as a key force in a much broader operation of imperialism domestically, and look beyond the police to understand how the 'internal' violence of the state is changing. In doing so, we also engage with the work of historian Robin Kelley, who situates contemporary anti-Black police violence in a much larger framework of transnational imperial war. We flag important transformations in the domestic face of empire within which the question of policing needs to be understood. More specifically, we trace shifting geographies of Black dispossession in order to bring more nuanced analyses to the relationship between repression and accumulation at work in the urban homeland of US imperialism. We suggest that the struggles over the suburban municipality of Ferguson are diagnostic of the remaking of American urban space more broadly. Police continue to do the violent work of the racial state in its internal colonies, but both the (racial) state and (racialised) space have been thoroughly transformed by the 'urban revolution'.

Black lives and internal colonies

In a lecture series entitled *Mike Brown's Body: Meditations on War, Race and Democracy*, Robin Kelley (2015) asserts, 'Mike Brown was a casualty of war'. Kelley argues that Michael Brown's murder should be understood as 'collateral damage in a perpetual war whose colonial roots are still alive'. In contrast to more limited historical claims emerging from security studies that locate unending warfare in the period post-2001, Kelley argues that the entire history of American racial oppression is a story of *perpetual war*. Kelley explicitly asserts that 'coloniality and capitalism are inseparable' and that the US racial state must be understood historically and geographically as transnational imperial violence, simultaneously reliant on indigenous dispossession and slave labour. 'War,' he argues, 'begins at the point of settlement with the shipment of kidnapped Africans across the Atlantic, the enslavement of indigenous people, with rape and pillage, forced removals, not to mention religious conversion and compulsory assimilation'.

Describing this violence in this way challenges given categories of the domestic and foreign that prevent us from seeing warfare as cutting through the national body politic, as it also crafts that national cartography. Kelley's arguments question the distinction between 'internal' and 'external' organised violence, and so, too, the supposedly foundational separation underpinning the modern state system between police and military force. In its place, Kelley offers a profoundly important rethinking of space. The 'domestic situation' becomes one site in a wider cartography of colonial violence that transformed, rather than expired, with emancipation.

While Kelley's claims may be bold, they also build on a long tradition in radical anti-colonial thought that analyses colonialism 'internally'. Bush (2008) and Gutiérrez (2004) trace the history of debates on 'internal colonialism' in some detail, highlighting its heyday in the 1960s Black Power movement. Its circulation expanded following the publication of Fanon's *Wretched of the Earth* (Bush 2008: 138), and the concept was taken up centrally in the work of Harold Cruse (1962) and Robert Allen (1969), who built on Carmichael and Hamilton's 1967 notion of 'colonial analogy'. Allen sought to develop the question of colonialism at home - not as analogy, but as ontology. Lively debates on the distinct nature of domestic colonialism unfolded through the 1960s and 1970s with key contributions from figures such as O'Dell (1967), Blauner (1969) and Bailey (1973). The use of the concept declined along with the Black Power movement in the late 1970s, although it has experienced revival more recently (Gutiérrez 2004, Allen 2005, Pinderhughes 2011).

The concept of internal colonialism allows for a recasting of national borders as tools of empire that enable its operation rather than bound its jurisdiction. If colonialism is no longer reserved for the domination of a people beyond national borders – if it can be conceptualised as a phenomenon with an 'internal' cartography as well - then it opens up the necessity of a geography of perpetual domestic war. For writers working in this tradition, the task was to analyse what makes colonial rule (and anti-colonial revolt) distinct when it is *internal* to the nation state at the centre of empire. In other words, this literature was centred in asking how empire oppresses, exploits, disciplines and dispossesses populations *internal* to its territorial core. At the centre of the answer lay the segregated cities of the United States. In fact, Gutiérrez (2004: 284) notes the urban basis for these debates, suggesting, 'internal colonialism as a theory grew out of the brutal urban conditions minorities faced in the United States'.

Writers in this tradition understand the Black 'ghetto' as a key space of discipline and extraction, and a terrain of police domination and violence. Crucial here is not only the violence executed on already constituted 'ghetto' spaces but the creation of these spaces by state policy, the fact of which Ta-Nehisi Coates (2014) and Rashad Shabazz (2015) remind us more recently. The Black 'ghetto' was formed through the social, economic, physical and epistemic violence of redlining, blockbusting, restrictive covenants, urban renewal and white flight, designed and enforced by the public private partnership of the state, the banks, residents associations and real estate firms (Rothstein 2014: 2).

While a variety of immigrant and ethnic enclaves constituted the American city, the Black 'ghetto' was distinct. Blauner (1969) highlights its forced rather than voluntary nature, through racially exclusionary policies, the durability of the Black 'ghetto' and the fact that land ownership remained concentrated in the hands of outsiders. Perhaps most importantly, Blauner and others writing in this tradition suggest that the aggressive nature of policing within was what distinguished Black ghettos from others. Police were understood in the violent terms of imperialism, where they 'are key agents in the power equation as well as the drama of dehumanization'. The Black 'ghetto' was of course a defining space of post-war America, generating such a deep association between race and space that terms like 'urban problems' and 'urban music' could circulate without even naming the Blackness that defined them. Indeed, in post-war America, 'urban' became code for 'Black' (Wacquant 1997, McKittrick 2006).

It is in this context of public policy that excluded Black people from access to the vast majority of urban space, contained Black people in small overcrowded areas with poor housing stock in the inner city, withheld credit from Black communities leading to the rapid deterioration of housing stock, aggressively policed these zones and frequently demolished them for the project of 'urban renewal' that inner cities saw the sustained disinvestment that gave rise to what Neil Smith termed the 'rent-gap'. For Smith, the inner city 'rent-gap' – the gap between actual and potential ground rent – was what primed the pump of gentrification (Smith 1996). It is perhaps his single best known contribution to debates about urban political economy, but Smith's rent-gap was always also about geo-political economy, where the specificities of who and where matter, and where the market relies on violent force. The rent-gap in American cities should be understood as a profoundly racial and violent effect of internal colonialism.

Gentrification and the suburbanisation of the 'internal colony'

For Smith, gentrification is a defining process of neoliberal urbanism, and one that he argued was 'globalising' decades ago (1996, see also Smith 2002). The rent-gap is not only a feature of local urban areas, but of an uneven global development that is increasingly contingent on the production of urban space. Of course, at the centre of Smith's work on gentrification is not only the circulation of capital but the social and spatial relations through which that circulation is secured. Smith's focus on revanchism highlights exactly that fact. Smith (2002) identified racialised police violence in New York as a key marker of the increasingly aggressive gentrification at work there. He flags the particular cases of Abner Louima and Amidou Diallo – both immigrant Black men violently attacked, Diallo murdered, by the NYPD in cases that were shocking in their brutality. Smith also cites the growing cost of payments to settle police brutality cases, noting that between 1994–1997, the NYPD paid $96.8 million to settle these lawsuits (Smith 2002). Crucial to this conception of gentrification is the displacement of working class, precarious and especially racialised peoples from inner city areas.

For Smith, gentrification unfolds not only through a racialised political economy but an imperial one. The question of 'the frontier' is not only posed in the title of his 1996 book, but is also subject to sustained analysis in its pages. Smith highlights the way that the vocabulary of gentrification is anchored in a colonial imaginary with pervasive references to, of course, the frontier, but also the 'Wild West', pioneering, cowboys and Indians and terra nullius. For Smith (1996), 'the frontier ideology continues to displace social conflict into the realm of myth, and at the same time to reaffirm a set of class-specific and race-specific social norms'. He describes the systematic evictions enacted through municipal policy as an 'effort to recolonize the city'. Smith (1996: 17) was insistent that the colonial frame is more than metaphor, that, 'the frontier imagery is neither merely decorative nor innocent... but carries considerable ideological weight'.

Black neighbourhoods across New York City – in Brooklyn, Manhattan and the Bronx – have been the focal point of some of the fiercest battles over gentrification and displacement. While the suburbanisation of Black communities and people 'might seem like mundane fact', ULTRA (2014) suggests that it 'actually haunts at a much more significant shift in the economic and racial geography of US metropolitan zones over the last 20 years'. The suburbanisation of Black communities emerges through the gentrification of inner city neighbourhoods and the displacement that ensues. This is a fact that is frequently gestured at in current debates about anti-Black state violence, but could be pried open much further. It is a striking feature of the geopolitical economies at work in a place like Ferguson, Missouri.

Over the last two decades. Ferguson has become a majority Black suburb as a result of gentrification and displacement in St. Louis and surrounding areas – although this shift is not reflected in the political leadership or police force, which remain almost entirely white. The transformation of local social geography has not only been dramatic, but rapid. From a population that was 85% white in 1980, Ferguson had become 69% Black by 2010 (Goldstein 2014). Gentrification of the inner city furthermore means that 'whites are now a solid majority in some neighbourhoods [of St. Louis] for the first time in decades' (Rothstein 2014). In a report outlining the public policies that created these conditions in Ferguson, Rothstein (2014) notes, 'Whereas 20th century segregation took the form of Black central cities surrounded by white suburbs, 21st century segregation is in transition – to whiter central cities with adjoining Black suburbs'. Indeed, writing at a very different moment, Neil Smith (1982: 145) would nevertheless assert the importance of these city-suburb processes and distinctions, suggesting that 'at the urban scale, the main pattern of uneven development lies in the relation between the suburbs and the inner city.'

Dispossession, displacement and crisis

If the redlining of the 20th century starved Black communities of capital, then subprime lending inverts this logic, and provides loans at exorbitant costs. The subprime crisis marks a shift from the bad risk of Black people and the withholding of capital, to the profitable risk of Black people via financial predation. Often referred to as 'reverse redlining', subprime lending refers to the high cost loans made to borrowers with impaired or limited credit histories, or those who have a higher ratio of debt to income. The 'reversal' of redlining was far from coincidental. It was in the wake of the Fair Housing Act and with the end of redlining that the mortgage industry began to target racialised groups with extraordinarily exploitative loans.

The subprime crisis has exacted one of the largest dispossessions of Black wealth in history – which some have conservatively estimated at $194 billion (Bocian, Li and Ernst 2010).

After decades of receiving little to no attention from traditional lenders, Black borrowers became the centre of attention for subprime institutions following legislation developed first under Reagan and then deepened under Clinton (Lewis forthcoming). Subprime lenders argued that they must charge higher fees to help protect them from higher risk borrowers defaulting on their loans, however, the effects were disastrous in their irony; it was precisely the onerous terms that led individual borrowers to default on their loans. Between 2007 and 2013, more than 12 million homeowners were evicted from their homes (Schwartz 2014). Entire communities were gutted as the number of foreclosures continued to soar. By the end of 2013, the American housing market had lost approximately $3.6 trillion in home equity (Schwartz 2014), leaving many with homes worth less than their mortgage debt. The effects of these processes have been the massive displacement of Black people from the inner city and the dramatic dispossession of Black wealth.

The shift of the Black population to Ferguson over the last two decades was also met by the subprime crisis, which hit the municipality – or rather the local Black population – hard. Today, half of Black homeowners in Ferguson have mortgage debt that exceeds the value of their homes, compared with the already high figure of 17% nationally (Goldstein 2014). While some markets around the country have showed signs of recovery, Ferguson presents a more disturbing trend. In the first half of 2015, foreclosure filings increased by 15.79% (Von Pohlmann 2015), and investment firms were buying up the town at highly discounted rates. Nearly a decade after the start of America's worst housing crisis, some of the same corporations responsible for the crash are now purchasing distressed properties and mortgages in bulk. At least 25% of all sales in Ferguson are to institutional investors. Most investors are out of state and absentee landlords, but even more insidious are the investor landlords that work to systematically gentrify neighbourhoods and displace residents, even in the suburbs. Perhaps more outrageous still, some residents in Ferguson (and elsewhere) now have to rent their own foreclosed homes back from firms that helped create the housing crisis.

During the height of the crisis, the federal government created a number of programmes they claimed would stem the rapid rise of foreclosure in communities. The Department of Housing and Urban Development (HUD) quietly created the Distressed Asset Stabilization Program (DASP) in 2012, designed to facilitate the sale of delinquent loans that had yet to go through the foreclosure mill. Since 2010, the programme has facilitated the sale of almost 100,000 mortgages to Wall Street Investors. The Federal Housing Finance Agency (FHFA oversees Fannie Mae and Freddie Mac) has also followed suit in selling tens of thousands of delinquent assets to hedge funds and private equity firms (Dreier and Sen 2014). Needless to say, the programme has been dismal in helping homeowners. A report produced by HUD showed that less than 6% of the 6,427 loans sold within the first three years of the programme were still in good standing (Bennett 2015).

The growth of the rental market also presents a unique opportunity for Wall Street, as millions of new renters were produced through foreclosures. Since 2007, the number of single-family homes for rent has increased by more than two million (Fernald 2013). While the rest of the economy went into a tailspin, a Wall Street giant, the Blackstone Group, was busy engineering a sophisticated product known as Rent Backed Securities (RBS), allowing them to profit again from the financial crisis. Rental-backed securities work very similar to mortgaged-backed securities, except monthly rental payments are used to pay bondholders rather than mortgage payments. Companies like the Blackstone Group, American Homes 4 Rent, Colony American Homes, and other Wall Street entities have raised close to $70 billion dollars to acquire distressed properties at steep discounts and charge as much as 180% of fair market rent (Dreier and Sen 2014). The Blackstone Group alone has spent almost $10 billion to purchase over 50,000 single-family properties (Dezember 2015).

Reports have revealed that some private equity firms are incorporating a mandatory 5% rent increase into their leases and even higher rate hikes for tenants with year-to-year leases (Call, Powell and Heck 2014). In some cases, institutional landlords have even required tenants to rent properties 'as is' and have written certain provisions into the lease to absolve the firm from any damages caused by their own negligence (Burns, Donley and Manzanet 2014). Economists and real estate professionals have already raised red flags with regard to this practice (Gottesdiener 2013), suggesting that tenants could be evicted not because they have missed a payment, but because their institutional landlord is unable to maintain a high occupancy rate in order to meet their bond obligations.

Ferguson as a flashpoint of empire at home

For good reason, there has been much focus on the particular nature of policing in Ferguson and across the United States by activists, the media and even the Department of Justice. The racism of the Ferguson police was highlighted in a recent U.S. Department of Justice report describing how a 'focus on revenue generation' helped foster a culture where 'many officers appear to see some residents, especially those who live in Ferguson's predominantly African-American neighbourhoods, less as constituents to be protected than as potential offenders and sources of revenue'. What is distinct about this moment is clearly not that police violently oppress Black people who are already imagined as either or both a threat to the social order or a resource for it. Rather, we insist on a more precise, but also much more diffuse set of transformations in the domestic violence of the state that operates centrally today through the financialisation of urban space.

How does 'internal colonial administration' proceed in the American city after successive waves of Black dispossession tied to gentrification and the subprime crisis? How can we understand the dislocation of Black communities and their immersion in debilitating debt as profoundly tied to the remaking of the American city, and so, too, the remaking of the US Empire? How do these shifts matter for the question of anti-Black police violence today?

Gentrification as a globalisation process is key to this transformation, and the force of the subprime crisis – erupting from the predatory financing of mortgages in US cities and unfolding into a global economic crisis – makes more sense in light of this thesis. That we now see some of the same firms that created the crisis profiting again through state subsidised and legally orchestrated dispossession perhaps supports a reading of these dealings in the violent ways we suggest here.

Suburban government is not only a site where the neoliberal policies of the racial state have been rolled out, but also a space of their genesis (Cowen 2005). There are direct connections to longer term experimentation with neoliberal forms of government in precisely these sorts of suburban spaces where the politics of property ownership underpinned by profoundly racialised notions of risk gave rise to distinctly segregated landscapes organised through finance (McGirr 2001).

If the 'Black ghetto' became a space of the administration of colonial rule domestically, then we have also seen a series of transformations in the government of urban space. Black communities have been displaced from these former 'ghetto' spaces through active and deliberate policies and processes of gentrification. As Smith (2002), Kelley (2007) and others highlight, gentrification is spurred by state policy and public dollars. This displacement is now layered with the theft of wealth we call the subprime crisis – a crisis that continues to unfold in communities across the United States, and which has led to a disastrous dispossession of Black wealth.

Translocal coalition and the internal colony

We have argued that a more fruitful engagement with the violence of the racial state can be had if we consider the police as one key actor in a much broader operation of empire at home. This approach allows us to map specific connections between repression and dispossession in and through US urban space. It asks that we recognise the violence of the US racial state as deeply entangled with the geo-political economies of the urban revolution and the financialisation of urban space. Black people and communities have not only survived persistent police violence since the formation of the American state, but a whole array of forms of state violence, including, centrally, through the government of urban land.

Putting Neil Smith's work together with the historical work on internal colonialism allows us to begin to map the way that contemporary forms of global accumulation anchored in the financialisation of urban space also occur domestically and target Black communities. We see this kind of cartographic practice at work in the Black Lives Matter movement today, most clearly in their transnational solidarity and coalition building work. It is not only important but also diagnostic that Ferguson activists understand their struggle as directly tied to that of others fighting the same violent empire, even if from halfway around the world. This is most visible perhaps in the explicit solidarity with Palestinian movements, and the coalitions being assembled with groups seeking justice for and in Gaza. In Ferguson, a 'Palestine Contingent' emerged and includes a wide range of groups such as the St. Louis Palestine Solidarity Committee, the Organization for Black Struggle, the U.S. Palestinian Community Network, Muslims for Ferguson, the U.S. Campaign to End the Israeli Occupation, the Council on American Islamic Relations, the St. Louis Palestinian BDS National Committee, National Students for Justice in Palestine, the Palestinian Youth Movement, American Muslims for Palestine and African Americans for Justice in the Middle East and North Africa. Initiatives such as 'Block the Boat' in Oakland, California also make explicit the transnational circulations that underpin empire. The release of a video highlighting specific connections between occupation in Palestine and anti-Black state violence in American cities and featuring high profile figures such as Lauryn Hill, Danny Glover and Rasmea Odeh, implicitly if not explicitly highlights this imperial frame. In fact, extensive efforts are underway to draw connections between state violence across transnational space – be it in Bahrain, Egypt, Miami or New York, among other sites and spaces. These efforts highlight the circulation of security tactics, technologies, and personnel, as well as the circulation of capital across empire, and help make the cartographies of the contemporary racial state and racial capitalism palpable, material and so more vulnerable to change.

References

Allen, R. L. (1969). *Black Awakening in Capitalist America: An Analytic History.* Garden City, NY: Anchor Books.

Allen, R. L. (2005). 'Reassessing the internal (neo) colonialism theory', *The Black Scholar*, 35, pp. 2–11.

Bailey, R. (1973). 'Economic aspects of the Black internal colony', *The Review of Black Political Economy*, 3, pp. 43–72.

Bennett, J. (2015). 'The government is selling thousands of homes to hedge funds without their owners' knowledge', *The Atlantic*. Available at: www.theatlantic.com/business/archive/2015/09/the-government-is-selling-thousands-of-homes-to-hedge-funds-without-their-owners-knowledge/406771/ [Accessed: 2 December 2015].

Blauner, R. (1969). 'Internal colonialism and ghetto revolt', *Social Problems*, 16, pp. 393–408.

Bocian, D. G., Li, W. and Ernst, K. S. (2010). *Foreclosures by Race and Ethnicity: The Demographics of a Crisis.* Durham, NC: Centre for Responsible Lending.

Available at: www.responsiblelending.org/mortgage-lending/research-analysis/foreclosures-by-race-and-ethnicity.pdf [Accessed: 3 December 2015].

Burns, R., Donley, M. and Manzanet, C. (2014). 'Game of Homes: the private-equity firm Blackstone could be your next landlord', *These Times*, 31 March Available at: http://inthesetimes.com/article/16424/game_of_homes [Accessed: 2 December 2015].

Bush, R. (2008). 'The internal colony hybrid reformulating structure, culture, and agency', in K. Smith and P. Leavy (eds.). *Hybrid Identities: Theoretical and Empirical Examinations*. Boston, MA: Brill, pp. 129–164.

Call, R., Powell, D. and Heck, S. (2014). *Blackstone: Atlanta's Newest Landlord. The New Face of the Rental Market*. Atlanta, GA: Occupy Our Homes Atlanta. Available at: http://homesforall.org/wp-content/uploads/2014/04/Blackstone ReportFinal0407141.pdf [Accessed: 9 November 2015].

Coates, T.-N. (2014). 'The case for reparations', *The Atlantic*. Available at: www.theatlantic.com/magazine/archive/2014/06/the-case-for-reparations/361631/ [Accessed: 1 December 2015].

Cowen, D. (2005). 'Suburban citizenship? The rise of targeting and the eclipse of social rights in Toronto', *Social and Cultural Geography*, 6(3), pp. 335–357.

Cruse, H. (1962). *Rebellion or Revolution?* New York: Morrow.

Dezember, R. (2015). 'Big landlords to merge, betting on rising rents', *Wall Street Journal*, 21 September. Available at: www.wsj.com/articles/big-landlords-in-talks-to-merge-betting-on-rising-rents-1442808003 [Accessed: 2 December 2015].

Dreier, P. and Sen, A. (2014). 'Hedge funds: the ultimate absentee landlords', *The American Prospect*. Available at: http://prospect.org/article/hedge-funds-ultimate-absentee-landlords-fall-preview [Accessed: 9 November 2015].

Fernald, M. (2013). *America's Rental Housing: Evolving Markets and Needs*. Cambridge, MA: Joint Center for Housing Studies of Harvard University. Available at: www.jchs.harvard.edu/sites/jchs.harvard.edu/files/americas_rental_housing_2015_web.pdf [Accessed: 5 November 2015].

Goldstein, M. (2014). 'Another shadow in Ferguson as outside firms buy and rent out distressed homes', *The New York Times*, 3 September. Available at: http://dealbook.nytimes.com/2014/09/03/another-shadow-in-ferguson-as-outside-firms-buy-and-rent-out-distressed-homes/?_r=0 [Accessed: 26 November 2015].

Gottesdiener, L. (2013). 'How Wall Street has turned housing into a dangerous get-rich-quick scheme—again', *Mother Jones*. Available at: www.motherjones.com/politics/2013/11/wall-street-buying-foreclosed-homes [Accessed: 9 November 2015].

Gutiérrez, R. A. (2004). 'Internal colonialism: an American theory of race', *Du Bois Review: Social Science Research on Race*, 1, pp. 281–295.

Kelley, R. D. G. (2007). 'Disappearing acts: Harlem in transition', in J. Hammett and K. Hammett (eds.). *The Suburbanization of New York: Is the World's Greatest City Becoming Just Another Town?* New York: Princeton Architectural Press, pp. 63–73.

Kelley, R. D. G. (2015). 'Mike Brown's body: an historical and political autopsy', *Princeton University African American Studies: Toni Morrison Lectures*, 13 April. Available at: http://livestream.com/aas21 [Accessed: 6 November 2015].

Lewis, N. K. (2017). *A Dream Denied: The Fight against the Mass Eviction of Families*. Doctoral thesis, Queen's University.

McGirr, L. (2001). *Suburban Warriors: The Origins of the New American Right*. Princeton: Princeton University Press.

McKittrick, K. (2006). *Demonic Grounds: Black Women and the Cartographies of Struggle*. Minneapolis: University of Minnesota Press.

O'Dell, J. (1967). 'A special variety of colonialism', *Freedomways*, 7(1), pp. 7–15.

Pinderhughes, C. (2011). 'Towards a new theory of internal colonialism', *Socialism and Democracy*, 25(1), pp. 235–256.

Rothstein, R. (2014). *The Making of Ferguson: Public Policies at the Root of Its Troubles*. Washington, D.C.: Economic Policy Institute. Available at: www.epi.org/files/2014/making-of-ferguson-final.pdf [Accessed: 26 November 2015].

Schwartz, A. F. (2014). *Housing Policy in the United States*, 3rd edition. New York: Routledge.

Shabazz, R. (2015). *Spatializing Blackness: Architectures of Confinement and Black Masculinity in Chicago*. Champaign, IL: University of Illinois Press.

Smith, N. (1982). 'Gentrification and uneven development', *Economic Geography*, 58(2), pp. 139–155.

Smith, N. (1996). *The New Urban Frontier: Gentrification and the Revanchist City*. London: Routledge.

Smith, N. (2002). 'New globalism, new urbanism: gentrification as global urban strategy', *Antipode*, 34, pp. 427–450.

ULTRA. (2014). *New Ghettos Burning*. August. Available at: www.ultra-com.org/project/new-ghettos-burning/ [Accessed: 12 September 2016].

Von Pohlmann, J. (2015). *How is the Ferguson, MO Housing Market Fairing One-Year after the Shooting of Michael Brown*. Irvine, CA: RealtyTrac. Available at: www.realtytrac.com/news/foreclosure-trends/how-is-the-ferguson-mo-housing-market-fairing-one-year-after-the-shooting-of-michael-brown/ [Accessed: 26 November 2015].

Wacquant, L. J. D. (1997). 'Three pernicious premises in the study of the American Ghetto', *International Journal of Urban and Regional Research*, 21(2), pp. 341–353.

PART VI

Neil Smith and beyond

24 Gentrification and the urban struggle

Neil Smith and beyond

Núria Benach and Abel Albet

Why does 'gentrification' deserve attention (today more than ever)?

Studies of urban transformation from the perspective of gentrification emerged in urban studies in the mid-1960s. Since then, they have increased in number, been enriched and diversified. Indeed, it is clear to everyone that the urban reality has changed enormously in the last fifty years. Yet, interpretations of gentrification associated with the dynamics of urbanisation under capitalism and the various applications of the concept to different geographic realities show that the term is still very much alive. It is perhaps inevitable to question how what started as a mere description of a process of urban change, or at most 'a valuable lens through which to examine a variety of intersecting phenomena in a city and/or neighbourhood context' (Lees, Shin and López-Morales 2016: xvi), has today become a very powerful theoretical tool to explain urban transformations in a much broader sense.

In 1991, when studies of gentrification had already filled space in the journals, thanks mainly to the impetus provided by Neil Smith's works on the subject from 1979 onwards (Smith 1979), Chris Hamnett identified five good reasons to account for the rise in gentrification studies observed at that time, a time when the global reach of the phenomenon was still not evident (Hamnett 1991: 173–174). First, and somewhat instrumentally, Hamnett considered that gentrification was a good topic to research in order to generate empirical and case studies of different types. Second, and rather more persuasively, he claimed that evidence of gentrification constituted a real challenge to traditional theories of residential location and urban social structure, which were dominated by the theorising of the Chicago school of urban sociology of the 1930s, introduced again under the reasoning of neoclassical economics that sustained neopositivist approaches of the 1960s to the urban land market. Third, gentrification had become a topic of great political relevance. Some stressed the positive economic effects of urban regeneration and renewal, whilst others focused on the social costs of these processes, and the people that were displaced by urban change. Fourth, gentrification was increasingly presented as the most visible face and the main front of contemporary metropolitan restructuring – a vision that owed much to Neil Smith's theoretical work on the topic. Finally, for Hamnett, gentrification had become one of the most important theoretical and ideological battlefields in urban geography and, by extension, in all human geography, the place where radically conflicting theoretical perspectives met: The more structuralist views, as represented by Neil Smith, came up against the more culturalist views, as defended by Hamnett himself, a position that was to meet with a sharp retort from Smith (Smith 1992).

In 2008, Lees, Slater and Wyly added three more reasons of weight as to why studies and debates on gentrification were still present after so many years, and why their relevance appeared renewed. First, gentrification is the leading edge of neoliberal urbanism, which, in ideological terms, is crudely reflected in the global neoliberal discourse on regeneration and urban renewal. Second, gentrification has gone global and can only be understood through its associations with processes that are also global. Third, globalisation is no longer confined, as it was in the past, to the inner city or to first world metropolises (Lees et al. 2008: xvii).

For these reasons, over the ensuing years and with the drastic changes that have occurred at all scales, the concept of gentrification has continued to be used and discussed. With all its complexities, casuistries and differences in meaning, gentrification interpreted as the result of uneven geographical development continues to be, in our opinion, the main theoretical contribution to urban studies that – inspired by Neil Smith – reveals the same uneven nature of capitalist urbanisation. In 2002, Neil Smith himself stated this in a clearer, more direct way: 'gentrification is the leading edge of global urbanism' (Smith 2002).

The evolution of a concept: definitions, applications and debates

In 1991, Hamnett drew attention to the very wide ranging literature on the subject of gentrification. Since then, the number of papers and texts on gentrification has increased almost exponentially, and excellent reviews have been published to systematise and clarify approaches (Lees et al. 2008, 2016).

Since Ruth Glass's pioneering study published in 1964 describing the process of urban change in some London neighbourhoods, the concept of gentrification has gradually been better defined, at the same time as it has been expanded to include new contexts. Initially, gentrification was the process of transformation of a working-class neighbourhood whereby the former residents were displaced by new middle-class residents. The result was an improvement in dwellings and urban infrastructure, and a consequent increase in value of these dwellings and rental prices. Over the last five decades, though, the definition of gentrification has expanded, and discussion of the term's meanings has gradually increased.

At the outset, gentrification only referred to the improvement of the old houses in some inner-city districts. However, the process expanded, and soon commercial properties began to be replaced and scholars began to speak of 'commercial gentrification', although relatively little attention has been paid to this process. Some areas underwent demolition and new construction in a process known as 'new-build gentrification', which led to the debate as to whether use could be made of the same term that some wanted to restrict solely to the process of residential renewal.

Over time, the political context in which the term 'gentrification' is used has also changed greatly. In the 1970s, the term was avoided. In its place, words with more positive connotations were used, such as 'reinvestment', 'revitalisation', 'renewal' or 'revival', that focused on the beneficial side of gentrification. However, since the 1980s, 'gentrification' has been used in a much more critical way. At the same time, organised resistance to gentrification has proliferated in a range of contexts (whilst politicians, property developers and even 'gentrifiers' continue to avoid a term that stresses the dark side of urban renewal).

In addition to the ongoing reflection on the conceptual difficulties and explanations of the phenomenon, academic papers have highlighted what Lees et al. (2016) have called the 'mutation of gentrification' to refer to temporal and spatial changes.

In the case of temporal changes, the evolution of gentrification has been described using a model of three stages or waves (Hackworth and Smith 2001):

1 The first stage occurred at the end of the 1960s and the start of the 1970s, when gentrification referred to the sporadic chaos induced by the state that affected small neighbourhoods in some of the biggest European and US cities.
2 The second stage took place at the end of the 1970s, when 'the process becomes implanted in hitherto disinvested central city neighbourhoods'. It also extended into medium-sized cities, and the role of the private sector became more important. In addition, from the middle of the 1980s, intensive political struggles broke out against the expulsion of the impoverished former residents.
3 The third stage occurred at the end of the 1980s, when the recession caused a fall in flows of capital towards neighbourhoods undergoing gentrification, and some scholars even began to speak of 'degentrification'.

The three-stage model fitted US and UK cities well and served to interpret what happened there. However, from the mid-1990s until the global financial crisis of 2007–2008, gentrification underwent a period of sustained growth, favoured by low interest rates. This was a period in which gentrification expanded and reached the non-industrialised world and in which entire neighbourhoods were renewed, often with state help.

This evolution shows the complexity of the urban phenomenon worldwide, which has been affected by various forms of gentrification. The process has expanded geographically (to reach metropolises such as São Paulo and Buenos Aires, the cities of Eastern Europe after the fall of the Berlin Wall, and many Chinese and Indian cities today), intensified (to reach impoverished areas that used to be of no interest to capital) and trickled down the urban hierarchy to reach medium-sized cities and old European industrial cities (with the paradigmatic case of Bilbao and the so-called Guggenheim effect).

The complexity of the changes has been so evident that many authors sought to modify the term to fit it better to the reality they were analysing, and justified its use by adding an adjective. Thus, the mutation of gentrification was signalled by new forms of the term, including 'rural gentrification', 'new-build gentrification' and 'super-gentrification'. Even more recently, the number of papers on new processes such as 'tourism gentrification' or 'green gentrification' has increased (Lees et al. 2008: 138 ff.).

'Rural gentrification' refers to the middle-class (re)settlement of rural areas (also known, in a play of words, as 'greentrification') and, so, gentrification was no longer an exclusively urban phenomenon. However, this statement is perhaps limited by too narrow a view of the current urban process, in which the boundary between rural and urban is increasingly blurred, particularly when the process in both contexts is extremely similar.

'New-build gentrification' presents a problem of a conceptual nature that has sparked a certain amount of debate. If gentrification was defined initially as a process of physical renewal or restoration of rundown areas of a city, could the construction of new buildings for the wealthier classes on old industrial land be analysed as a similar process? In this case, there were no old buildings to renovate, and the process did not lead to the expulsion of the former residents. However, those who defended the existence of new-build gentrification claimed that there were certain forms of indirect, almost symbolic expulsion if the process led to the arrival of the middle-upper classes, if the landscape was completely renewed and, above all, if the capital was reinvested in urban areas that had undergone a long process of disinvestment.

'Super-gentrification', or gentrification that is superimposed on an already gentrified neighbourhood, is a sign of the broad scope and maturity of the phenomenon. It represents an intensification of the process, as it involves a level of financial or economic investment beyond that which triggered the initial stage of gentrification. To some extent, it could lead us to question the classic explanations based on the process of disinvestment-reinvestment, because here there is no rent gap (Lees et al. 2008: 149).

In recent years, an increasing number of papers have examined other forms of gentrification. 'Green gentrification' refers to gentrification facilitated by the creation or restoration of an environmental amenity. The greening of former industrial areas has been a widespread trend in urban redevelopment and includes waterfront planning, green infrastructure creation, brownfield redevelopment and other local and state green initiatives. This process has major social implications, as it can lead to an increase in the price of housing that can outweigh the benefits of a clean or green neighbourhood, as described by Miller in his empirical study of Gowanus Canal in Brooklyn (Miller 2016). The same conclusion was reached by Wolch et al., who stated that 'Ultimately, this can lead to gentrification and displacement of the very residents the green space strategies were designed to benefit' (Wolch et al. 2014). In a very different sense, Anguelovski used the case of the historical centre of Barcelona to show that social movements can 'use their environmental endeavours as tools to address stigmas attached to their place, control the land and its boundaries, and build a more transgressive form of democracy' (Anguelovski 2013).

Finally, we should mention gentrification linked to tourism. The connection between tourism and the urban economy has become a major topic of debate in recent decades. The increasing demand for tourism has been matched by the parallel creation of tourism supply in many cities, which have found their economic salvation in this industry. Nevertheless, until relatively recently, little attention had been paid to the connection between urban tourism and gentrification, which is doubtless one of the most visible manifestations of tourism in many urban landscapes (García Herrera, Smith, Mejías Vera 2007). What is known as 'tourism gentrification' is a highly representative form of the third wave of gentrification. In this case, the process is no longer limited to certain countries, but rather it is global and closely connected to circuits of capital and global culture in which the state plays a crucial role (Smith 2002). The issue here is that the use of tourism as a government strategy for urban regeneration leads almost inevitably to gentrification through the massive inflow of capital into the real estate market (Gotham 2005: 1100). Cities that have become tourist cities, such as New Orleans (particularly through the touristification of the Vieux Carre, as analysed by Gotham), Venice and Barcelona, with their essential commercial component, show the narrow connection between local institutions, the real estate industry and the global economy (Sorando and Ardura 2016).

Over the years, the situations that could be described or interpreted from the perspective of gentrification have changed and the perspective itself has broadened. Moreover, Neil Smith played a key role in this process. Initially, the concept of gentrification was used to try to understand changes in inner city neighbourhoods. Today it contains the keys to understanding global urbanisation.

Initially, some saw the changes as the solution to the problem of urban decline at the end of the 1950s, although over time this view has gradually dissipated. Brian Berry, for example, approached the changes in this way. He described them as an ephemeral, temporary phenomena resulting from the entry into an over stretched housing market of the baby boomer generation, who saw an opportunity in the cheaper dwellings available in declining inner city areas. However, reality disproved this hypothesis when gentrified areas continued to expand in the 1980s, and the focus shifted from demand to supply. It was at this point that the work of Neil Smith had a great impact. Drawing on the Marxist urban theory being developed by David Harvey, he described gentrification as the expression of an urban restructuring process driven by the demands of advanced capitalism.

The interpretation was based on movements of capital (the cycle of depreciation or destruction of value to which all urban land is subject), expressed in the rent gap hypothesis in Neil Smith's 1979 article 'Toward a theory of gentrification', significantly subtitled 'A back to the city movement by capital, not people' (Smith 1979).

After that first hard-hitting study, Neil Smith continued to develop his theory of gentrification and was undoubtedly, as Lees et al. stated, the first scholar of gentrification to stress the relationship between globalisation, neoliberalism and the role played by the state. His two central arguments were that the neoliberal state 'becomes a consummate agent of – rather than a regulator of – the market, the new revanchist urbanism that replaces liberal urban policy in cities of the advanced capitalist world increasingly expresses the impulses of capitalist production rather than social reproduction' (Smith 2002: 427), and that gentrification had definitively become a global process. These two central arguments have ensured that gentrification has become one of the key perspectives in contemporary urbanism.

One of the ideas derived from this political perspective of gentrification was that the expulsion of poor people and the homeless from the urban public space was used as a 'necessary' measure to make the city suitable for middle class consumer habits. Of course, the 'revanchist city', a term coined by Neil Smith (Smith 1996), is supported by neoliberal policies for cleaning up cities, as can be seen in the numerous cases of urban policies designed to 'clean' and 'order' the urban public space.

Finally, the idea of gentrification as a new urban colonialism is also derived directly from Neil Smith's proposals on 'the new urban frontier'. According to Atkinson and Bridge, 'the geographical spread of gentrification over the last 20 years has been reminiscent of earlier waves of colonial and mercantile expansion' (Atkinson and Bridge 2005: 2). For these authors, gentrification has many characteristics that could be considered colonial: forms of deregulation, corporate forms of urban management that promote revanchist measures and the experiences of the poorest and most vulnerable residents. Clearly, the conquest of the poorest spaces in the city by the most privileged shares many similarities with colonialism: the same processes of dispossession, with the same processes of resistance, ideological imposition, symbolic dispossession and dispossession of memory, invisibilisation and even stigmatisation, deligitimisation of alternative discourses and, when required by the circumstances, the use of force and repression.

Gentrification beyond the 'usual suspects'

Gentrification has gone global. This has become a mantra-like sentence in many recent works on gentrification to show that examples of the process can be found beyond the 'usual suspects' of the USA and Europe (Lees et al. 2008). Lees herself, this time in collaboration with authors from a wider geographic area, has qualified gentrification as 'planetary' (Lees et al. 2016). In clear agreement with studies of the urban phenomenon and of the scales of urban development (Brenner and Schmid 2011, Merrifield 2013), gentrification is now presented as a process that occurs at the planetary scale, with an increasing emphasis on the range of contexts in which it occurs. This shift is also justified as it incorporates studies that have both criticised the English-speaking world's hegemony of critical urban theory, which acritically assumes that gentrification follows a similar path worldwide and highlights the need to 'decolonize' gentrification. By integrating the the studies of 'comparative urbanism' undertaken by key authors such as Jennifer Robinson (Robinson 2011) and the reflections on urbanisation in the Global South by Ananya Roy (Roy 2009), a new direction has been taken, in which urban theory produced by Western countries, and even theories on neoliberal urbanism, are being called into question (Parnell and Robinson 2012).

However, we should avoid drawing a division between the 'usual suspects' and the Global South. As Roy has indicated, all cities have processes like those of the Global South (poverty, unsustainability and an informal economy), and all have processes that are similar to those of New York or Tokyo (a concentration of wealth and massive real estate investments of global capital), but all of them also have their specific features (Roy 2009). What is required is perhaps a new approach that contributes new material to the theory formulated in the West.

This new approach brings with it studies on other, non-Western cities and on processes that cannot be labelled as 'gentrification'. This has led some to consider whether the theory of the Western hemisphere can really be useful for non-Western cities and if it can account for everything that happens in them, or whether perhaps the time has come to abandon the concept. According to this perspective,

> The limitation of gentrification, as an analytic, is that it fails to grasp transformations in the peri-urban and outer areas of post-socialist and post-colonial cities where the most violent displacement is taking place and where non-fully privatised tenure endures. It is precisely these areas that much of the new, Southern gentrification literature has sought to subsume within its remit, and it is in precisely these areas that nobody cares about gentrification.
>
> (Ghertner 2015: 560)

Similarly, some authors opt to focus on the specific characteristics of their own contexts. For example, in Spain and Latin America, processes of gentrification have been found to vary substantially from what is observed in the United States and the United Kingdom. The emerging nature of these processes (both as an urban reality and a critical analysis), the range of cities in which they take place, the diversity of national realities and the lack of specific language to describe them are factors that justify establishing a certain distance from Anglo-American critical discourse. However, the aim is not to attack this discourse, but rather to complement or nourish it (Janoschka, Sequera and Salinas 2014).

The main question is whether the concept of gentrification helps or hinders understanding of urban processes in other contexts. Most studies do not doubt the validity of the concept itself, but rather they are chary of its capacity to explain contexts other than those of Anglo-American countries. Clearly, there is a concern that the indiscriminate use of gentrification could have counterproductive effects, which could hamper the analysis required in each context.

It is also important to avoid a concept that has

> substance of its own, has transgressed the limits of academia and has become a sociopolitical issue and a stake in itself. We should, therefore, constantly challenge it by revealing/recalling its implicit contextual assumptions and by comparing them to the contextual realities of the analysis each time at hand.
>
> (Maloutas 2012: 44)

The good news is that everybody will benefit from this: Case studies will proliferate, and the theory will be enriched and made more robust.

The recent volume edited by Lees, Bang Shin and López-Morales addresses a range of situations, positions and reservations about critical urban theory, and even includes some support for the old rent gap theory, countering other theories that are allegedly more sophisticated, but do not reflect the political vocation with which rent gap was created. Tom Slater was emphatic when he stated that rent gap is 'fundamentally about class struggle, about the structural violence visited upon so many working class people in contexts these days that are usually described as "regenerating" or "revitalising"' (Slater 2015: 9). This idea was echoed by Lees, Bang Shin and Lópes-Morales: 'the rent gap is a simple idea that anti-gentrification activists can read, understand and debate, and then apply to their everyday realities' (Lees et al. 2016: 74).

Here, it is interesting to note that anti-gentrification has also gone global: Gentrification has gone from being an academic and, therefore, elitist concept, to a word that is fully integrated into the discourse of social movements and that of many citizens with only the slightest interest in the problems caused by neoliberal urbanism. Giving a name to what is happening in your neighbourhood helps to better understand the phenomenon and contextualise it, to understand its origins and predict its consequences and, thus, promote collective mobilisation.[1] Naming the phenomenon also helps people to identify with similar problems in other cities of the country or continent, which in turn helps to better assimilate the magnitude and connotations of the phenomenon, to better identify those who tend to be its main promotors and its victims and, all in all, help create feelings of solidarity that extend beyond their own reality. Anyway, contradictions not only affect those who persecute benefit from the exploitation of cities, but also threaten the resistants to gentrification. Maintaining spaces of hope that resist the siege of the capital is, then, a task of maximum difficulty but, also, of an extremely creativity, as Sorando and Ardura (2016: 125–155) clearly show following examples from several cities.[2]

Resistance to gentrification is resistance to neoliberal urbanism

We live in an age characterised by neoliberal policies implemented with increasing force and intensity. In this context, neoliberal urbanism no longer differentiates between places, but rather it extends itself across the entire planet, adapting itself selectively to different realities and casuistries. Gentrification, understood as a strategy for dispossession and the consequent appropriation of land rents by a minority, has become a planetary phenomenon and, as discussed, can be found in many different territories, quite distinct from the impoverished centres of developed metropolises that were initially its preferred area of action.

Today, resistance to neoliberal urbanism necessarily means resistance to gentrification. This resistance involves two essential factors. First, there is an urgent need to further our critical interpretations of the urban processes that serve neoliberal capitalism. Due to the global, but also local, scope of gentrification, studies carried out in different parts of the world should no longer be considered as 'case' studies, but treated as contributions that are essential to a full coherent understanding of the phenomenon. This book seeks to pay tribute to Neil Smith by also highlighting the importance of research on gentrification that is carried out outside of the Anglo-American world and the main Western metropolises.

Second, resistance to gentrification has to be fought on another equally important front: We need to remain alert to the dangers of the adaptation of theoretical debates and the distortion of conceptual discourses. 'Gentrification' continues to be a dirty word that is used, above all, to reveal the dark side of regeneration, renewal, revitalisation and an infinite number of other terms that refer to the supposed advantages of transforming the city though its radical restructuring. Hence, it is essential we address the evidence of increasing inequality, segregation and eviction, in the face of the hegemonic, fatalistic and economistic discourses (Massey 2011) that consider gentrifying dynamics – however much they might be hidden behind other names – to be good, beneficial and/or to have no alternatives. In response to shocking claims that 'some of these large-scale urban renewal (read gentrification) schemes are supported by the very populations they dispossess' (Lees et al. 2016: 17), we must condemn the ideological manipulations of actions that serve only a minority. When the state no longer steps in to control the voracity of capital and public investments diminish (or disinvestments become widespread), as often happens in the first stages prior to gentrification, the devastation of poverty and lack of solidarity spread, and injustice and hate are exacerbated. The association between the 'dictates of the markets' and neo-conservatism leads to an urban cultural policy (which today many like to call *governance*) in which political rhetoric serves the interests of neoliberalism, questions the proposals of participative and grassroots democracy and promotes strong hostility of citizens to forms of egalitarianism and solidarity that do not fit the logic of the market. Again, some collateral actions are typical of gentrifying processes: rejection of immigrants and working classes, criminalisation of alternative initiatives (squats and cooperatives, among others), undermining of protests and social movements and stigmatisation of entire neighbourhoods to legitimise their renewal, among others.

We need to break with the common argument that urban problems are simply the result of 'natural' economic forces (market inertia), against which it is almost impossible to act. The immediate consequence of this is the depoliticisation of urban thought and of the debate about the city, given that socio-spatial inequalities (which are intensified in the current neoliberal phase) have been emptied of all political meaning and content (Hall and Massey 2010). The city is constantly presented not so much as a place in which to live, but as a place in which to do business. Thus, the objectives and effects of urbanism become the acquisition of immediate individual gains rather than the redistribution of these benefits in acts of solidarity; urban appearances and urban marketing instead of citizens' welfare; competition and corruption instead of social and spatial justice.

Resistance to gentrification and to neoliberal urbanism, therefore, involves remembering (and acting in accordance with these ideas) that the solution to the main problems of the contemporary city and the devastating effects of neoliberal urban intervention lies solely in seeking to change the system that, out of its own need, creates these problems. Neil Smith called for a revolution. He considered that we should not give up the real possibility of revolution, as this represented the first and last guarantee of social and spatial transformation and regeneration and the only way to overcome the inequalities and exploitation that are implicit in capitalism. Knowing Neil's dynamic, passionate, enthusiastic nature, it is easy for us to imagine him marching decisively through the streets, stirring the masses to action, raising a red flag, ready to occupy any new Winter Palace. In 'The Revolutionary Imperative' Neil warns that, despite everything, the greatest 'tragedy' that holds back the revolution today 'is less the political onslaught by the right than the political non-response of the left' (Smith 2009: 51).

Neil's stubborn realism did not cloud his profound optimism: He recognised that the left was highly divided, but also that the range of civil movements and protests was changing the situation rapidly and effectively (and, when he left us in 2012, Neil could not have imagined to what extent!). It is true that the power of the media and the manipulation of politicians is almost infinite, but it is also true that they cannot hide the fact that we find ourselves in a time of great crisis in the capitalist and neoliberal system and, so, at an opportune moment to make profound changes. It is also true that many revolutions have failed, but the revolution may arrive at any time 'like a thief in the night'... and stay: the revolution will arrive, said Neil, if it brings tools with it: some tools are intellectual ideas, other are our own bodies, organised politically and socially.

The necessary two-way relationship between theory and action was discussed by Andy Merrifield, who rightly said that 'nobody is mobilised by a theory', but he also stressed that theories should stay close to where they can be useful (Merrifield 1997). It is true that when theories are more than academic exercises and are indeed useful, they can and should be applied and mobilised in specific cases. Time seems to have shown that Henri Lefebvre was right when he stated that resistance to global processes is only possible from the space of everyday life. For him, revolution should be based on the possibility that the quotidian acts over the urban, and the urban over the global, although they should not be seen as separate scales but simultaneous ones. The big episodes of social fights against capitalism have not only been political events but revolutions of the urban space and of daily life; changing the world has meant changing the way that every day we live our daily life' (Goonewardena 2012). Although he accepted this idea (in theory and practice), Neil Smith did not stop analysing strategies of capital accumulation at the global scale nor, of course, did he stop working to promote global justice.

Today, Neil Smith continues to be a point of reference both for theory (for academia and the conceptual debate) and for action (for social movements and for urban activism). Today, the legacy of Neil Smith continues to express criticism of the geography of capitalism in a way that can stir the imagination, contribute to the design of liberating lifestyles and help recover the revolutionary project that shows us that other worlds are possible.

All is not lost: The neoliberalisation of the city will remain incomplete while there exists the chance to break with the discourse that justifies urban space as a good and the city as an exchange value rather than as a use value. For this reason, by means of the conference held in Barcelona in 2015 and through the pages of this book, we wish to pay tribute to the citizens in those areas of the city that are under greatest attack from the processes of gentrification. We hope that the contents of this book will be of use in the struggle of those who, despite the uncontrolled voracity of neoliberal urbanism seen in an increasingly wide range of contexts, continue to resist (Figure 24.1).

Figure 24.1 Neil Smith at Tompkins Square Park (New York) after its renewal (2012).
Source: Deborah Cowen collection.

Notes

1 *Staying Put: The Anti-Gentrification Handbook* is a good example of a succesful cooperative initiative promoted by directly affected citizens, social movements, and university researchers. As it is mentioned in its statement, the handbook

> explains why the regeneration of council estates often results in established communities being broken up and moved away, and housing becoming more expensive. It is designed to help local communities learn about gentrification and the alternatives they can fight for. Through the experiences of council tenants, leaseholders and the wider community in London, it contains ideas, stories, tools and resources.

It brings 'the tools necessary to halt the social cleansing of council estates and developing community-led alternatives for sustaining existing communities.' It can be downloaded at https://southwarknotes.files.wordpress.com/2014/06/staying-put-web-version-low.pdf

2 Among many others, see actions taking place in San Francisco through 'The Anti-Eviction Map Project' (www.antievictionmap.com), in New York with 'the Gentrifier Calculator' (Kirk 2015, deMause 2015), in Berlin see 'The Gentrification Blog' (https://gentrificationblog.wordpress.com/about/), in Rome the 'Rete Anti-Sfratto Roma Est' (https://fucina62.noblogs.org/files/2013/12/Opuscolo-Antisfratto.pdf) and the 'Libera Repubblica di San Lorenzo' (www.liberarepubblicadisanlorenzo.it). For worldwide actions, see 'International Alliance of Inhabitants' (www.habitants.org/zero_evictions_campaign) and Blanco and Bosoer (2016). For the Spanish case see 'Vivero de inciativas

ciudadanas' (http://viveroiniciativasciudadanas.net/2015/03/10/madrid-desahuciado/), 'Gentrificación no es un nombre de señora' by Left Hand Rotation (www.lefthandrotation.com/gentrificacion/), 'Plataforma d'Afectats per la Hipoteca' (https://pahbarcelona.org) (see also Colau and Alemany 2013), 'Assemblea de Barris per un Turisme Sostenible' (https://assembleabarris.wordpress.com). For an overview see Observatorio Metropolitano (2007), Montaner, Álvarez and Muxí (2012), Montaner et al. (2013), Aricó, Mansilla and Stanchieri (2016), Observatorio Metropolitano de Madrid (2015), Subirats and García Bernardos (2015).

References

Anguelovski, I. (2013). 'Beyond a livable and green neighborhood: asserting control, sovereignty and transgression in the Casc Antic of Barcelona', *International Journal of Urban and Regional Research*, 37(3), pp. 1012–1034.

Aricó, G., Mansilla, J. A. and Stanchieri, M. L. (eds.) (2016). *Barrios corsarios. Memoria, luchas urbanas y cambio social en los márgenes de la ciudad neoliberal.* Barcelona: Pol·len.

Atkinson, R. and Bridge, G. (eds.) (2005). *Gentrification in a Global Context. The New Urban Colonialism.* New York: Routledge.

Blanco, J. and Bosoer, L. (2016). *Ciudades en disputa: Gentrificación. Propuestas de abordaje didáctico.* Madrid: Contested Cities.

Brenner, N. and Schmid, C. (2011). 'Planetary urbanisation', in M. Gandy (ed.). *Urban Constellations.* Berlin: Jovis, pp. 10–13.

Colau, A. and Alemany, A. (2013). *¡Sí se puede! Crónica de una pequeña gran victoria.* Barcelona: Destino.

deMause, N. (2015). 'The gentrifier's guide to not being an asshole', *The Village Voice.* August 25. Available at: www.villagevoice.com/news/the-gentrifier-s-guide-to-not-being-an-asshole-7529761

García Herrera, L. M., Smith, N. and Mejías Vera, M. A. (2007). 'Gentrification, displacement, and tourism in Santa Cruz de Tenerife', *Urban Geography*, 28(3), 276–298.

Ghertner, D. A. (2015). 'Why gentrification theory fails in "much of the world"', *City*, 19(4), pp. 552–563.

Glass, R. (1964). 'Introduction' in Centre for Urban Studies (ed.). *London, Aspects of Change.* London: Macgibbon & Kee, pp. xiii–xlii.

Goonewardena, K. (2012). 'Space and revolution in theory and practice: eight theses', in N. Brenner, P. Marcuse and M. Mayer (eds). *Cities for People, Not for Profit. Critical Urban Theory and the Right to the City.* New York: Routledge, pp. 86–101.

Gotham, K. F. (2005). 'Tourism gentrification: the case of New Orleans' Vieux Carré (French Quarter)', *Urban Studies*, 42(7), pp. 1099–1121.

Hackworth, J. and Smith, N. (2001). 'The changing state of gentrification', *Tijdschrift voor Economische en Sociale Geografie*, 92(4), pp. 464–477.

Hall, S. and Massey, D. (2010). 'Interpreting the crisis', *Soundings: A Journal of Politics and Culture*, 44, pp. 57–71.

Hamnett, C. (1991). 'The blind men and the elephant: the explanation of gentrification', *Transactions of the Institute of British Geographers*, 16(2), pp. 173–189.

Janoschka, M., Sequera, J. and Salinas, L. (2014). 'Gentrification in Spain and Latin America – a critical dialogue', *International Journal of Urban and Regional Research*, 38(4), pp. 1234–1265.

Kirk, C. (2015). 'Are you a gentrifier?', *Slate*. July 30. Available at: www.slate.com/articles/business/moneybox/2015/07/gentrification_calculator_are_you_a_neighborhood_gentrifier.html
Lees, L., Shin, H. B. and López-Morales, E. (2016). *Planetary Gentrification*. Cambridge: Polity Press.
Lees, L., Slater, T. and Wyly, E. (2008). *Gentrification*. London: Routledge.
Maloutas, T. (2012). 'Contextual diversity in gentrification research', *Critical Sociology*, 38(1), pp. 33–48.
Massey, D. (2011). 'Ideology and economics in the present moment', *Soundings: A Journal of Politics and Culture*, 48, pp. 29–39.
Merrifield, A. (1997). 'Between process and individuation: translating metaphors and narratives of urban space', *Antipode*, 29(4), pp. 417–436.
Merrifield, A. (2013). 'The urban question under planetary urbanization', *International Journal of Urban and Regional Research*, 37(3), pp. 909–922.
Miller, J. T. (2016). 'Is urban greening for everyone? Social inclusion and exclusion along the Gowanus Canal', *Urban Forestry & Urban Greening*, 19, pp. 285–294.
Montaner, J. M., Álvarez, F. and Muxí, Z. (eds.) (2012). *Arxiu Crític Model Barcelona 1973–2004*. Barcelona: Ajuntament de Barcelona/ETSAB-UPC.
Montaner, J. M., Álvarez, F., Muxí, Z. and Casanovas, R. (eds.) (2013). *Reader Modelo Barcelona 1973–2013*. Barcelona: Ajuntament de Barcelona/Comanegra.
Observatorio Metropolitano (ed.). (2007). *Madrid: ¿la suma de todos? Globalización, territorio, desigualdad*. Madrid: Traficantes de Sueños.
Observatorio Metropolitano de Madrid. (2015). 'Introducción', in Observatorio Metropolitano de Madrid (ed.). *El mercado contra la ciudad. Globalización, gentrificación y políticas urbanas*. Madrid: Traficantes de Sueños, pp. 17–26.
Parnell, S. and Robinson, J. (2012). '(Re)theorizing cities from the Global South: looking beyond neoliberalism', *Urban Geography*, 33(4), pp. 593–617.
Robinson, J. (2011). 'Cities in a world of cities: the comparative gesture', *International Journal of Urban and Regional Research*, 35(1), pp. 1–23.
Roy, A. (2009). 'The 21st-century metropolis: new geographies of theory', *Regional Studies*, 43(6), pp. 819–830.
Slater, T. (2015). 'Planetary rent gaps', *Antipode*, pp. 1–24. doi:10.1111/anti.12185.
Smith, N. (1979). 'Toward a theory of gentrification. A back to the city movement by capital not people', *Journal of the American Planning Association*, 45, pp. 538–548.
Smith, N. (1992). 'Blind man's buff, or Hamnett's philosophical individualism in search of gentrification', *Transactions of the Institute of British Geographers*, 17(1), pp. 110–115.
Smith, N. (1996). *The New Urban Frontier. Gentrification and the Revanchist City*. London: Routledge.
Smith, N. (2002). 'New globalism, new urbanism: gentrification as global urban strategy', *Antipode*, 34(3), pp. 434–457.
Smith, N. (2009). 'The revolutionary imperative', *Antipode*, 41, pp. 50–65.
Sorando, D. and Ardura, J. (2016). *First We Take Manhattan. La destrucción creativa de las ciudades*. Madrid: Los libros de la catarata.
Subirats, J. and García Bernardos, Á. (eds.). (2015). *Innovación social y políticas urbanas en España*. Barcelona: Icaria.
Wolch, J. R., Byrne, J. and Newell, J. P. (2014). 'Urban green space, public health, and environmental justice: the challenge of making cities "just green enough"', *Landscape and Urban Planning*, 125, pp. 234–244.

Index

For Product Safety Concerns and Information please contact our EU representative GPSR@taylorandfrancis.com
Taylor & Francis Verlag GmbH, Kaufingerstraße 24, 80331 München, Germany

www.ingramcontent.com/pod-product-compliance
Ingram Content Group UK Ltd.
Pitfield, Milton Keynes, MK11 3LW, UK
UKHW020018200926
475702UK00014B/1957
9780367219123